T0092646

Synthesis Lectures on Engineering, Science, and Technology

The focus of this series is general topics, and applications about, and for, engineers and scientists on a wide array of applications, methods and advances. Most titles cover subjects such as professional development, education, and study skills, as well as basic introductory undergraduate material and other topics appropriate for a broader and less technical audience.

Amal Banerjee

Semiconductor Devices

Diodes, Transistors, Solar Cells, Charge Coupled Devices and Solid State Lasers

Amal Banerjee
Analog Electronics
Kolkata, West Bengal, India

ISSN 2690-0300 ISSN 2690-0327 (electronic)
Synthesis Lectures on Engineering, Science, and Technology
ISBN 978-3-031-45749-4 ISBN 978-3-031-45750-0 (eBook)
https://doi.org/10.1007/978-3-031-45750-0

This Springer imprint is published by the registered company Springer Nature Switzerland AG
The registered company address is: Gewerbestrasse 11, 6330 Cham, Switzerland

Paper in this product is recyclable.

This work is dedicated to

My late father Sivadas Banerjee
My late mother Meera Banerjee

The two professors who taught me all about solid state physics and semiconductor devices:

Dr. Sanjay Kumar Banerjee
late Dr. James C Thompson

A dear friend and guide: Dr. Andreas Gerstlauer

Acknowledgements

The author is offers his heartfelt thanks to his past thesis advisor Dr. Sanjay Kumar Banerjee. Dr. Banerjee checked the table of contents and two sample chapters to provide valuable feedback on the initial book proposal. Dr. Banerjee is the Director, Microelectronics Research Center, Department of Electrical and Computer Engineering, University of Texas at Austin.

Contents

1 Fundamental Quantum and Statistical Mechanics of Crystalline Solids

… by nature I am a conservative man, but a theoretical explanation had to be found, whatever the cost … Max Planck at his Nobel Prize acceptance speech.

1.1 Energy Bands

Quantum mechanics [1–8] explains how and why all fundamental physical entities (electrons, protons, atoms, molecules, etc.,) exist and their properties. Quantum mechanics is based entirely on *experimentally established* fact that **energy levels of all physical entities (electrons, protons, neutrons, atoms and molecules) are discrete and quantized**. That is, the allowed energy levels of even the simplest atom—the hydrogen atom are E, 2E, 3E, …. **but never** E, 1.5E, 2.67E … etc., *As the numerical values of these discrete energy levels are small, from a macroscopic view, these energy levels appear smeared out into one continuous energy band.* **Nothing would exist without quantum mechanics, as everything at the core is made of atoms, electrons, protons etc.**

One key foundation of quantum mechanics is the particle|wave duality of matter, proposed by Louis De-Broglie, which states that the momentum and wavelength of a particle (e.g., electron) are related as:

$$p = \frac{\hbar}{\lambda_{DB}} \quad where\ p, \hbar, \lambda_{DB} \tag{1.1}$$

A. Banerjee, *Semiconductor Devices*, Synthesis Lectures on Engineering, Science, and Technology, https://doi.org/10.1007/978-3-031-45750-0_1

are respectively the momentum, Planck's constant and De-Broglie wavelength. *The behaviour of a fundamental particle (and entities made up of these fundamental particles—atoms, molecules etc.,) can be explained either by invoking its wave properties or particle properties.*

The Schrodinger's wave equation explains the behaviour of elementary physical entities, e.g., the details of an electron's motion in a solid, by exploiting that entity's wave-like properties. *This equation, derived in 1925, was first successfully used to explain the experimentally observed, discrete frequencies of light emission from an excited hydrogen atom.* The simplest form of this equation is the time-independent version, i.e., in one dimension given as:

$$\frac{-\hbar^2 d^2\psi}{2mdx^2} + U(x)\psi(x) = E\psi(x) \tag{1.2}$$

Invoking conservation of energy, the Scrodinger's wave equation in one dimension can be re-written in terms of the Hamiltonian operator as:

$$H\psi(x) = E\psi(x) \tag{1.3}$$

In both the above expressions,

$$\psi(x),\ U(x),\ m,\ E$$

are respectively the particle's (e.g., electron's) wavefunction, potential energy, rest mass and total energy. *The fundamental physical importance of the wavefunction is that the expression*

$$\psi(x)\psi^P(x) = |\psi(x)|^2 \tag{1.4}$$

is the probability of finding the particle (e.g. electron) in the 1-dimensional region bounded by (x, x + dx). *The superscript P represents the complex conjugate of the wave function.* This is the Niels Bohr interpretation of the wavefunction. Clearly,

$$\int |\psi(x)|^2 dx = 1 \quad -\infty \le x \le \infty \tag{1.5}$$

The solutions to the Schrodinger's equation are the possible discrete energy levels that can be occupied by the particle (e.g., electron). These discrete energy levels give rise to energy bands. Once a potential has been defined, the Schrodinger's wave equation can be solved. Of the many possible potential functional forms that might be used, the one very attractive for solid state physics purposes, is the periodic Dirac delta function potential. The Dirac delta function, is defined as:

$\delta(x - a)$ has value infinity at x = a, and *is zero for all* other values of x.

The periodic (period N) Dirac delta function is:

$$U(x) = \beta \sum \delta(x - la)\ 0 \leq l \leq (N-1)a\ \beta\ constant \tag{1.6}$$

The periodic Dirac delta function, applied to the Schrodinger's wave equation, transforms it to:

$$\frac{d^2\psi(x)}{dx^2} + U(x)\psi(x) = 0 \tag{1.7}$$

whose general solution, with arbitrary constants A, B is listed below. The constants are evaluated using boundary conditions:

- The wavefunction is continuous at a boundary.
- The first derivative of the wavefunction is continuous infinitesimally close to the boundary, but **not at** the boundary, where the potential is infinite (Dirac delta function):

$$\psi(x) = A\cos\left(\frac{\sqrt{2mE}}{\hbar}\right)x + B\sin\left(\frac{\sqrt{2mE}}{\hbar}\right)x \tag{1.8}$$

A periodic potential has a huge number of boundaries, at each of which the potential energy U tends to infinity. This problem is easily circumvented by invoking the periodic potential's unique properties as embedded in the Bloch's Theorem. *This theorem states that for a periodic potential U(x + a) = U(x), the solutions to Schrödinger's equation satisfy:*

$$\psi_k(x) = u_k(x)e^{jkx}\ u_k(x)\quad lattice\ periodicity \tag{1.9}$$

The subscript k indicates that u(x) has different functional forms for different values of the Bloch wavenumber k. In case if u(x) is not periodic but a constant, the Bloch wave becomes a plane wave. *Therefore a Bloch wave is a plane wave modulated by a function that has the periodicity of the lattice.* An alternative form of Bloch's Theorem is:

$$\psi_k(x + a) = e^{jka}\psi_k(x) \tag{1.10}$$

If there is a discontinuity in the first derivative of the wavefunction, then the kinetic energy term tends to infinity, but the equation is still satisfied if the potential energy tends to infinity $U \rightarrow \infty$. When the *effective mass Schrödinger wave equation*, is examined the boundary condition for the derivative of ψ must include the effective mass, where $\psi_k(x), |\psi_k(x)|^2$ are periodic.

This periodicity property is because an electron has an equal probability of being at any of the identical sites in the linear array, and breaks down at the edges of the lattice. The electrons deep within the lattice are unaware that the periodicity property is broken at the edges, since the array is very long compared to the separation. between atoms, i.e., *when N (the number of lattice nodes|sites), is very large*. The generalized periodic boundary condition becomes:

$$\psi_k(x + Na) = \psi_k(x) \tag{1.11}$$

which leads to

$$\psi_k e^{jkNa} = \psi_k(x) \quad k = \frac{2\pi n}{Na} \quad n : \text{integer} \tag{1.12}$$

where k is the *Bloch wavenumber*. Now, the constants A, B can be determined. In the linear region −a < x < 0. Schrodinger's wave equation is re-written as:

$$\psi_k(x) = e^{-jka}\left(\left(A\cos\left(\frac{\sqrt{2mE}}{\hbar}(x+a)\right) + B\sin\left(\frac{\sqrt{2mE}}{\hbar}\right)(x+a)\right)\right) \tag{1.13}$$

The expressions for the constants A, B are then (evaluated at x = 0):

$$B = e^{-jka}\left(\left(A\cos\left(\frac{\sqrt{2mE}}{\hbar}a\right) + B\sin\left(\frac{\sqrt{2mE}}{\hbar}a\right)\right)\right) \tag{1.14}$$

As the derivative of the Dirac delta function is discontinuous at x = 0, the derivative of ψ is also discontinuous at x = 0. Therefore the discontinuity must be evaluated to get another expression linking A and B. For U(x) = βδ(x) the discontinuity condition implies:

$$\Delta\frac{d\psi}{dx} = \frac{2m\beta}{\hbar^2}\psi(0) \tag{1.15}$$

Therefore, from the derivatives of ψ at x = 0:

$$\frac{\sqrt{2mE}}{\hbar}\left(A - e^{-jka}\right)\left(A\cos\left(\frac{\sqrt{2mEa}}{\hbar}\right) - B\sin\left(\frac{\sqrt{2mEa}}{\hbar}\right)\right) = \frac{2m\beta B}{\hbar^2} \tag{1.16}$$

Equation 1.16 is obtained by integrating the Schrödinger's equation over a tiny interval spanning x = 0. The integral of the first derivative term is the required discontinuity and is equal to the integrals over the Eψ and Uψ terms. In the former term E is a constant and ψ is finite, so integrating over an infinitesimal interval gives zero. Identical arguments hold for the Uψ term. However, as U = ∞ at x = 0, a special property of the Dirac delta, i.e., its integral over all space is 1, is exploited. **After some careful manipulation of these expressions, the fundamental expression that governs the physics of energy bands is obtained, independent of the constants A, B.**

$$\cos(ka) = \cos\left(\frac{\sqrt{2mE}}{\hbar}a\right) + \frac{a\beta m\sin\left(\sqrt{2mEa}\right)}{\hbar^2\frac{\sqrt{2mEa}}{\hbar}} \tag{1.17}$$

This key equation embodies the secret of energy bands: the right-hand side is a function of the energy E, but the left-hand side strictly enforces that f(E) must be bounded by ± 1. Energy is quantized. The energy bands corresponding to the allowed values of $-1 \leq f\left(\frac{\sqrt{2mE}a}{\hbar}\right) \leq +1$, and the forbidden regions (bandgaps—separating the permitted bands), are displayed on a plot of energy E versus Bloch wavevector k (extended zone plot)—Fig. 1.1a. The first zone extends over $\frac{-\pi}{a} < k < \frac{\pi}{a}$; the second zone is split into two: $\frac{-2\pi}{a} < k < \frac{-\pi}{a}$ $\frac{\pi}{a} < k < \frac{2\pi}{a}$. Range of k in each zone is $\frac{2\pi}{a}$. The total number of lattice sites is N. *As N is very large in semiconductor devices, the separation of neighbouring k values (=* $\frac{2N\pi}{a}$*), is so small that the E-k relation appears continuous within a band.* $k = \frac{2\pi}{\lambda}$ is the general relationship between wavelength and wavevector. For the specific case of a Bloch wavevector, $\vec{P}_{CRYSTAL} = \hbar\vec{k}$ is called the *crystal momentum. This is the momentum of an electron (for a specific Bloch wavevector) in a crystal as a consequence of applied forces.*

A reduced zone [8–16] plot (Fig. 1.1b) also displays the E-k relationship by compressing all of its information into the first zone, by horizontally shifting each of the curves from the higher order zones in the extended-zone plot by an appropriate multiple of $\frac{2\pi}{a}$. For the positive wavevectors in the 4th and 5th zones, i.e., $\frac{3\pi}{a} < k < \frac{5\pi}{a}$, the wavevector can be re-written as:

$$k = \frac{4\pi}{a} + k^P \quad \frac{-\pi}{a} \leq k^P \leq \frac{\pi}{a} \tag{1.18}$$

So the wavefunction becomes:

$$\psi_k(x) = u_k(x)e^{\frac{j4\pi x}{a}}e^{jk^P x} \quad = \psi_{k^P}(x) \tag{1.19}$$

The terms $e^{\frac{j4\pi x}{a}}$ and u have the same period a, so can be combined into a new periodic function. Also, as the changes to u and k are complementary, the wavefunction is

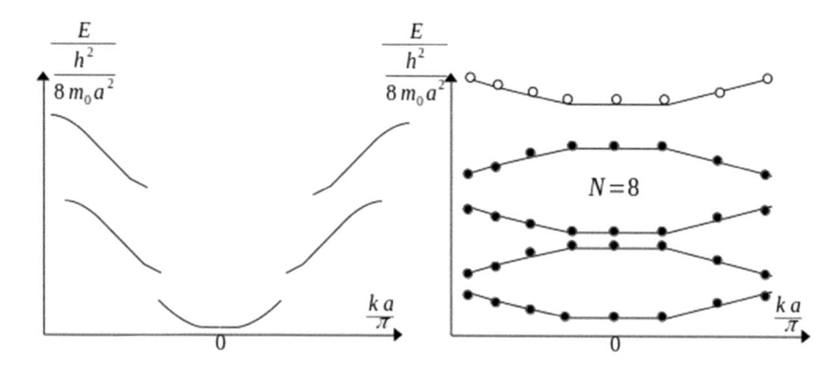

Fig. 1.1 **a** E-k extended zone (k -Bloch wavevector) plot **b** Reduced zone plot for N = 8 crystal momenta

unchanged. The shift in k of 4π/a takes the band of the 4th zone (positive k) to the range $\frac{-\pi}{a} < k < 0$, and the 5th band to $0 < k < \frac{\pi}{a}$. The bands in the new scheme are completed by similar operations on the corresponding, negative-k portions of the 4th and 5th bands from the extended-zone plot. Identical translations of appropriate multiples of 2π/a, bring all of the other bands into the first zone. The resulting plot is the reduced zone plot, Fig. 1.1. The first zone, which now contains all the bands, is first Brillouin zone, or Brillouin zone. In the reduced-zone plot the crystal momentum is renamed the *reduced crystal momentum.*

1.1.1 Physics of Energy Bands

The underlying physics of energy bands is simple [1–16]. Let a beam of electrons of wavelength λ propagate a one dimensional lattice. Some of the electrons scatter off two neighbouring lattice sites. The two portions of the reflected beam would reinforce constructively if the Bragg condition for normal incidence is satisfied.

$$2a = \beta\lambda \tag{1.20}$$

where a is the spacing between lattice sites and b = 1, 2, 3, ... is an integer. As the number of lattice sites is very large, multiple Bragg reflections would make the beam bounce around (i.e., reflected back–forth) in the crystal resulting in a standing wave. The wavevectors at which this occurs are:

$$k = \frac{\pm 2\pi}{\lambda} = \frac{\pm b\pi}{\lambda} = \frac{G_b}{2} \tag{1.21}$$

As a result, energy bandgaps, within which there are no propagating waves, arise at Brillouin zone boundaries because of the strong Bragg reflection. In the above expression:

$$G_b = \frac{2\pi b}{a} \tag{1.22}$$

is a set of multiples of 2π and the reciprocal of the lattice spacing a. ***The multiples are called reciprocal lattice numbers, and become vectors in two|three dimensional lattices. The translation numbers used to obtain the reduced zone plot from the extended zone plot are the reciprocal lattice.***

1.1.2 Material Classification Using Quantum States

The reduced zone plot shown above (Fig. 1.1b) is valid for very small number of lattice sites or primitive unit cells each with a single electron (monovalent). For a reduced zone plot:

$$|k_{MAX}| = \frac{\pi}{a} \quad |n_{MAX}| = \frac{N}{2} \quad \frac{k\pi}{a} = 0, \pm 0.2, \pm 0.4, \pm 0.6\ldots \tag{1.23}$$

Thus n is a state of reduced crystal momentum that can be occupied by an electron. As the end values n = ±N/2 are the same point, the total number of distinct n numbers in the reduced zone is equal to N, the number of lattice sites of primitive unit cells.

From Pauli's Exclusion Principle (no two fundamental physical entities (e.g., electrons) can have an identical set of quantum numbers), each reduced crystal momentum state can be occupied by two electrons, if and only if they have opposite spin. The quantum number for electron spin is ±1/2 and there is one quantum number—n, for the crystal momentum. In the reduced zone scheme, n is restricted to values between −N/2 and N/2, so that another number is essential to separate states with the same value of reduced wavevector, but different values of energy. This number is the band index. Summarizing,

- Each band contains 2N states, where N is the number of primitive unit cells that form the crystal lattice.
- For the special case of a monovalent single atom primitive cell there will be N valence electrons, which T = 0 K will occupy the bottom half of the first band.
- If there were 2 valence electrons per primitive cell the entire first band would be occupied at 0 K. This means that bands will be either completely filled or completely empty if there is an even number of electrons in the primitive unit cell.
- The highest fully occupied band at 0 K is the valence band, and the lowest unfilled band at 0 K is the conduction band. The energy gap between these bands is called the bandgap.
- For a full band, a filled state with crystal momentum is matched by a filled state with crystal momentum.
- Crystal momentum is the electron momentum due to external forces, such as an applied electric field. So, there can be no net motion of charge carriers, as long as the electrons stay in the full band.

When thermal energy is added to the system by increasing T, then the monovalent case the electrons can respond to this stimulus by moving into allowed states of higher energy and crystal momentum within the half full first band. If in addition an electric field is applied, electrons will be accelerated into states of higher crystal momentum, and there would be current flow—as in metals. In the divalent case, the only way to get a net gain in crystal momentum would be if some electrons could be sufficiently energized to cross the forbidden energy bandgap and populate some of the states in the empty conduction band, in which they would gain crystal momentum from an applied field. If this bandgap is very large, the probability of electrons acquiring sufficient energy to overcome the bandgap is very small, and this is an insulator. If the bandgap is not too large, some electrons can be excited into the conduction band, and this is a semiconductor—useful semiconductors have a bandgap in the range 0.5–3.5 eV.

1.2 Crystal Structure and Semiconductor Energy Bands

For one dimensional lattice, the reciprocal lattice number is:

$$G_b = \frac{2\pi b}{a} \tag{1.24a}$$

Thus the reciprocal lattice number is associated with the reciprocal space, which in the one dimensional lattice, consists of a linear array of points separated by 2π/a, where a is the spacing of primitive unit cells in the direct lattice. *In real three space, the primitive unit cell is a volume, and have reciprocal lattice vectors which have a magnitude of some multiple of 2π divided by the spacing between planes of atoms. The direction of the reciprocal lattice vector in reciprocal space is orthogonal to that of the planes in real space.* The primitive unit cell in reciprocal space for the real space face-centred cubic (FCC) lattice is a truncated octahedron Fig. 1.2. The Cartesian axes corresponds to directions of the Bloch wavevector k. These directions are orthogonal to planes in the direct lattice. Therefore these are denoted in the same way as the normals to crystal planes, using Miller indices Fig. 1.3. In the Cartesian coordinate system of the direct lattice, the (100) plane intersects the x,y,z axes at a, ∞, ∞, respectively. The latter set becomes (100) by using reciprocal of each intercept and reducing to integer values. The normal to this plane is specified by the same set of numbers, but with a different parenthesis, i.e., [100]. Axis labelling is arbitrary, so that the (−1,0,0) and (0,1,0) surfaces have exactly the same properties as (100) surface, denoted as a set {100}: the equivalent normals directions are designated as 100.

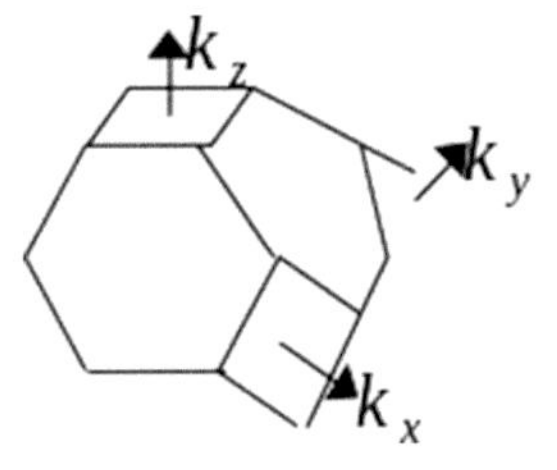

Fig. 1.2 Part of Brillouin zone|reciprocal lattice for face centered cubic (FCC) direct lattice

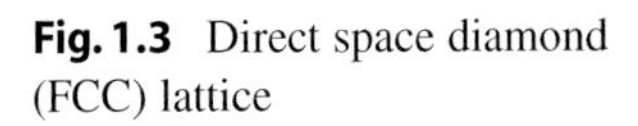
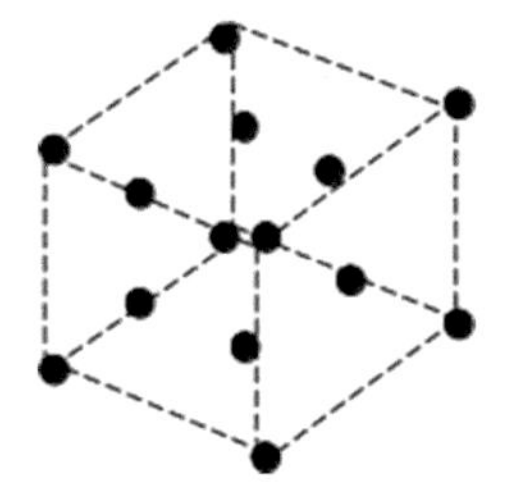

Fig. 1.3 Direct space diamond (FCC) lattice

The set [100], denotes the normal to the k_x, k_y, k_z surface that has intercepts in reciprocal space of G_1, 1, ∞ ∞. In Fig. 1.2 this direction is from the origin of k-space—point out through the center of the square surface at the X-point. Another key direction of interest is [111], which passes from to L at the center of the hexagonal faces of the reciprocal lattice unit cell. Given the complicated interaction of the direct and reciprocal lattices and their corresponding vectors the band structure for three dimensional lattice can be computed only special computer programs.

1.2.1 Crystal Momentum, Effective Mass, Negative Effective Mass

An electron's crystal momentum is due to an applied external force. A key property of n electron moving in a crystal with an applied external stimulus is its effective mass.

An electron in a one dimensional lattice is either in the conduction band or the partially filled valence band/When an external force $F_{x,ext}$, is applied, the electron gains energy from the field according to:

$$\frac{dE}{dt} = F_{x,EXTERNAL} \frac{dx}{dt} = F_{x,EXTERNAL} v_x \quad (1.24b)$$

To estimate the appropriate velocity of the electrons in a crystal using Bloch wavefunctions (whose key underlying concept is that the probability of finding an electron at some point in a primitive unit cell is the same for all of the primitive unit cells of the crystal), an accurate estimation of the electron's location in the crystal, is required. **A single wavefunction gives the electron's crystal momentum (with the wavevector k), but no precise spatial information about the electron. Linear combination of waves of slightly different k creates a *wavepacket*. A wide range of wavenumbers (k's) translates to a tightly constrained spatial wavepacket. Consequently, the electron will appear to have mass at a point, i.e., to be particle-like**. *The electron can then be considered as a classical, Newtonian mechanics like particle with a trajectory*. Thus, the velocity of the center of the wavepacket—**group velocity** is used to describe the motion of the wvepacket. Given that $E = \hbar\omega$ and using the definition of group velocity:

$$v_{GROUP} = \frac{d\omega}{dk} \quad v_x = \frac{dE}{\hbar dk_x} \quad \frac{dE}{dt} = \frac{dE}{dk_x}\frac{dk_x}{dt} \quad (1.25)$$

Combining Eqs. 1.24–1.25 and some manipulation gives:

$$F_x = \frac{d(\hbar k_x)}{dt} \quad (1.26)$$

Therefore the time rate of change of the crystal momentum is the force acting on the electron, traversing a periodic structure. *No knowledge of the mechanical momentum of the electron, (which changes periodically in response to the crystal field) is needed.* **The**

response to an external field is calculated by estimating the time dependence of the crystal momentum. The acceleration acting on the electron is:

$$a_x = \frac{dv_x}{dt} = \frac{d^2 E dk_x}{\hbar dk_x^2 dt} = \frac{d^2 E d(\hbar k_x)}{\hbar^2 dk_x^2 dt} \tag{1.27}$$

Now using Newton's force law F = ma the effective mass of the electron is then:

$$m^{EFFECTIVE}(E) = \frac{1}{\frac{d^2 E}{\hbar^2 dk_x^2}} \tag{1.28}$$

The effective mass depends on the direction. In three dimensions, systems it is a tensor. It also depends on the band structure, which in turn depends on the potential energy of the crystal. *The effective mass is not equal to the free electron mass m.*

The effective mass of the electron is positive at the bottom of bands,—where the E-k relation is concave upwards, and negative at the top of bands, where the E-k relation is convex upwards.

At the bottom of the conduction band the effective mass is positive, so a positive force causes a positive change in crystal momentum, forcing the electron to accelerate in the direction of the applied force. During this traversal, as the electron moves up the band, it passes through a crystal momentum state at which its effective mass becomes infinite and so negative. **The transition from positive to negative effective mass point occurs where the acceleration due to the external force is overcome by the increasing Bragg reflection of the Bloch waves as the Brillouin-zone boundary is approached**. *The momentum transfer from the applied force to the electron becomes less than the momentum transfer from the lattice to the electron.* However the conduction band electrons do not enter this part of the zone, and so stay near the bottom of the band, and accelerate in the direction of the applied force.

The top of the valence band is most important when analyzing the motion of charge carriers. For a net change in crystal momentum of the electrons in the valence band, empty states in the band must exist, so that the electrons can move into them. These empty states exist near the top of the band because the electrons will want to stay in their lowest possible energy states, in absence of external stimuli. The empty states near the top of the valence band are called holes. If an electron is excited into one of these empty states, an empty state will appear lower down in the band. *This energy exchange provides energy to the hole, i.e., the hole energy increases moving downwards the E-k diagram.* The hole effective mass is positive near the top of the band. So, holes accelerate in the same direction as the applied external force.

Electrons and holes have positive effective mass and charge with equal magnitude and opposite polarity. An intrinsic semiconductor with a full valence band and an empty conduction band, is electrically neutral as the electron|hole charges balance eachother. From charge balance:

$$\int -qn_{INTRINSIC,VALENCE\ BAND} + qn_{ATOM}d\Omega = 0 \tag{1.29a}$$

where q is the electronic charge and Ω the total volume. If the semiconductor is now perturbed with an with an external stimulus, e.g., thermal excitation, then, using the same charge balance concept:

$$\int -qn^{P}_{INTRINSIC,VALENCE\ BAND} + qn_{ATOM} - qn_{INTRINSIC}d\Omega = 0 \tag{1.29b}$$

$$\int q\left(-n^{P}_{INTRINSIC,VALENCE\ BAND} + n_{ATOM} - n_{INTRINSIC}\right)d\Omega = 0 \tag{1.29c}$$

$n_{INTRINSIC}$, $p_{INTRINSIC}$ are respectively the number of electrons|holes in the conduction|valence bands.

$$\int q(p_{INTRINSIC} - n_{INTRINSIC})d\Omega = 0 \tag{1.30}$$

Clearly, the polarity of electron and hole charges are different.

From the previous discussions, it is clear that the bottom|top of the E-k diagram are very important. Specifically, if the effective mass can be approximated as a constant, rather than being energy dependent, several important results can be extracted:—this occurs only when the E-k expression is parabolic. The E-k relationship for a free electron is parabolic. By analogy, for electrons near the bottom of the conduction band, and for holes near the top of the valence band the kinetic energy is:

$$E - \hat{E} = \frac{(\hbar k)^2}{2m^P} \tag{1.31}$$

where $\hat{E} > 0$ is the energy of the extremity of the approximate parabolic band, and m^P is the constant *parabolic effective mass*. The result obtained after applying this approximation to a sample E-k diagram is shown in Fig. 1.4. *The parabolic band approximation is useful for quick analysis because charge carriers in real world semiconductor devices are found in band extremity regions.*

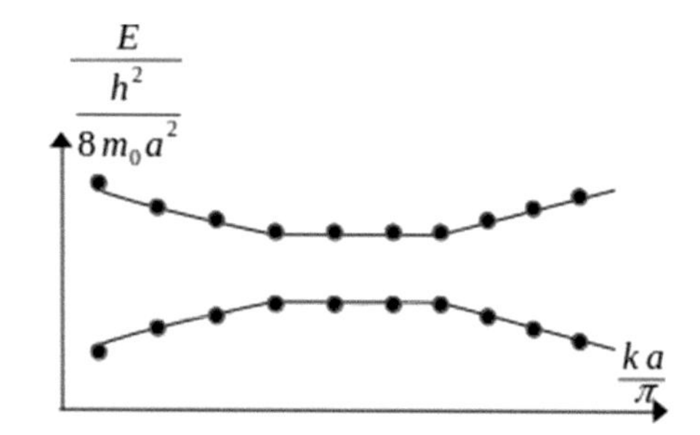

Fig. 1.4 **a** Parabolic curve fit to conduction band bottom and valence band top **b** Band structure and energy band diagram relationship for a homogeneous semiconductor in uniform electric field

Fig. 1.4 (continued)

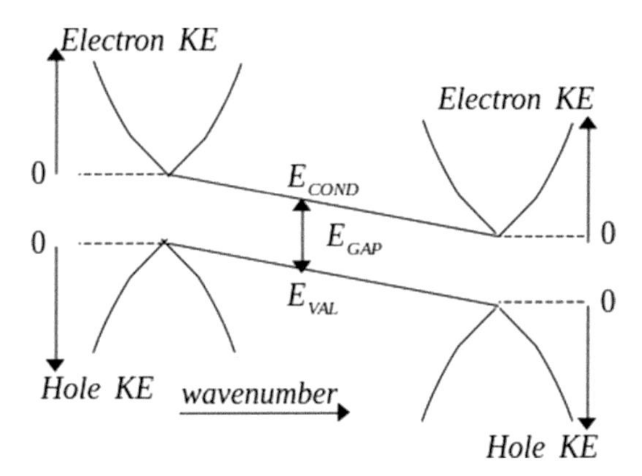

The constant effective mass idea can be extended to three dimensions leading to constant energy surfaces, where for the bottom of the conduction band:

$$E - \hat{E}_{CONDUCTION,0} = \frac{\hbar^2}{2}\left(\frac{k_x^2}{m_x^P} + \frac{k_y^2}{m_y^P} + \frac{k_z^2}{m_z^P}\right) \tag{1.32}$$

where m_x^P, m_y^P, m_z^P are the longitudinal effective electron mass, and transverse effective electron masses in the x, y and z directions respectively.

In case of the valence band, for common semiconducting materials as Si and GaAs, there are degenerate energy states, and therefore interaction between electrons and holes cannot be ignored.

$$E(k) - \hat{E}_{VALENCE,0} = \frac{\hbar^2}{2m}\left(Ak^2 \mp \sqrt{B^2k^4 + C\left(k_x^2k_y^2 + k_y^2k_z^2k_z^2 + k_z^2k_x^2\right)}\right) \tag{1.33}$$

where the negative sign corresponds to the heavy hole band and the positive sign to the light hole band. A, B and C are parameters to be determined experimentally.

1.3 Effective Mass Schrodinger's Wave Equation

Schrodinger's wave equation can be re-formulated in terms of the effective mass, when the potential energy is a superposition of individual values, due to the periodic lattice U_L and some external macroscopic, potential energy U_M, e.g., applied electric field; the potential energy due to a variation in ionized impurities in the crystal, e.g., p–n junction.

Now, Schrödinger Wave Equation becomes:

$$\frac{-\hbar^2 d^2\psi(x)}{2mdx^2} + (U_L(x) + U_M(x))\psi(x) = E\psi(x) \tag{1.34}$$

under the assumption that the conduction band potential has a parabolic form:

$$E_{\nu}(k) = E_{CONDUCTION,0} + \frac{\hbar^2 k^2}{2m^P} \tag{1.35}$$

where ν is the band index and $E_{\mathrm{CONDUCTION},0}$ is the energy of the lowest level in the conduction band. The single band effective mass conduction band Schrodinger's wave equation is:

$$\frac{-\hbar^2 d^2 F(x)}{2mdx^2} + U_M(x)F(x) = EF(x) - E_{CONDUCTION,0}F(x) \quad \psi(x) = u_{k0}F(x) \tag{1.36}$$

where F is the envelope function of the original wavefunction, and $u_k(0)$ is the periodic part of the Bloch wavefunction, evaluated at the bottom of the conduction band.

Several conditions must be satisfied for this equation to generate meaningful results.

- This holds for a single band and will have to be relaxed for the valence band, at the top of which both heavy and light holes are present in separate bands.
- u is independent of k in the neighborhood of k_0.
- F(x) varies slowly with x, i.e., when compared to the spatial variation of the lattice potential energy due to the periodicity of the crystal.
- The parabolic-band effective mass is applicable when electron energies are restricted to near the bottom of the conduction band.
- Information on the atomic scale variation of the electron concentrations is not needed, as the sum of the probability densities of all the electrons involves the envelope functions, which produce a smoothed out version of the true electron concentrations.

The effective mass Schrodinger's wave equation can be re-written as:

$$\frac{-\hbar^2 d(dF(x))}{2dx\left(m^P(x)dx\right)} + E_{CONDUCTION}(x)F(x) = EF(x) \tag{1.37}$$

where $E_C(x) = E_C(0) + U_M(x)$. The boundary condition for the derivative of F that conserves current is:

$$\left.\frac{dF_{CONDUCTION}}{m^P_{CONDUCTION}dx}\right|_{x=0} = \left.\frac{dF_{VALENCE}}{m^P_{VALENCE}dx}\right|_{x=0} \qquad F_{CONDUCTION}(0) = F_{VALENCE}(0) \tag{1.38}$$

The band structure of a semiconductor embodies information about the energy in k-space, which is conveyed by the expression for parabolic energy bands combined with the external potential energy.

$$E = \frac{\hbar^2 k^2}{2m^P} + E_{CONDUCTION,0} + U_M(x) = E_{CONDUCTION} + \frac{\hbar^2 k^2}{2m^P} \quad (1.39)$$

$E_{CONDUCTION}$ is the position dependent conduction band potential energy—*conduction band edge. These expressions contain information about the spatial variation of the lowest conduction band energy level, obtained by drawing $E_{CONDUCTION}(x)$, assuming that energies above it at any position x are the kinetic energies of electrons at that location.* The resulting plot is an energy band diagram. The relationship between it and the parabolic dispersion relationship is shown in in Fig. 1.4. The applied electric field uniform, which causes a linear change in $U_M(x)$.

1.3.1 Electron Excitation from Valence to Conduction Band—Hole Creation

An electron in the valence band can be excited to the conduction band in a number of ways. Consequently a *hole* is created in the valence band.

1.3.1.1 Thermal Excitation

For temperatures T > 0 K, the lattice atoms vibrate about their mean positions. *Sometimes, the ambient temperature is high enough to excite the electrons to a level such that the local amplitudes of vibration are sufficient to break a valence bond—the released electron enters the conduction band.* This transition occurs because the electron absorbs a phonon. **A phonon is the quantum of lattice vibrational energy**. As atoms in the lattice are bonded to their neighbours, these lattice vibrations travel through the material as waves because the random vibrations of each atom are coupled to its neighbors. The crystal lattice, being a mechanical structure, can only vibrate in specific modes.

- The atomic displacements in the direction of the atomic chain create longitudinal modes. *The associated phonons are acoustic or optic, depending on whether the displacements of neighboring atoms are in phase or out-of-phase.*
- If the atoms vibrate in the two directions perpendicular to the direction of the atomic chain, the displacements create transverse modes.

In real three dimensional crystals, the lattice periodicity results in band structure of the phonons that have properties similar to that of electrons. For cubic lattice or cubic lattice like materials the two transverse modes are the same—doubly degenerate. Other factors as atomic mass etc., also influence phonon band structure.

1.3.1.2 Optical Excitation

Electron–hole pairs in a crystal lattice can be generated when the crystal absorbs optical energy at appropriate frequencies. The energy and momentum balance equations are:

$$E_{electron}\left\langle 1^P\right\rangle - E_{hole}\langle 1\rangle = \hbar\omega_{photon} + \sum\left(\hbar\omega_{photon} - \hbar\omega_{electron}\right) \tag{1.40a}$$

$$\hbar\left(k_{electron}\left\langle 1^P\right\rangle - k_{hole}\langle 1\rangle\right) = \hbar\left(k_{photon} + \sum\left(\beta_{phonon} - \beta_{electron}\right)\right) \tag{1.40b}$$

where $\beta_{phonon}, \beta_{electron}$ are the wavenumbers of the generated phonons and electrons respectively. **Phonons need to be generated to excite the electrons from the valence to the conduction band**. The photon wavenumber is:

$$k_{photon} = \frac{En_r}{c\hbar} \tag{1.41}$$

where c, E, n_r are the speed of light in vacuum, energy and refractive index of the crystalline material. In a direct bandgap (*maximum of valence band lies vertically below minimum of conduction band*) material, interband electron transitions at energies near the bandgap need not involve any momentum change, so phonons need not be involved. So photon absorption in direct bandgap materials occur more readily than in indirect bandgap materials. But as silicon is widely used in solar cells and photodetectors indicates that absorption can be strong in indirect bandgap materials as well. For photon energies near the bandgap, the phonons do not need to have much energy, and there are many acoustic phonons of this type available.

1.3.1.3 Electrical Excitation

Electrical generation of electron–hole pair is achieve when a highly energized electron from an external source hits an atom in the lattice. Energy transfer from the incident electron to the target atom occurs, and this extra gained energy sometimes energizes some of the valence band electrons to acquire sufficient energy to enter the conduction band. This is called impact ionization. The energy and momentum balance equations are:

$$\begin{aligned} E_{electron}\left\langle 1^P\right\rangle + E_{electron}\left\langle 2^P\right\rangle &= E_{electron}\langle 1\rangle + E_{hole}\langle 2\rangle \\ k_{electron}\left\langle 1^P\right\rangle + k_{electron}\left\langle 2^P\right\rangle &= k_{electron}\langle 1\rangle + k_{hole}\langle 2\rangle \end{aligned} \tag{1.42}$$

Electrons are confined to the lower energies of the conduction band by frequent collisions with atoms, defects, and any impurities in the lattice. If an electron is rapidly accelerated by an applied electric field, it could gain sufficient energy between collisions for impact ionization to occur. If the applied electric field is very strong, it could trigger avalanche breakdown.

1.3.1.4 Chemical Excitation

Unlike all the previous electron–hole pair excitation methods, **the chemical excitation method is permanent and is the basis for all semiconductor devices**. *If some of the lattice atoms in a homogeneous material lattice are substituted with some impurity atoms so* *that the resulting lattice has an excess of electrons the resulting material is a n type (electron rich) type semiconductor. Conversely, substitution of intrinsic lattice atoms with impurity atoms such that the resulting material has an excess of holes results in a p type semiconductor.* Removing an atom from the lattice is difficult, so that high temperature diffusion or most widely used ion implantation of the impurities is needed. The impurity atom must have morelless electrons in its valence band than the atom it is replacing. The extra electrons left over after the impurity atom is embedded in the lattice are free to enter the conduction band—n type semiconductor. Similar arguments hold for a p type semiconductor. *A donor impurity atom 'donates' an electron, while an acceptor atom removes an electron from the intrinsic lattice.*

1.4 Recombination

An external stimulus can create an electron–hole pair. Similarly, the excited electron in the conduction band can lose energy by radiating it and then recombine with holes in the valence band—*recombination.*

1.4.1 Band-Band Recombination

Band-band recombination is the inverse of optical generation, for a direct bandgap material. The energy lost by the electron is usually radiated as a light photon. No phonons are generated, so that the lattice is not heated up This type of recombination is called radiative recombination. While the electron in the conduction band is waiting for a large number of phonons to be simultaneously emitted, as would be required for a direct band-to-band transition (e.g., gallium arsenide), a single photon is created instead. The recombination rate is:

$$R_{RADIATION} = Bnp \tag{1.43a}$$

where B, n, p are respectively the radiative recombination coefficient, the concentrations of electrons and holes. For an indirect bandgap material such as Si, $B \approx 10-14\ \frac{cm^3}{s}$. Radiative recombination is much less likely to occur in indirect bandgap materials because phonons need to be generated as well, to take up the change in crystal momentum. The energy and momentum relations are:

$$\begin{aligned} E_{electron}\langle 1\rangle - E_{electron}\left\langle 1^P\right\rangle &= \hbar\omega_{photon} \\ k_{electron}\left\langle 1^P\right\rangle &= k_{electron}\langle 1\rangle \end{aligned} \tag{1.43b}$$

1.4.2 Recombination-Generation Center Recombination

Recombination-generation (RG)-center recombination is a result of crystal imperfections. These imperfections include missing atoms from their regular lattice sites, lattice distortions in the vicinity of impurity atoms. As a perfect periodicity gives rise to energy bands and bandgaps, the presence of crystalline defects disturbs the periodicity of the structure, resulting in localized energy levels within the bandgap. In contrast to the localized levels of donors and acceptors, the energy levels of defects and unwanted impurities are not confined to energies close to the band edges. These defect induced localized energy levels are distributed throughout the bandgap, which facilitate the recombination process by providing temporary (metastable) states for electrons. Each transition involves fewer phonons than required for band-band recombination in one step. Clearly, these crystal imperfection induced energy levels occur near the middle of the bandgap. *These are most effective as recombination centers because the probability of electron capture from the conduction band is the same as the probability of hole capture from the valence band—Shockley–Read–Hall recombination process.*

The rate of recombination depends on the presence of electrons, holes and traps. However, the rate limiting step will be the capture of the minority carrier by the trap: the capture of the majority carrier is far more probable. Thus, for p-type material the recombination rate is:

$$R_{GENERATION,\,RECOMBINATION} \simeq nN_T r \equiv An \quad (1.43c)$$

where r is the temperature dependent rate constant, N_T is the trap concentration and A is the trap dependent recombination coefficient. The energy and momentum balance equations are:

$$\begin{aligned} E_{electron}\langle 1\rangle - E_{hole}\left\langle 1^P\right\rangle &= \sum \hbar\omega_\beta \\ \hbar\left(k_{electron}\langle 1\rangle - k_{hole}\left\langle 1^P\right\rangle\right) &= \hbar \sum \beta \end{aligned} \quad (1.43d)$$

1.4.3 Auger Recombination

This is the reverse of the impact ionization electron–hole creation process. The energy and momentum produced by the electro, hole recombination is transferred to either a second electron or hole. These excited carriers then lose their energy and momentum by emitting phonons. The emission process occurs as there an abundance of available states in the bands. The charge carrier can easily transfer with changes in momentum and energy that are compatible with the phonon dispersion relationship. When the excited carrier is an electron, the energy and momentum-balance equations are:

$$E_{electron}\langle 1 \rangle - E_{hole}\left\langle 1^{P} \right\rangle = \sum \hbar\omega_{\beta}$$
$$\hbar\left(k_{electron}\langle 1 \rangle - k_{hole}\left\langle 1^{P} \right\rangle\right) = \hbar \sum \beta \tag{1.44a}$$

This recombination occurs when one of the carrier concentrations is very high, yet both carriers are of importance, such as in solar cells. Similar carrier concentrations exist in the space charge region of LEDs, and in the base of heterogeneous junction bipolar transistor(HBT)s. Each recombination event, involves one carrier of one type and one carrier of the complementary type. The recombination rate for the sum of the process. is the corresponding process involving the excitation of a hole:

$$R_{AUGER} = np(Cn + Dp) \tag{1.44b}$$

where C, D are Auger recombination coefficients.

1.4.4 Recombination Lifetime

The environment inside a semiconductor device is dynamic. Recombination is always occurring, and at equilibrium, recombination rate must equal electron–hole pair generation rate. The net rate of recombination out of equilibrium is given, by the difference between the non-equilibrium and the equilibrium-recombination rates. This net rate is often denoted as U and has units of per unit time. For radiative recombination,

$$U_{RAD} = B(np - n_0 p_0) \tag{1.45a}$$

where n_0, p_0 are thermal equilibrium values. For a set of Δn generated electron hole pairs, the recombination rate is proportional to:

$$np - n_0 p_0 = (n_0 + \Delta n)(p_0 + \Delta n) \ - \ n_0 p_0 \tag{1.45b}$$

which can be re-written as:

$$np - n_0 p_0 = \Delta n^2 \ + \ \Delta n(n_0 \ + p_0) \tag{1.45c}$$

Equation 1.45c, modified for p-type semiconductor and low|high injection cases respectively is:

$$np - n_0 p_0 = \Delta n^2 \ + \ \Delta n p_0 \ = \ \Delta n p_0 \tag{1.45d}$$

The above set of expressions show how the general case can be modified to the p type semiconductor ($\Delta n = \Delta p, \quad n_0 \ll p_0$), and then low level|high injection. The net electron recombination rate is:

$$U_{electron,RADIATION} = \frac{\Delta n}{\tau_{electron,RADIATION}} \tag{1.45e}$$

where $\tau_{e,RAD}$ is the electron minority carrier radiative recombination lifetime. The expressions for the various recombination process lifetimes are:

$$\tau_{eletron,RADIATION} = \frac{1}{Bp_0} \vee \frac{1}{B(p_0 + \Delta n)} \tag{1.45f}$$

recombination-generation center:

$$\tau_{electron,GENERATION-RECOMBINATION} = \frac{1}{A} \tag{1.45g}$$

For Auger recombination in p type material high level injection:

$$\tau_{electron,AUGER} = \frac{1}{C(\Delta n p_0 + n_0 p_0 + \Delta n^2)} + \frac{1}{D(2\Delta n p_0 + p_0^2 + \Delta n^2)} \tag{1.46h}$$

overall

$$\frac{1}{\tau_{electron}} = \frac{1}{\tau_{electron,GENERATION-RECOMBINATION}} + \frac{1}{\tau_{electron,RADIATION}} + \frac{1}{\tau_{electron,AUGER}} \tag{1.46i}$$

1.5 Carrier Concentrations [9–16]

The electron concentration at position $\vec{r}$ and inside a volume Ω is:

$$n(\vec{r}, t) = \frac{1}{\Omega} \sum filled\ states \tag{1.47a}$$

To convert this sum into an integral two key points must be noted.

- Separation over all momentum space k between states in one dimension is 2π/L, where L is the length of one dimensional crystal lattice.
- Each momentum state in k space can contain two electrons with opposite spins (Pauli Exclusion Principle).
- An integral over all k states, including empty ones requires a distribution function to account for the probability of a particular state being occupied

$$n(\vec{r}, t) = \frac{1}{\Omega} \sum filled\ states = \frac{1}{\Omega} \frac{1}{\frac{2\pi}{\Omega}} \int f\left(\vec{r}, \vec{k}, t\right) d\vec{k} \tag{1.47b}$$

This general equation is not useful. So, the electron concentration in the three dimensional, time-independent case and to simplify the notation, the position identifier is removed. The above equation becomes:

$$n = \frac{1}{4\pi^3} \int f\left(\vec{k}\right) d\vec{k} \tag{1.47c}$$

The electrons fill up states from the origin to higher values of momentum (corresponding to higher values of wavevector $\vec{k}$) and as the number of available states is huge, the approximate surface of the actual volume of k-space being filled is a sphere of radius k, i.e., *the distribution function f depends only on the magnitude of k*. Therefore, in spherical coordinates:

$$n = \frac{1}{4\pi^3} \iiint f(\vec{k}) k^2 \sin(\theta) d\theta d\phi dk \quad 0 \le \theta \le \pi \quad 0 \le \phi \le 2\pi \tag{1.47d}$$

This integral evaluates to:

$$n = \frac{1}{4\pi^3} \int 4\pi^2 f(k) dk = \frac{k^2 f(k)}{\pi^2} = g(k) f(k) \tag{1.47e}$$

here **g(k)** us the density of states in the k space, with units of $\frac{1}{km^3}$. As it is easier to measure energy than wavenumbers, the above integral for the density of states is re-written as:

$$\int g(E - E_{CONDUCTION}) d(E - E_{CONDUCTION}) = \int g(k) d(k) \tag{1.47f}$$

To evaluate closed form analytical expressions, the parabolic band approximation with the density of states effective mass is used, for which the kinetic energy of an electron is:

$$E - \hat{E}_{CONDUCTION} = \frac{\hbar^2}{2}\left(\frac{k_x^2}{m_x^P} + \frac{k_y^2}{m_y^P} + \frac{k_z^2}{m_z^P}\right) = \frac{\hbar^2 k^2}{2m^P_{DENSITY\ OF\ STATES}} \tag{1.47g}$$

For materials as gallium arsenide, the constant-energy surfaces are spherical for electrons, the density of states effective mass is the parabolic-band effective mass. In such cases. the above equations, after some manipulation give the following equations for the energy dependent conductance and valence band density of states expressions.

$$g_{CONDUCTION}(E - E_{CONDUCTION}) = \frac{8\sqrt{2}}{\hbar^3}\sqrt{E - E_{CONDUCTION}} \left(m^P_{electron,DENSITY\ OF\ STATES}\right)^{\left(\frac{3}{2}\right)} \quad E \ge E_{CONDUCTION}$$

$$g_{VALENCE}(E_{FERMI} - E) = \frac{8\sqrt{2}}{\hbar^3}\sqrt{E_{FERMI} - E}$$

$$\left(m^{P}_{electron,DENSITY\ OF\ STATES}\right)^{\left(\frac{3}{2}\right)} \quad E_{FERMI} \geq E \qquad (1.47\text{h,i})$$

In the densities of states in three dimensional real space increase parabolically with energy away from the band edges. After some manipulation of the above equations, the electron and hole concentrations (unit volume) are:

$$n = \int g_{CONDUCTION} f(E)dE \quad p = \int g_{VALENCE}(1 - f(E))dE \qquad (1.47\text{j})$$

For silicon, the energy of electrons in the conduction band is:

$$E - E_{CONDUCTION,0} = \frac{\hbar^2}{2}\left(\frac{k_x^2}{m^{P}_{electron,longitudinal}} + \frac{k_y^2 + k_z^2}{m^{P}_{electron,transverse}}\right) \qquad (1.47\text{k})$$

where $m^{P}_{electron,longitudinal}$, $m^{P}_{electron,transverse}$ are longitudinal|transverse electron effective masses. The electron density of states effective mass is:

$$m^{P}_{electron,DENSITYOFSTATES} = 6.0^{0.66}\left(m^{P}_{electron,longitudinal}\left(m^{P}_{electron,transverse}\right)^2\right)^{0.66} \qquad (1.47\text{l})$$

For holes, for both silicon and gallium arsenide, the density of states effective mass includes both light|heavy hole effective masses.

$$m^{P}_{hole,DENSITYOFSTATE} = \left(\left(m^{P}_{hole,heavy}\right)^{1.5} + \left(m^{P}_{hole,light}\right)^{1.5}\right)^{0.66} \qquad (1.47\text{m})$$

1.6 Thermal Equilibrium and Fermi–Dirac Statistics

The expression for the density of states for the valence band contains the term for the Fermi level. A semiconductor under idealized thermal equilibrium is analyzed to understand the Fermi level. *Under thermal equilibrium there are no external sources of energy to excite electrons except for a constant temperature heat source.*

1.6.1 Collisions and Scattering

Generation and recombination processes control the carrier concentrations in the conduction and valence bands. Under thermal equilibrium, the thermally activated band-to-band and chemical generation processes are active, with possibly one|all of the others: (radiative, recombination-generation center, Auger). **For thermal equilibrium, the net sum of these various generation|recombination rates must be zero. This means that as energy**

(typically heat) is taken from the device to create an electron–hole pair, so that energy must be returned to the lattice on recombination—energy balance. Under thermal equilibrium charge carriers in must remain in their lowest energy states within each band. *That is, charge carriers continuously collide with atoms|ions|crystalline defects and dissipate energy obtained from the external heat source.*

If an electron is a wave a collision between it and a lattice atom would be the interaction of the electron wave with a phonon. *As the atoms vibrate about their mean positions with a temperature dependent amplitude, lattice periodicity is disturbed*. Then the potential in the system non-periodic. But, if the potential perturbations are small, the electron wavefunctions can be approximated with Bloch wavefunctions, with different values from those for a stationary, periodic lattice—both eigenvalues|eigenfunctions change. **Physically the state of the electron will change on interaction with a vibrating atom i.e., the electron scatters to a new state on interacting with an atom|electron|ion|another|lattice defect**.

At low temperatures, when the atomic thermal vibrations are small, **ionized impurity scattering** is the dominant scattering mechanism. If there are many carriers present, they can screen other carriers from the attractive or repulsive effect of the impurity ion's charge, leading to **screened ionized-impurity scattering**. If the concentration of carriers is extremely large, carrier-carrier scattering becomes important. In doped semiconductors at higher temperatures, **phonon scattering** dominates. Longitudinal acoustic phonons alternately compress and dilate the lattice; the resulting strain deforms the band edges, leading to **deformation potential scattering**. In semiconductors with no crystal inversion symmetry, e.g.,, gallium arsenide, the strain may generate a potential via the piezoelectric effect, leading to **piezoelectric scattering**. Silicon and GaAs both have two atoms per unit cell, so when these atoms move in opposite directions, optic phonons are generated, leading to **optical phonon scattering**.

1.6.2 Fermi–Dirac Statistics and the Fermi Level

The Fermi Dirac statistics is the most probable distribution of electrons among the available states in the conduction and valence bands. This probability distribution function is evaluated by considering one of the types of collisions (elastic, electron–electron) responsible for maintaining thermal equilibrium.

Two electrons have respectively energies $E_1,\ E_2,\ E_1^P,\ E_2^P$ before and after an elastic collision. Then:

$$E_1^P + E_2^P = E_1 + E_2 \tag{1.48a}$$

This collision occur only when there are filled states at energies $E_1,\ E_2$, and empty states at $E_1^P,\ E_2^P$. The second requirement is essential to satisfy Pauli's Exclusion Principle. The probability of occupancy is denoted as f(E), and the rate of this collision:

$$r_{1,1^P:22^P} = cf(E_1)f(E_2)\left(1 - f\left(E_1^P\right)\right)\left(1 - f\left(E_2^P\right)\right) \tag{1.48b}$$

where C is the rate constant. **Thermal equilibrium is preserved anywhere in the lattice only when electrons with starting energies E_1^P, E_2^P, and ending energies E_1, E_2 collide with the collision rate stated in Eq. 1.48. No collision occurs without filled states at the starting energies and empty states at the ending energies**. This means that:

$$r_{1^P,1:2^P,2} = cf\left(E_1^P\right)f\left(E_2^P\right)(1 - f(E_1))(1 - f(E_2)) \tag{1.48c}$$

These two collision rates must be the same, by equating the two expressions and dividing the result by the product is the probability distribution functions at the four energy values gives:

$$\left(\frac{1}{f(E_1)} - 1\right)\left(\frac{1}{f(E_2)} - 1\right) = \left(\frac{1}{f\left(E_1^P\right)} - 1\right)\left(\frac{1}{f\left(E_2^P\right)} - 1\right) \tag{1.48d}$$

A solution to this expression is of the form:

$$\frac{1}{f(E)} - 1 = Ae^{\beta E} \tag{1.48e}$$

To calculate β, a large energy value E is examined. *Selecting a a large value of E ensures that there is a very small probability of quantum states at this energy being filled by electrons.* Therefore:

$$f(E) = \frac{e^{-\beta E}}{A} \ll 1 \tag{1.48f}$$

Then the probability of two electrons occupying the same state, is infinitesimally small (Pauli's Exclusion Principle)—i.e., *these electrons are non-interacting*. Exploiting some basic thermodynamics concepts, the electron distribution function under conditions of thermal equilibrium is:

$$f_{FERMI,DIRAC}(E) = \frac{1}{1 + e^{\frac{E - E_{FERMI}}{k_B T}}} \tag{1.48}$$

where E_F, k_B, T are respectively the Fermi energy level, Boltzmann constant and absolute temperature in Kelvin.

The Fermi Dirac distribution is the probability of an electron occupying a state of energy E at thermal equilibrium because it varies in value from 0 to 1. The probability of occupancy of states with energy E falls off rapidly as E exceeds the Fermi level E_F. In n-type semiconductor material there are many electrons that must be accommodated in the conduction band. So states will be filled-up to higher energies than in the case of p-type material. In p-type material, there are few electrons. For higher energy states to have a higher probability of being filled, the Fermi level must be raised, i.e., *the Fermi level is higher in energy for n-type material than it is for p-type.*

1.6.3 The Fermi Level and Equilibrium Carrier Concentrations

Using the Fermi Dirac distribution, the thermal equilibrium intrinsic electron, hole concentrations are:

$$n_0 = \int g_{CONDUCTION}(E) f_{FERMI,DIRAC}(E) dE \quad E_{CONDUCTION} \leq E \leq BANDTOP$$
$$p_0 = \int g_{VALENCE}(E)\left(1 - f_{FERMI,DIRAC}(E)\right) \quad BANDBOTTOM \leq E \leq E_{FERMI}$$
(1.49a,b)

The limits of integration can be taken to be $\pm$ infinity, without adding any error. This is because the Fermi Dirac distribution function evaluates to zero very high in the conduction band and likewise very large in the valence band at the top, so that $1 - f_{FD}$ goes to zero. Then the equilibrium electron concentration becomes:

$$n_0 = N_{CONDUCTION} F_{\frac{1}{2}}(a_F) \tag{1.49c}$$

where $F_{\frac{1}{2}}(a_F)$ is the Fermi–Dirac integral of order half. The values of this integral is tabulated in standard look up tables. In special cases, when e.g., a< -2, the integral can be approximated as:

$$F_{\frac{1}{2}}(a_F) \rightarrow e^{a_F} \tag{1.49d}$$

In the above integral, the material specific constant terms are combined into $N_C N_C$

$$N_{CONDUCTION} = 2\left(\frac{2m^P_{electron,DENSITY\ OF\ STATES} k_B T}{\hbar^2}\right)^{\frac{3}{2}} \tag{1.49e}$$

Now combining all these results together, the equilibrium electron concentration is:

$$n_0 = N_{CONDUCTION} e^{\frac{E_{FERMI} - E_{CONDUCTION}}{k_B T}} \tag{1.49f}$$

This is the Maxwell Boltzmann equation. The corresponding expression for the holes is:

$$p_0 = N_{VALENCE} e^{\frac{E_{VALENCE} - E_{FERMI}}{k_B T}} \tag{1.49g}$$

The above expressions can be modified to provide information about intrinsic carrier concentrations in a semiconducting material.

$$n_{INTRINSIC} = N_{CONDUCTION} e^{\frac{E_{FERMI,INTRINSIC} - E_{CONDUCTION}}{k_B T}}$$
$$p_{INTRINSIC} = N_{VALENCE} e^{\frac{E_{VALENCE} - E_{FERMI,INTRINSIC}}{k_B T}} \tag{1.49h}$$

where $E_{GAP} = E_C - E_V$/ In an intrinsic semiconductor, $n_{INTRINSIC} = p_{INTRINSIC}$ so that the above two expressions can be equated, to give the intrinsic electron concentration:

$$n_{INTRINSIC} = \sqrt{N_{CONDUCTION} N_{VALENCE} e^{\frac{-E_{FERMI}}{2k_B T}}} \tag{1.49i}$$

The intrinsic Fermi level and the valence band energy are related as:

$$E_{FERMI,INTRINSIC} - E_{VALENCE} = \frac{E_{GAP}}{2} + \frac{3k_B T}{4} \ln\left(\frac{m^P_{hole,DENSITY\ OF\ STATES}}{m^P_{electron,DENSITY\ OF\ STATES}}\right) \tag{1.49j}$$

From the above expressions it is clear that the intrinsic concentration is a temperature-dependent material constant, and that the intrinsic carrier concentration decreases as the bandgap increases, independent of the Fermi energy:

$$n_0 p_0 = n_{INTRINSIC} p_{INTRINSIC} = n^2_{INTRINSIC} \tag{1.49k}$$

These expressions hold for moderately doped, non-degenerate semiconductors only. The charge balance expression is:

$$q(p_0 + N_{DONOR} - n_0 - N_{ACCEPTOR}) = 0 \tag{1.49l}$$

The mean velocity of an equilibrium distribution can be estimated from information presented above. The analysis exploits the "velocity-sphere" approach to estimate the average x-directed velocity. The radius v_{th} is the mean thermal speed. Electron velocities are randomly distributed in direction and the number of electrons with velocities in the range θ to θ + dθ is related to the area of a strip of length $2\pi v_{th} \sin(\theta)$ and width $v_{th} d\theta$. To express this as a fraction of the total number of electrons, divide by $4\pi v^2_{th}$ to get $\frac{\sin(\theta) d\theta}{2}$ and then integrate over $0 \le \theta \le \frac{\pi}{2}$. The result is the mean unidirectional velocity.

The equilibrium carrier concentration for electrons is:

$$n_0(E) = g_{CONDUCTION}(E) f_0(E) \tag{1.49m}$$

which after simplification with the Maxwell–Boltzmann distribution is:

$$f_{MAXWELL-BOLTZMANN} = \frac{n_0}{N_E} e^{\frac{E_{CONDUCTION} - E}{k_B T}} \tag{1.49n}$$

The result for $n_0 = \frac{10^{19}}{cm^3}$ is shows that the distribution is split into two. The distribution on the right is for the electrons with positive crystal momentum, i.e., velocity component in the positive x-direction. The part on the left is its oppositely directed or complementary part. Each half of the distribution contains exactly $\frac{n_0}{2}$ electrons, and is termed a *hemi-Maxwellian* (**hemispherical Maxwellian**). At any plane in the material, the flow of electrons in the positive going distribution will be opposed by the flow of negative going electrons from a neighboring distribution. So there is no net current. However, if the conditions in a device were near-equilibrium, due to a non-uniform doping density, then the opposing charge flows at an intermediate plane would not cancel, resulting in current-diffusion. Current flow due to a near-equilibrium hemi-Maxwellian or hemi-Fermi–Dirac (**hemispherical Fermi–Dirac**) can arise due to injection of carriers into the base of an HBT(Heterogeneous Bijunction Transistor) or into the channel of a FET. To compute the current, the mean speed of the electrons in the total distribution in thermal equilibrium is needed—given by:

$$v_{mean,thermal\ equilibrium} = \frac{-\int n_0(v)v dv}{\int n_0(v) dv} \quad 0 \le v \le \infty \tag{1.49o}$$

where $n_0(v)$ is the *number of electrons per volume per unit velocity*. The denominator is the total electron concentration. Recognizing that:

$$E - E_{CONDUCTION} = \frac{m^P_{CONDUCTIVITY} v^2}{2} \tag{1.49p}$$

the expression for the equilibrium thermal mean velocity is, after some manipulation:

$$v_{mean,thermal\ equilibrium} = \sqrt{\frac{8k_B T}{m^P_{CONDUCTION}} \frac{F_1}{F_{\frac{1}{2}}}} \tag{1.49q}$$

Mean drift velocity and drift current relation Is simple:

$$\vec{J}_{electron,PLUS} = -qn_0 v_R \hat{x} \quad \vec{J}_{electron,MINUS} = -qn_0 v_R(-\hat{x}) \tag{1.49r}$$

where ‘PLUS’ and ‘MINUS’ refer to the positive and negative x directions.

References

1. Griffiths, D. J., & Schroeter, D. F. (2019). *Introduction to quantum mechanics.* Cambridge University Press.
2. Feynmann, R. P., Leighton, R. B., & Sands, M. L. (1989). *The Feynman Lectures on Physics* (Vol. 1). Addison Wesley.
3. Susskind, L., & Friedman, A. (2015). *Quantum mechanics: The theoretical minimum.* Amazon Books.
4. Sakurai, J. J., & Napolitano, J. (2020). *Modern quantum mechanics.* Cambridge University Press.
5. Nouredine, Z. (2009). *Quantum mechanics: Concepts and applications.* Wiley.

6. Shankar, R. (2012). *Principles of quantum mechanics*. Springer Science & Business Media.
7. Matthews, P., & Venkatesan, K. (1978). *A textbook of quantum mechanics*. Tata McGraw Hill.
8. Huang, K. *Introduction to statistical physics* (2nd ed.). ISBN 10 9781420079029; 13 978–1420079029.
9. Reichl, L. E. (2016). *A modern course in statistical physics*. Wiley VCH. ISBN 10 3527413499; 13 978-3527413492.
10. Landau, L. D., & Lifschitz, E. M. *Statistical physics, Part 2, Course of Theoretical Physics* (3rd ed., Vol. 5). ISBN 10 0750633727; 13 978-0750633727.
11. Beale, P. D. *Statistical mechanics* (3rd ed.). ISBN 10 0123821886; 13 978–0123821881.
12. Kittel, C., & Kroemer, H. *Thermal physics* (2nd ed.). ISBN 10 0716710889; 13 978-0716710882.
13. Griener, W., Neise, L., Stocker, H., & Rischke, R. *Thermodynamics and statistical mechanics (Classical theoretical physics)*. ISBN 10 0387942998; 13 978-0387942995.
14. Sze, S. M., & Ng, K. K. *Physics of semiconductor devices*. Copyright © 2007 John Wiley & Sons, Inc. All rights reserved. Print ISBN 9780471143239, Online ISBN 9780470068328. https://doi.org/10.1002/0470068329
15. Streetman, B., & Banerjee, S. K. (2013). *Solid state electronic devices: Pearson new international edition*. Pearson Education Limited. ISBN 10 1292025786; 13 9781292025780.
16. Pullfrey, D. (2013). *Understanding modern transistors and diodes*. Cambridge University Press. ISBN 13 978-0521514606.

Charge Transport (Current Flow) in Semiconductors

2

2.1 General Concepts

From the previous chapter, electrons and holes move through the semiconductor in momentum states. These carriers are both continually generated, annihilated and scattered to new momentum states (collisions with vibrating atoms, ionized impurities, inter carrier scattering etc.,). In thermal equilibrium large charge carrier fluxes are present, but no net current flows. *This equilibrium is broken to get a non-zero net flow of charge* ***only when*** *a driving force within the semiconductor is introduced, in the form of an energy gradient. This energy gradient is related to the electrostatic potential* ψ, *as well as a kinetic energy gradient.* If each of the n electrons per unit volume have a kinetic energy u, the total kinetic energy density W is *nu*. *This is dissipative transport, i.e., the directed momentum of electrons injected into a region from a hemi-Maxwellian distribution, is dissipated by scattering events.* If the region is so short that the injected electrons can traverse it without scattering, then it is ballistic transport. Ballistic transport occurs in tunnelling nanoscale devices.

2.2 Relation Between Charge, Current and Energy [1–8]

The fundamental definitions of charge, current and energy densities can be obtained using very simple physical concepts. To simplify the analysis, parabolic bands with an isotropic effective mass, is assumed. The electron charge density is:

$$-qn = \frac{-q}{4\pi^3}\int f\left(\vec{k}\right)d\vec{k} \tag{2.1}$$

Defining the x direction velocity as $v_x = \frac{dE}{\hbar dk_x}$ the x direction current flux is:

A. Banerjee, *Semiconductor Devices*, Synthesis Lectures on Engineering, Science, and Technology, https://doi.org/10.1007/978-3-031-45750-0_2

$$\vec{J}_{electron,x} = \frac{-q}{4\pi^3} \int v_x f\left(\vec{k}\right) d\vec{k} \tag{2.2}$$

and the corresponding energy density is:

$$W_{electron,x} = \frac{1}{8\pi^3} \int m^P_{electron} v_x^2 f\left(\vec{k}\right) d\vec{k} \tag{2.3}$$

which includes the kinetic energy term $(\frac{m^P v_x^2}{2})$. The distribution function $f\left(\vec{r}, \vec{k}, t\right)$, that accurately describes the non-equilibrium conditions necessary for a net current, is a difficult to determine analytically. So an equation is formulated that describes how the distribution function changes in both real space and in k-space.

2.3 The Boltzmann Transport Equation

The full six dimensional $\vec{r}, \vec{k}$ space is called phase space [1–7]. The following analysis uses the two dimensional sub-space x, k_x. The crystal momentum $\hbar k$ is the momentum due to forces other than the periodic crystal forces. These external forces determine the functional form of $f(x, k_x, t)$ as the probability of an electron at position x having a momentum $\hbar k_x$ at time t. The subsequent analysis uses the semi-classical model of the electron, since by Heisenberg's Uncertainty Principle, it is impossible to determine the position and momentum of an electron simultaneously. The classical electron is compatible with the effective-mass formalism in which the group velocity of a wave packet is involved. Group velocity introduces elements of quantum mechanics.

An applied force F_x acts on an electron which moves a distance $v_x \Delta t$ in time Δt where the x-direction velocity is $v_x = \frac{dE}{\hbar dk_x}$. The electron's momentum changes, $F_x = \hbar \frac{\Delta k_x}{\Delta t}$. As the electron moves from one state in to another state in the same phase space, the probability of occupancy of these two states must be equal.

$$f(x, k_x, t) = f\left(x + \Delta x, k_x + \frac{F_x \Delta t}{\hbar}, x + \Delta x\right) \tag{2.4}$$

In the limit $\Delta t \to 0$ Eq. 2.4 is expanded in a Taylor's series and re-written as:

$$\frac{\partial f}{\partial t} = -\left(\frac{v_x \partial f}{\partial x} + \frac{\partial F_x \partial f}{\partial x \partial k_x}\right) \tag{2.5}$$

The external force is due to an applied electric field E_x which makes the electrons collide with a force $F_{x,coll}$, so that:

$$\left.\frac{\partial f}{\partial t}\right|_{collision} = \frac{-\partial f F_{x,collision}}{\hbar \partial k_x} \tag{2.6}$$

Now, combining all the forces gives the Boltzmann Transport equation:

$$\frac{\partial f}{\partial t} = \frac{-v_x \partial f}{\partial x} + \frac{q \partial E_x \partial f}{\hbar k_x} + \left.\frac{\partial f}{\partial t}\right|_{collision} \tag{2.7}$$

The method of, moments is used to solve this equation indirectly, so that useful expressions for charge density, current density and kinetic-energy density can be obtained without very complicated mathematical manipulations. *The method of moments consists of multiplying each term of the above equation by some factor Υ and integrating over $\frac{d\vec{k}}{4\pi^3}$. The velocity v_x is included in Υ and the resulting moments are labelled according to the power to which v_x is raised. The goal is to solve for: $-qn$ using the 0th order moment, $\vec{J}_{electron}$ using the first order moment, followed by solving for $W_{elextron}$ using the second order moment.*

Introducing a dummy variable Φ that can represent any of the quantities $-qn$, $\vec{J}_{electron}$, $W_{electron}$ etc., depending on moment order, the Boltzmann Transport equation is transformed to:

$$\frac{\partial \Phi}{\partial t} + \nabla \cdot \vec{J}_\Phi = G_\Phi - R_\Phi \tag{2.8}$$

where the flux of Φ is $\vec{J}_\Phi$, and G_Φ, R_Φ respectively the generation|annihilation rates of Φ. The order of the terms above is different from that in the original equation, and is in three dimensional form to distinguish between divergence and gradient. The reformulated transport equation has a clear physical interpretation: the difference between generation and loss of Φ in a volume increases it with time, or outward flow.

2.3.1 Method of Moments Solution of Boltzmann Equation—Continuity Equations [1–11]

The first three generated moments, from the Boltzmann Transport equation contain the required key information and are examined in detail. *For charge density, there is no generation or loss term. The electric field does not generate any charge, and there is no scattering loss: the energy loss from collision is redistributed amongst other allowed momentum states. The charge density* **continuity equation** *is*:

$$\frac{\partial n}{\partial t} = \frac{\partial J_{electron}}{q \partial x} \tag{2.9}$$

The current density continuity equation exploits the physical fact that current is generated by the external applied electric field, and lost due to scattering—momentum loss. The momentum loss is characterized by the momentum relaxation time τ_M. The time dependency of the current, after removal of the external applied field, must be of the form $\vec{J}(t) = \vec{J}(0)e^{\frac{-t}{\tau}}$, so that the current density variation with time is:

$$\frac{d\vec{J}}{dt} = \frac{-\vec{J}(t)}{\tau_M} = -R_\Phi \tag{2.10}$$

The expression for τ is obtained by replacing R_Φ in the first order moment term involving $\frac{\partial f}{\partial t}$:

$$\langle\langle \tau_M \rangle\rangle = \frac{-\int k_x f d\overrightarrow{k}}{\int k_x KK d\overrightarrow{k}} \quad KK = \left.\frac{\partial f}{\partial t}\right|_{collision} \tag{2.11}$$

where the double averaging term is very important. $\vec{J}_\Phi$ in the first-involves v_x^2, which is related to the x-direction kinetic energy density. Therefore, $\nabla \cdot \vec{J}_\Phi$ is a flux of kinetic energy density. **Thus, the first order moment equation is that if the loss of x direction momentum due to scattering is greater than the gain of x direction momentum due to the field, then to maintain the steady state, there must be a net inflow of kinetic energy.** In the kinetic energy–density continuity equation the generation term is the product of the current density and the field, i.e., the electrons in $\vec{J}_{electron,x}$ are accelerated by E_x, thereby gaining kinetic energy. This kinetic energy gain is dissipated by scattering events. The total kinetic energy is randomized. The scattering events are embodied in an averaged energy relaxation time τ_E.

2.3.2 Drift Diffusion Equations

The Drift Diffusion equation pair is used by semiconductor design engineers to do the first-cut design. *The Drift equation applies when there is an* ***electric field and zero charge carrier concentration gradient***. On the other hand, *the Diffusion equation describes the situation when there is* ***a charge carrier concentration gradient and zero electric field gradient***. *Normally charge carrier movement in a semiconductor device has both drift and diffusion contributions, with one dominating the other.* Electron kinetic-energy density is a result of concentration of n electrons, in where each electron possesses some kinetic energy. In one dimension for a total of n electrons:

$$W_{electron,x} = n \cdot x_{KINETIC\,ENERGY} = \frac{1}{8\pi^3}\int m^P_{electron} v_x^2 f d\overrightarrow{k} \tag{2.12}$$

defining $\langle u_x \rangle$ as the averaged kinetic energy of the electron in the x direction:

$$\langle u_x \rangle = \frac{1}{8n\pi^3}\int m^P_{electron} v_x^2 f d\vec{k} \tag{2.13a}$$

The derivative of the total electron kinetic energy with respect to x is:

$$\frac{dW_{electron,x}}{dx} = \left(\langle u_x \rangle \frac{dn}{dx} + n\frac{d\langle u_x \rangle}{dx}\right) \tag{2.13b}$$

and so the divergence term is $\frac{-2qdW_{electrom,x}}{m^{P}dx}$.

The Drift Diffusion equation is based on the notion that the contribution to the kinetic energy density gradient from the spatial change in the electron concentration is much larger than the contribution from any spatial change in the average kinetic energy of each electron. Then in steady state, the the Drift Diffusion equation becomes:

$$\vec{J}_{electron,x} = q\left(n\mu_{electron}\vec{E} + D_{electron}\frac{dn}{dx}\right) \tag{2.14}$$

where the electron mobility and diffusivity terms are:

$$\mu_{electron} = \frac{q\langle\langle\tau_M\rangle\rangle}{m^{P}_{electron}} \quad D_{electron} = \frac{2\mu_{electron}\langle u_x\rangle}{q} \tag{2.15a,b}$$

The expression for diffusion Is also called the Einstein Relation, and reduces to $\frac{D_{electron}}{\mu_{electron}} = \frac{k_BT}{q}$ when the mean, x-directed kinetic energy is $u_x = \frac{k_BT}{2}$. These expressions hold for low doping densities under near-equilibrium conditions. **Electron mobility is a statistically estimated macroscopic property obtained by aggregating the microscopic scattering of individual electrons specified by the average momentum relaxation time.** The approximation underlying the above Drift Diffusion equation is justified in situations where the regions of a device is long enough for any gradients in u_x due to injection of energetic carriers from a neighboring region. These assumptions are inapplicable to state-of-the-art ultra small dimension (e.g., sub-micron gate length MOSFET) devices.

2.3.3 Hydrodynamic Transport Equations

Moving from the idealized one dimensional situation to a more realistic configuration, when component of the gradient in kinetic energy density cannot be ignored, the second order of continuity equation must be used. To fully exploit the continuity equation for W the flux of kinetic energy density S must be estimated first. S is the product of W and some velocity that describes the electron ensemble. This velocity has a component due to the applied field, and a component due to the scattering processes. The ideal electron gas model is used and components due to the external applied field are neglected. The electrons' kinetic energy is related to their random motion. In analogy with the Kinetic Theory of Gases, electron kinetic energy is characterized by a temperature—electron temperature $T_{ekectron}$. At equilibrium, the electron and lattice temperatures are the same. Then the electron kinetic energy, as a junction of the temperature is:

$$W_{electron,x} = n\langle u_x\rangle \approx \frac{nk_BT_{electron}}{2} \tag{2.16}$$

The drift component of the electron energy is ignored, so that the kinetic energy density flux or power density S is a diffusive flow, which depends on the gradient of the electron temperature:

$$S_{electron,x} \approx \frac{-\kappa_{electron} \partial T_{electron}}{\partial x} \tag{2.17}$$

where $\kappa_{electron}$ is the thermal conductivity of the electron gas.

Applying the above expression to the previous first order expressions, in combination with the current continuity equation, gives a revised expression for the steady-state current density:

$$J_{electron,x} = \mu_{electron}\left(nE_x q + k_B\left(T_{electron}\frac{dn}{dx} + n\frac{dT_{electron}}{dx}\right)\right) \tag{2.18}$$

Some manipulation with the above expressions gives the kinetic energy density continuity equation:

$$\frac{\partial W_{electron,x}}{\partial t} - \kappa_{electron}\frac{\partial^2 T_{electron}}{\partial x^2} = J_{electron,x}E_x - \frac{k_B(T_{electron} - T_{lattice})}{2\langle\langle \tau_E \rangle\rangle} \tag{2.18a}$$

Combining the above equations and concepts in an unified set of equations are the *hydrodynamic continuity equations* based on the concept of electron gas—*hydrodynamic*:

$$\frac{\partial n}{\partial t} = \frac{\partial J_{electron}}{q\partial x} \tag{2.18b}$$

$$\frac{\partial J_{electron,x}}{\partial t} = \frac{q}{m^P_{electron}}\left(\frac{\partial W_{electron,x}}{\partial x} + nqE_x\right) - \frac{J_{electron,x}}{\langle\langle \tau_M \rangle\rangle} \tag{2.19}$$

2.3.4 Semiconductor Device Design Equations

The hydrodynamic equations are the basis for a master set of equations to design and analyse semiconductor devices devices with a *semiclassical* approach. The first and third equations in the set of hydrodynamic equations describe the conservation of charge density and kinetic energy density of electrons only in the conduction band. To generalize these equations for real world semiconductor devices, additional terms that allow for interband transfer of electrons are required, to account for recombination and generation processes. The charge continuity equation must include generation of charge carriers from optical and impact-ionization processes. The sum of the effects of these processes is denoted as $G_{optical,inpactionisation}$. Thermal generation and recombination mechanisms are combined in a *net recombination rate* U. This introduces an additional term in the right hand side of the kinetic energy density equation given by $G_{optical,impact\ ionisation} - U$. The continuity equation for kinetic energy density must also include recombination and generation as these processes change the number of carriers in a band and the kinetic energy density. Recombination of one electron removes $\frac{3k_N T_{electron}}{2}$ of kinetic energy from the electron ensemble in the conduction band. Impact-ionization generation involves a loss of kinetic energy by one electron, accompanied by the boosting of another electron to the conduction band, so now two electrons can gain energy from the field. The overall change in kinetic energy density due to recombination and generation is

$$H = \frac{3k_B T_{CHARGECARRIER} U}{2} + E_{GAP} G_{IMPACTIONIZATION} \tag{2.20}$$

Recombination, generation processes involve electrons and holes, so a separate set of equations is needed to express conservation of hole charge, and kinetic energy density. Like the second equation in the hydrodynamic set(continuity of electron flow|current density), a corresponding equation for holes is essential. Also, the concentrations of electron and hole charge densities, −qn|qp respectively could lead to local space charge, perturbing the electrostatic potential ψ. Thus, Poisson's Equation must be added to the master set.

$$\begin{aligned} \frac{-\vec{\nabla} \cdot \vec{J}_{electron}}{q} &= G_{OPTICAL\,IMPACT\,IONIZATION} - U_{electron} \\ \frac{-\vec{\nabla} \cdot \vec{J}_{hole}}{q} &= G_{OPTICAL\,IMPACT\,IONIZATION} - U_{hole} \end{aligned} \tag{2.21a,b}$$

$$\begin{aligned} \vec{J}_{electron} &= n\mu_{electron}\left(\frac{k_B T_{electron} \nabla n}{n} + k_B T_{electron} - q\nabla\psi\right) \\ \vec{J}_{hole} &= p\mu_{hole}\left(\frac{k_B T_{hole} \nabla p}{p} + k_B T_{hole} + q\nabla\psi\right) \end{aligned} \tag{2.22a,b}$$

$$\begin{aligned} &-\vec{\nabla} \cdot (\kappa_{electron} \nabla T_{electron}) = E_{GAP} G_{electron,IMPACTIONIZATION} \\ &+ \frac{3k_B}{2}\left(U T_{electron} - \frac{n(T_{electron} - T_{lattice})}{\langle\langle \tau_E \rangle\rangle}\right) - \vec{J}\nabla_{electron} \cdot \nabla\psi \\ &-\vec{\nabla} \cdot (\kappa_{hole} \nabla T_{hole}) = E_{GAP} G_{hole,IMPACTIONIZATION} \\ &+ \frac{3k_B}{2}\left(U T_{hole} - \frac{n(T_{hole} - T_{lattice})}{\langle\langle \tau_E \rangle\rangle}\right) - \vec{J}_{hole} \cdot \nabla\psi \end{aligned} \tag{2.23a,b}$$

$$\nabla^2 \psi = \frac{q}{\varepsilon}(p + N_{ACCEPTOR} - n - N_{\mathrm{DONOR}}) \tag{2.23c}$$

Generation|recombination terms G,U in the charge continuity equations depend on the carrier concentrations n,p. The parameters μ and energy relaxation time τ_E are material properties, so can be determined experimentally. Energy relaxation time and the carrier thermal conductivities are dependent on carrier temperature, and the mobility is field dependent, since the momentum relaxation time τ_M is dependent on the applied force.

These dependencies can be expressed empirically, or derived from detailed theoretical calculations. The two assumptions underlying the above set of design equations.

- Carrier transport is semi-classical, as Newton's Laws were used to formulate the Boltzmann Transport equation.
- Kinetic energy density contribution to W of the average velocity that the carriers might have is neglected.

The set of equations consists of five independent equations in five unknowns, n, p, $T_{electron}$, T_{hole}, and ψ. The set has to be solved numerically. Even though the carrier temperatures $T_{electron}$, T_{hole} are functions of position, the lattice temperature is constant—isothermal equation set. Spatial variation in the lattice temperature a lattice energy balance equation is needed. A real semiconductor has physical heat dissipation mechanisms A simple expression for heat balance in the steady state is:

$$-\nabla \cdot \left(\kappa_{MATERIAL} \nabla \vec{T}_{lattice}\right) = \vec{E} \cdot \left(\vec{J}_{electron} + \vec{J}_{hole}\right) \tag{2.24}$$

2.3.4.1 Electron and Hole Mobility

Mobility is defined in terms of the average momentum relaxation time τ_M. An applied force changes momentum, and since some scattering mechanisms are momentum dependent, mobility is be field dependent—electrons are accelerated by an electric field, and move to new momentum states. In the parabolic band approximation, the field moves electrons to new velocities. The average of these field related velocity changes is the drift velocity. From the current density expressions, the component of current that is directly related to the electric field is, for the electron current, $qn\mu_{electron}E$. The electron drift velocity is:

$$v_{\mathrm{electron,DRIFI}}(E) = \left|v_{\mathrm{electron,\ DRIFT}}(\vec{E})\right| \stackrel{\mathrm{def}}{=} \mu_{\mathrm{electron}}\, E \tag{2.25}$$

The drift velocity for electrons is in the opposite direction to the applied electric field.

2.3.4.2 Conductivity Electron, Hole Effective Masses

The definition of mobility includes effective mass. For a single, spherical constant energy surface, as utilized in the drift diffusion and hydrodynamic equations, the relevant effective mass is the band effective mass. For multiple spherical surfaces in the valence band, and multiple prolate spheroids in the conduction band of Si, a modified definition for mobility is needed. To derive this modified definition of mobility, the expression for the material conductivity must be re-written Its value is given by the ratio of the drift current density to the electric field:

$$\sigma = \frac{1}{\rho} = q(n\mu_{electron} + p\mu_{hole}) \tag{2.26}$$

The unit is Mho.

In both Si and GaAs, there are two (light|heavy) hole bands. Both are degenerate at the top of the band. If the average momentum relaxation times are the same for holes in both of these bands, then the hole conductivity is:

$$\begin{aligned}\sigma_{\text{hole}} &\stackrel{\text{def}}{=} \frac{\vec{J}_{\text{hole, DRIFT}}}{\vec{E}} = q^2\langle\langle\tau_M\rangle\rangle\left(\frac{p_{\text{heavy hole}}}{m^p_{\text{heavy hole}}} + \frac{p_{\text{light hole}}}{m^p_{\text{light hole}}}\right)\\ &= q^2\langle\langle\tau_M\rangle\rangle\left(\frac{p_{\text{heavy hole}} + p_{\text{light hole}}}{m^p_{\text{hole,CONDUCTION}}}\right)\end{aligned} \tag{2.27}$$

The corresponding conductivity effective masses for holes is:

$$\frac{1}{m^P_{hole,CONDUCTION}} = \left(\frac{\sqrt{m^P_{heavyhole}}}{\left(m^P_{heavyhole}\right)^{1.5}} + \frac{\sqrt{m^P_{lighthole}}}{\left(m^P_{lighthole}\right)^{1.5}}\right) \tag{2.28}$$

Using similar reasoning as above, the conductivity and conductivity effective mass for electrons in silicon is:

$$\begin{aligned}\sigma_{\text{electron}} &\stackrel{\text{def}}{=} \frac{\vec{J}_{\text{electron, DRIFT}}}{\vec{E}}\\ &= nq^2\langle\langle\tau_M\rangle\rangle\left(\frac{1}{3m^p_{\text{electron LONGITUDINAL}}} + \frac{2}{m^p_{\text{electron, TRANSVERSE}}}\right)\\ &= nq^2\langle\langle\tau_M\rangle\rangle\left(\frac{1}{3m^p_{\text{electron, CONDUCTION}}}\right)\end{aligned} \tag{2.29}$$

2.3.4.3 Drift, Diffusion and Thermal Currents

The electron|hole drift current densities based on mobility arguments, are as below. Although the current densities are in the same direction, the electrons and holes move in different directions.

$$\vec{J}_{electron,DRIFT} = -qn\mu_{electron}(E)\nabla\psi = qn\mu_{electron}(E)\vec{E} = -qn \tag{2.30}$$

The actual distribution of the carriers in k-space is not explicit in these equations: it is implicitly taken into account in either the expressions for the mobility or the drift velocities, both as functions of the applied electric field. The drift current expressions can be used when the distribution function is different from its equilibrium form.

At low fields, e.g., 100 V/cm, the distribution of carriers is slightly perturbed from its equilibrium value. Each electron in a velocity state (parabolic bands) of could be shifted to a state $\frac{\hbar k}{m^P} + v_{DRIFT}$ due to acceleration by the field between collisions. Denoting the forward| backward-going parts of the distribution by 'RIGHTDIR|LEFTDIR'

$$\begin{aligned} \vec{J}_{electron,DRIFT} &= \vec{J}_{electron,DRIFT,RIGHTDIR} + \vec{J}_{electron,DRIFT,LEFTDIR} \\ &= nq\vec{v}_{electron,DRIFT} \end{aligned} \tag{2.31}$$

where n, not n_0, is used to denote small perturbation from equilibrium. The electron distribution is described in these conditions by a quasi Fermi energy. The distribution in this low field case is called a displaced Maxwellian. The velocity distribution function is:

$$f(v) = \frac{n_0}{N_{CONDUCTION}} e^{\frac{m^P_{electron,CONDUCTION}(v - v_{electron,DRIFT})^2}{2k_B T}} \tag{2.32}$$

The carrier concentration in this small perturbation case is function of drift velocity and is approximately:

$$n_0(v) = \frac{8\pi\left(m^P_{electron,DENSITYOFSTATES} m^P_{electron,CONDUCTION}\right)^{1.5} v^2}{h^3 e^{\frac{m^P_{th} v^2}{2k_B T}} e^{\frac{E - E_{CONDUCTION}}{k_B T}}} \tag{2.33}$$

The diffusion electron and hole current densities are:

$$\vec{J}_{electron,DIFFUSION} = k_B T_{electron} \Delta n \quad \vec{J}_{hole,DIFFUSION} = k_B T_{hole} \Delta p \tag{2.34}$$

where the quantity $k_B T_{electron}$, $k_B T_{hole}$ measures the mean kinetic energy of each carrier. This means that a kinetic energy gradient and therefore a kinetic energy and charge flow is set up at constant electron|hole temperature. For a given concentration gradient electrons|holes diffuse in the same direction, but the current densities have opposite signs.:Like the drift velocity current, the distribution function is not the same as for the equilibrium case. Likewise, the diffusion currents differ from equilibrium or near equilibrium currents. For quick analysis, diffusion currents can be considered to originate from equilibrium distributions. Let two Maxwellian distributions with different electron concentrations n_1, n_2 be separated by a distance *2 l*. The current density at an intermediate plane at x = 0, is:

$$\vec{J}_{electron} = -qn(n_1 - n_2) \tag{2.35}$$

The two distributions have to be close enough so that the electrons from each can cross the plane at x = 0 **without** collisions. Collisions change the distributions. The mean distance between collisions is called the **mean free path**. The concentration difference $\frac{n_1 - n_2}{2}$ is calculated by a first order Taylor's series expansion about the plane x = 0.

$$n_2 = n(0) + \frac{ldn}{dx} \tag{2.36}$$

Now, using the expression for $\frac{n_1}{2}$ the final expression for the diffusion current density is:

Here two Maxwellian distributions have produced a net current, apparently inconsistent as Maxwellian distributions are equilibrium distributions, and a net current is not allowed. This means that if there were a diffusion current at equilibrium, then it would be negated by a current due to some other mechanism, e.g., np junction at equilibrium.

$$\vec{J}_{electron,DIFFUSION} = qD_{electron}\frac{d\vec{n}}{dx} \tag{2.37a}$$

The thermal currents in a semiconductor are

$$\vec{J}_{electron,THERMAL} = k_B T_{electron}\mu_{electron}\overrightarrow{\nabla}T_{electron} \tag{2.37b}$$

$$\vec{J}_{hole,THERMAL} = -k_B T_{hole}\mu_{hole}\vec{\nabla}T_{hole} \tag{2.37c}$$

These currents exist when neighbouring carrier ensembles have different average kinetic energies—the injection of high energy electrons into a p-type region in which the existing carriers are at near equilibrium. Gradually the hot electrons cool towards the near equilibrium distribution by scattering and recombination. The newly created electron temperature gradient drives the injected carriers forwards. Thermal current occurs in HBTs and HEMTs operating at high currents.

2.4 Ballistic Transport of Charge Carriers

Previously considered drift, diffusion and thermal currents are a consequence of scattering that dissipates any momentum imparted to charge carriers by an applied force. **Ballistic|collisionless transport occurs in regions of a semiconductor that are shorter than the mean free path l**. The mean free path in near equilibrium conditions is:

$$D_{electron} = \frac{k_B T_{lattice}\mu_0}{2qv_R} = \frac{k_B T_{electron}\mu}{q} \tag{2.38}$$

where the subscript '0' indicates near equilibrium with a low applied field. The electron gas is in equilibrium with the lattice.

For very low doping densities μ_0 is about 1300 $\frac{cm^2}{Vs}$ for Si, and about six times larger for GaAs. For typical values of $2v_R$ at low doping and lattice temperature 300 K is approximately $10^7\frac{cm}{s}$ for Si, and about twice this value for GaAs. Then l_0 is about 30 nm for Si and about 100 nm for GaAs, and are inversely proportional to the doping density.

Ballistic charge transport is a consequence of quantum mechanical tunnelling. A real world semiconductor device (e.g., MOSFET) is analyzed to explain how ballistic transport works. A potential profile consists of two regions of constant potential energy, U1 and U3, which are separated by a thin barrier, with a triangular top, as e.g., the potential profile in the oxide of a MOSFET. **Quantum mechanics permits an electron in region 1 to tunnel through the barrier to region 3—*wave particle duality*.** For an electron with wavefunction Ψ the corresponding probability density is $\Psi\Psi^P$ and the time rate of change of this probability density as the electron tunnels from region 1 to region 3 is (using the effective mass form of the Schrodinger equation and neglecting envelope function)

$$\frac{dP}{dt} = \frac{\psi \partial \psi^P}{dt} + \frac{\psi^P \partial \psi}{dt}$$
$$= -j\hbar\left(\psi^P\left(U(x) - \frac{\hbar^2\nabla^2\psi}{2m^P_{electron}}\right) + \psi\left(U(x) - \frac{\hbar^2\nabla^2\psi^P}{2m^P_{electron}}\right)\right) \tag{2.39a}$$

The probability density current $\vec{J}_P$ is related to the time rate of change of probability as:

$$\frac{dP}{dt} = \frac{-j\hbar}{2m^P_{electron}}\nabla \cdot \left(\psi^P \nabla\psi - \psi\nabla\psi^P\right) = -\nabla \cdot \vec{J}_P \tag{2.39b}$$

The key metric is the transmission probability that controls whether an electron in region 1 will tunnel into region 3. This is the same as computing the probability density current. Schrodinger's wave equation, using the effective mass, and valid from regions 1 to 3 is:

$$(U - E)\psi - \frac{\hbar^2 d^2\psi}{2m^P_{electron} dx^2} = 0 \tag{2.39c}$$

The two boundary conditions are that the wavefunction be continuous at each boundary and the product of the reciprocal of the effective mass times the first derivative of wave function is continious across each boundary/ The potential U has no time dependency. The solutions in the three regions are:

$$\psi_1(x) = Ae^{jk_1x} + Be^{-jk_1x} \quad \psi_2(x) = Ce^{-jk_2x} + De^{jk_2x} \quad \psi_3(x) = Fe^{jk_3x} \tag{2.40}$$

where the parameters are:

$$k_1 = \frac{\sqrt{2m_1^P(E - U_1)}}{\hbar} \quad k_2 = \frac{\sqrt{2m_2^P(E - U_2)}}{\hbar} \quad k_2^P = \frac{\sqrt{2m_2^P(U_2 - E)}}{\hbar}$$
$$k_3 = \frac{\sqrt{2m_3^P(E - U_3)}}{\hbar} \quad k_1, k_2^P, k_3 \; real \quad k_2 \; imaginary \tag{2.41a,b}$$

The constants A, B, C, D and F are evaluated using the boundary conditions. The probability density in the first region is purely oscillatory, inside potential barrier region exponential and in region 3 oscillatory. **Physically, a wave enters region 1 and propagates sinusoidally. When it enters the potential barrier region. its magnitude is reduced exponentially as it traverses the barrier region—an evanescent wave. If the barrier region is thin and some fraction of the probability density wave exits the barrier into region 3.** So relative amplitudes can be calculated. The probability density currents are

$$J_{P,A} = \frac{\hbar k_1 |A|^2}{2m_1^P} \quad J_{P,F} = \frac{\hbar k_3 |F|^2}{2m_3^P} \tag{2.42}$$

The transmission coefficient is:

$$T = \frac{k_1 m_1^P |A|^2}{k_3 m_3^P |F|^2} \tag{2.43}$$

The analyticallfunctional form of the transmission coefficient is very difficult to compute. Using simplifying assumptions, the following simpler expressions are:

$$T = \frac{16e^{\frac{-2d}{\hbar}\sqrt{2m_2^P(U_2-E)}}}{4 + \left(\frac{k_2^P m_1}{k_1 m_2} - \frac{k_3 m_2}{k_2^P m_3}\right)^2} \tag{2.44a}$$

This expression is simplified further, using more simplifying assumptions:

$$T = e^{\frac{-2d}{\hbar}\sqrt{2m_2^P(U_2-E)}} \tag{2.44b}$$

As expected, these simplified expressions for the transmission coefficient are inaccurate, and the best estimates for the transmission coefficient are obtained by carefully analyzing experimental data. It is left for the interested reader to determine what simplifying assumptions were used to derive the above expressions.

The above concepts are easily extended to transitions from a continuum of states. The electrons in region 1 are not confined and E_C is flat. The fundamental expression for the x-directed current density in region 1 multiplied by the transmission probability gives the expression for the x-directed electron current density in region 3.

$$\vec{J}_{electron,x} = \frac{-q}{4\pi^3}\int T_{tunnelling}(k_x) v_x f\left(\vec{k}\right) d\vec{k} \tag{2.45a}$$

Assuming that:

$$m_{1,y}^P = m_{1,z}^P = m_{ORTHOGONAL}^P \tag{2.45b}$$

the energy can be simplified as:

$$E = E_x + \frac{\hbar^2\left(k_y^2 + k_z^2\right)}{2m^P_{ORTHOGONAL}} \tag{2.45c}$$

Transforming to polar coordinates, the expression for the electron current density in the x direction becomes:

$$J_{electron,x} = \frac{-q}{2\pi}\int T_{tunnelling}(k_x)v_x dk_x \int\left(\frac{m^P_{ORTHOGONAL} f(E)dE}{\pi\hbar^2}\right)$$
$$0 \le k_x \le \infty \quad E_{ORTHOGONAL} \le E \le \infty \tag{2.46}$$

The second integral represents an areal density. It is the density of electrons n_{2D} in the two dimensional sheet of electrons at the insulator semiconductor interface having an energy E_x i.e., each of the electrons spread across the surface that has energy E_x has a probability of tunnelling $T(k_x)$ with a velocity v_x. The integral solved by variable substitution, gives:

$$n_{2D}(E_{FERMI} - E_x) = \frac{m^2_{ORTHOGONAL} k_B T \ln\left(1 + e^{\frac{E_{FERMI}-E_x}{k_B T}}\right)}{\pi\hbar^2} \tag{2.47}$$

Applying another appropriate change of variables, (left as an exercise for the interested reader) the current density is:

$$J = \frac{-q}{\hbar}\int T_{tunnelling}(E_x)n_{2D}(E_{FERMI} - E_x)dE_x \quad U_1 \le E_x \le \infty \tag{2.48}$$

These arguments can be extended to quasi bound states. This is left as an exercise for the reader.

References

1. https://doi.org/10.1007/978-3-7091-6963-6
2. https://doi.org/10.1007/978-3-030-35993-5
3. Jerome, J. W. *Analysis of charge transport: A mathematical study of semiconductor devices.* ISBN 10 3642799892 ISBN 13 978–3642799891
4. Ferry, D. *Semiconductor transport* (1st ed.). Published March 16, 2000 by CRC Press ISBN 9780748408665.
5. El-Saba, M. *Transport of information-carriers in semiconductors and nanodevices.* ISBN 13 9781522523123 ISBN10 152252312X EISBN13: 9781522523130
6. Scholl, E. *Nonlinear spatio-temporal dynamics and chaos in semiconductors.* Online ISBN: 9780511524615.
7. Blokhin, A. M., et al. *Qualitative analysis of hydrodynamical models of charge transport in semiconductors.* Nova Science Publishers Inc ISBN: 9781617617911, 9781617617911.
8. Askerov, B. M. (1994). *Electron transport phenomena in semiconductors.* World Scientific.

9. Brennan, K. F., & Ruden, P. P. *Topics in high field transport in semiconductors.* ISBN 10 9810246714 ISBN 13 978-9810246716.
10. Seeger, K. H. *Semiconductor physics.* ISBN: 978-3-662-02576-5 https://doi.org/10.1007/978-3-662-02576-5
11. Pullfrey, D. (2013). *Understanding modern transistors and diodes.* Cambridge University Press. ISBN 13 978-0521514606.

Homogeneous (np) and Heterogeneous (Np) Semiconductor Junction Fundamentals

3

3.1 Homogeneous and Heterogeneous Semiconductor Junctions [1–7]

- A homogeneous semiconductor junction is created when the base material on the two sides of a semiconductor junction is the same—e.g., n doped silicon in contact with p doped silicon.
- A heterogeneous semiconductor junction is created when the base material on two sides of the junction are different—e.g., galliun nitride (GaN) in intimate contact with aluminum arsenide (AlAs). The doping type on the two sides of the junction could be either n or p.
- A homogeneous semiconductor junction is denoted np and a heterogeneous semiconductor junction as e.g., Np. **For a heterogeneous semiconductor junction, the uppercase letter denotes *the doping type of the side with the larger bandgap:*** *i.e., for the Np heterogeneous semiconductor junction, the n doped region has higher bandgap than the p doped region.*

Semiconductor junctions are the building blocks of solar cells, LEDs, homogeneous|heterogeneous bipolar transistors (BJT, HBT), homogeneous|heterogeneous field effect transistors (JFET, MOSFET, MESFET, HEMT), IGBT, solid state lasers etc.

3.1.1 Homogeneous Semiconductor Junction—np

The equilibrium energy levels and the corresponding energy band diagram for a homogeneous np junction is shown in Fig. 3.1a, b. In Fig. 3.1a shows the separated n|p type regions, including the conduction|valence band edges, related to the potential energies

A. Banerjee, *Semiconductor Devices*, Synthesis Lectures on Engineering, Science, and Technology, https://doi.org/10.1007/978-3-031-45750-0_3

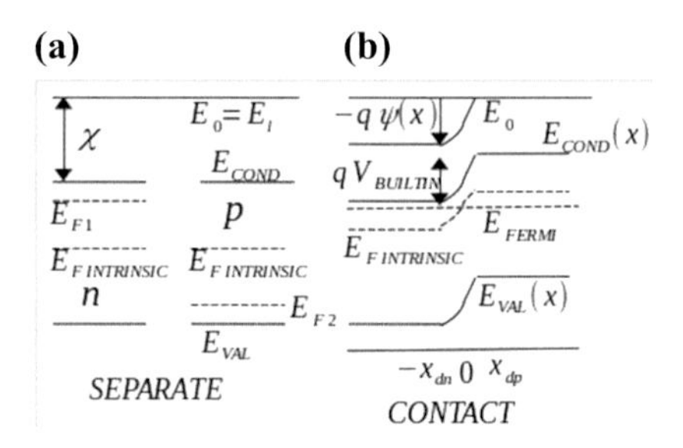

Fig. 3.1 **a**, **b** Energy band diagram for separate n|p type material (**a**) and n|p material in intimate contact with band bending (**b**)

of the electrons|holes respectively. The force free vacuum energy *reference* level E_0 is conventionally assigned the value 0. All other energy levels are defined with reference to it, and are negative. **The vacuum reference energy level is the energy that an electron would have if it were removed from the semiconductor, in absence of potential energy gradients (electrostatic forces) in the semiconductor**. The positive energy required to remove the electron $(E_0 - E_{COND})$ is the *electron affinity* of the material, denoted as χ. Starting with the expression for the equilibrium carrier concentration:

$$n_0 = N_{COND} e^{\frac{E_{FERMI} - E_{COND}}{k_B T_{lattice}}} \tag{3.1a}$$

where E_{COND}, N_{COND} are respectively the conduction band energy bottom and effective density of states. The Fermi energy is:

$$E_{FERMI} = E_{COND} + k_B T_{lattice}(\ln(n_0) - \ln(N_{COND})) \tag{3.1b}$$

$$E_{FERMI} = E_0 - \chi + k_B T_{lattice}(\ln(n_0) - \ln(N_{COND})) \tag{3.1c}$$

$$E_{FERMI} = k_B T_{lattice}(\ln(n_0) - \ln(N_{COND})) \; - \; \chi = \mu \tag{3.1d}$$

The final equation indicates that the Fermi energy is equal to the chemical potential energy μ in field free conditions. As χ and N_{COND} are material constants, the Fermi energy increases as more electrons are added to the material. The difference between the Fermi energies in the two separate materials of Fig. 3.1a is the difference in chemical potential energy:

$$E_{FERMI1} - E_{FERMI2} = k_B T_{lattice} \frac{n_{01}}{n_{02}} = \Delta\mu \tag{3.2}$$

When the two materials are put in contact (Fig. 3.1b), ***this difference in chemical potential energy between the two differently doped regions cannot be maintained any more, and the system equilibrates***. *The minimum energy state of thermal equilibrium is reached when the difference in the chemical potential energies vanishes, i.e., when sufficient electrons have been transferred from the n side to the p side so that* $E_F = E_{F1} = E_{F2}$.

*The transferred electrons at start increase the np product on the p side near the junction, increasing recombination rate, and the annihilation of many electron–hole pairs. This creates a region of negative space charge (acceptor ions) on the p side of the junction. A region consisting of equal magnitude of positive space charge (donor ions) appears on the n side from the departure of the transferred electrons. This space-*charge *region is a source of internal potential energy, with an internal electrostatic potential difference called the built-in potential|voltage V_{BI}, expressed as a positive quantity.* The actual, position dependent electrostatic potential $\psi(x)$ is shown in Fig. 3.1b. In this graph, the potential energy $-q\psi(x)$ is a positive potential. So the local vacuum level $E_{INTRINSIC}$ is negative with respect to the reference energy E_0. The reference potential ($\psi = 0$) is conventionally set to zero at the end of the p region: $x = x_P$. Otherwise:

$$E_{FERMI} = \mu(x) - q\psi(x) \tag{3.3}$$

The Fermi energy is the electrochemical potential energy, based on thermodynamics of a system with both chemical potential energy and electrostatic potential energy.

Expressions for the built-in potential and its corresponding profile are obtained easily. At equilibrium, the drift and the diffusion currents cancel each other resulting in:

$$\vec{J}_{electron} = \mu_{electron}(k_B T_{lattice} \nabla n - qn\nabla\psi) = 0 \tag{3.4}$$

$$\int d\psi = \frac{k_B T_{lattice}}{q} \int \left(\frac{dn}{n}\right) \quad n_{0p} \le n \le n_{0n} \tag{3.5}$$

Equation 3.5, after integration gives the expression for the built-in potential:

$$V_{BI} = \frac{k_B T_{lattice}}{q} \ln\left(\frac{n_{0n}}{n_{0p}}\right) \overset{\text{det}}{=} V_{THERMAL} \ln\left(\frac{N_A N_D}{n_i^2}\right) \tag{3.6}$$

With reference to Fig. 3.1a, b, simple expressions for the potential profile for a bias free homogeneous np semiconductor junction, are obtained:

$$-q\psi(x) = E_I(x) - E_0 \quad -q\psi(x) = E_{COND}(x) - E_{COND}(x_P) \tag{3.7a, b}$$

Under equilibrium conditions, applying Maxwell–Boltzmann statistics, the one dimensional electron|hole concentrations are:

$$n_0(x) = n_0(x_P)e^{\frac{\psi(x)}{V_{THERMAL}}} \tag{3.8a}$$

$$p_0(x) = p_0(x_P)e^{\frac{-\psi(x)}{V_{THERMAL}}} \tag{3.8b}$$

Now applying these expressions to Poisson's equation (equilibrium conditions) gives the following non-linear differential equation, to be solved numerically:

$$\frac{-d^2\psi}{dx^2} = \frac{q}{\varepsilon}(p_0(x) + N_{DONOR} - n_0(x) - N_{ACCEPTOR}) \tag{3.9a}$$

where $N_{ACCEPTOR}$, N_{DONOR} are the acceptor and donor concentrations.

The space charge region inside a homogeneous np junction is very important for its operation and therefore its properties must be determined accurately. But (3.9a) is a nonlinear equation and can be solved accurately only using numerical techniques.

The *depletion approximation scheme* fully linearizes (3.9a) and so the results so obtained with the linearized version are not very accurate. According to the depletion approximation, the width of the space charge region is:

$$W = \sqrt{\left(N_{DONOR}x_{dn}^2 + N_{ACCEPTOR}x_{dp}^2\right)\left(\frac{1}{N_{ACCEPTOR}} + \frac{1}{N_{DONOR}}\right)} \tag{3.9b}$$

3.1.2 How to Create|Draw np Junction Equilibrium Energy Band Diagram

The energy band diagram of a homogeneous np semiconductor junction is created using the following step-by-step method, starting with **the n, p region are not in contact**.

- Draw a horizontal solid line to represent the reference energy level E_0.
- The electron affinities for the n, p regions are different. *With respect to the reference energy level, draw two horizontal lines, below the reference energy level.* The vertical distance with respect to the reference level of each of the horizontal lines is **equal to that region's (n|p) electron affinity**. *This horizontal line is the bottom of the conduction band for that region—E_{COND}.*
- *Using the horizontal line for the bottom of the conduction band for each of the regions (n|p) as reference, draw two new horizontal lines, below the horizontal line that denotes the bottom of the conduction band for that region.* **For each of the n|p regions the vertical distance between the bottom of the conduction band and the new horizontal line is the bandgap.** *This new horizontal line for each region is the top of the valence band E_{VAL} for that region.*
- Draw a horizontal line in between the horizontal lines for the conduction and valence bands to indicate the approximate doping level—e.g., for a highly doped n region, this line must be close to the bottom of the conduction band. This is the Fermi level.

Now the n, p regions are put in contact—np junction is created.

- Draw a horizontal line across the length of the entire n, p regions combined. This is the Fermi level.

- Add the conduction band bottom and valence top for both the n, p regions **keeping a gap in the middle where there would be a discontinuity between the conduction, valence band levels of the n, p regions.**
- Select one side of the device as a reference side and draw a horizontal line to indicate $E_I = E_0$ for that side only. The line must not extend into the interfacial region.
- For the non-reference side, insert a horizontal line **a vertical distance equal to the electron affinity, above the conduction band bottom**. This is $E_I = E_0$ for the non-reference side.
- Connect the $E_I = E_0$ horizontal lines for reference|non-reference regions with straight lines. Repeat for the conduction and valence band bottom|top horizontal lines for the reference|non-reference regions respectively.

3.2 Homogeneous np Junction with External Bias

An external bias (voltage) V_{APP} is applied to the two extremities of the np junction, via metal contacts at each extremity. With reference to the p side, as the zero reference for electrostatic potential, the applied bias is negative and the total potential difference across the device is:

$$E_{FERMI}(-x_N) - E_{FERMI}(x_P) = V_{BI} + V_{APP} \tag{3.10}$$

In metals with an odd number of valence electrons per primitive unit cell, the highest occupied band is half-filled with electrons. As a metal has a huge number of free electrons, any electron exchange between a metal contact and semiconductor occurs without disturbing the metal from its thermal equilibrium state. External bias applied to an np-junction is a difference in electrochemical potential between the metal end contacts and is:

$$E_{FERMI}(-x_N) - E_{FERMI}(x_P) = -qV_{APP} \tag{3.11}$$

$E_{FERMI}(x_P)$ is taken as the reference energy for the electron potential energy $-qV_{APP}$ due to an applied voltage V_{APP}. **A forward bias** *condition is when the applied potential on the n side is less than or negative compared to the p side. The opposite case when the n side is more positive than the p side is* **reverse bias.**

The applied voltage is dropped across the junction, because **conductivity is much lower in the space-charge region compared to either in the n or p regions of the device**. *With miniscule number of charge carriers (electrons|holes) in the space charge region it acts as a resistor*. So most of the voltage is dropped across the resistive space charge region. For the forward bias case the potential barrier at the junction would be lowered by $|V_{APP}|$. This lowering enables more electrons to pass from the n side to the p side of

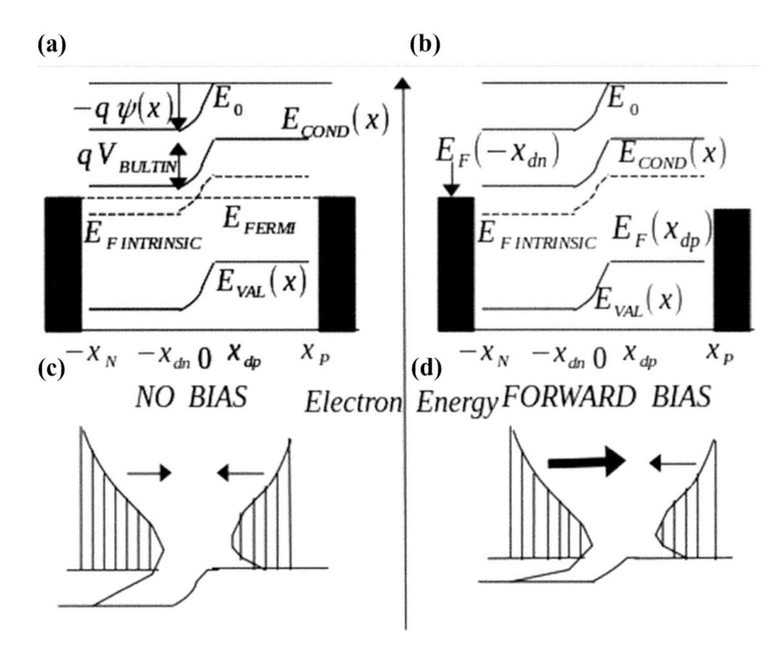

Fig. 3.2 **a**, **b** Energy band diagram for separated (equilibrium) and in contact (forward biased) homogeneous n, p semiconductors. **c**, **d** Electron concentration (hatched area) and flow (arrow) across separated (equilibrium) and in contact (forward biased) homogeneous n, p semiconductors. Thick arrow shows direction of maximum electron flow in (**d**)

the junction, resulting in a net current. When the current density is large, e.g., $\frac{10^4\,\mathrm{A}}{\mathrm{cm}^2}$, the flow of electrons extends into the n side the device starting from the p side. For a test case doping density of $\frac{10^{19}}{\mathrm{cm}^3}$ the field needed to support the chosen current density, the field is approximately $\frac{6\,\mathrm{mV}}{\mu\mathrm{m}}$. Thus, in microelectronics, the actual voltage dropped across this region will be very small compared to the value for the applied voltage of a few volts. Therefore the applied voltage is dropped entirely across the space charge region. The energy band (equilibrium separated, contact forward biased) and related (electron concentration) diagrams are shown in Figs 3.2a–d.

3.2.1 How to Create Energy Band Diagram for Homogeneous np Junction at Non-equilibrium

The scheme elaborated on earlier to create the energy band diagram for a homogeneous np junction at equilibrium is extended to address the common non-equilibrium case. Right at start, identify the selected reference side (n|p) for the homogeneous case. Then:

- Copy the selected reference portion of the unbiased homogeneous np junction band diagram **without the Fermi level.**
- On the non-reference side of the new diagram, a new horizontal line is added for E_I at a depth of $-qV_{APP}$ below the previous position of E_I. For the forward bias np junction this will raise E_I on the n side above its equilibrium position.
- **The non-reference side (of the original homogeneous np junction) is now completed by adding E_{COND} at a vertical position χ below E_0 and similarly E_{VAL} is added at a vertical position E_{GAP} below E_{COND}.**
- *E_I must join up smoothly across the junction.* Forward bias reduces the potential drop across the junction, so less charge is needed to support the potential difference, and the space charge region shrinks. *Thus the horizontal sections of E_I are extended further towards the external metal contacts before E_I segments from the two sides of the non-equilibrium np junction are joined by a smooth curve.* This process is repeated for the conduction and valence band edges. In reverse bias, the procedure is the same, but the larger voltage drop across the junction means that the depletion region is widened relative to its equilibrium value.

As the electric field in regions of a np junction outside the space charge region is very weak, there is little deviation from charge neutrality. The response of the majority carrier (holes) to the injection of electrons from the n side shows the presence of new negative charge beyond the edge of the depletion region on the p side which attract holes. These electrons flow in from the metal contact. Finally, holes drawing energy from the tiny electric field in this region, reach the site of the excess electrons. Their positive charge is negated by the negative charge of the contact site electrons. Thus, the region of length L between the contact and the excess electrons is a parallel plate capacitor of capacitance C. The corresponding resistance is R. The associated RC time constant defines the dielectric relaxation time as $\tau_D \stackrel{\text{def}}{=} RC_{QNR} = \frac{\sigma}{\varepsilon}$ and ε, σ are the electrical permittivity and material conductivity respectively. So the majority carriers respond to any changes in a time of the order of the dielectric relaxation constant. **These regions are called quasi neutral region (QNR).** Under low-level injection of minority carriers, the majority carrier concentration remains very close to its equilibrium value, and is subjected to a very small field and the distribution of the majority carriers is quasi Maxwellian.

In reverse bias (Fig. 3.3a, b) a positive potential is applied to the n side contact with and a negative bias is applied to the p side contact—i.e., *the n side is node is more positive than the p side.* Equivalently, with the p side grounded, and the n side connected to a positive bias, the np junction diode is reverse biased. The barrier height of the junction is $q(V_{BI} + V_{APP})$. The space charge layer becomes wider:

$$W = \sqrt{\frac{2q}{\varepsilon}(V_{BI} + V_{APP})\left(\frac{1}{N_{ACCEPTOR}} + \frac{1}{N_{DONOR}}\right)} \tag{3.12}$$

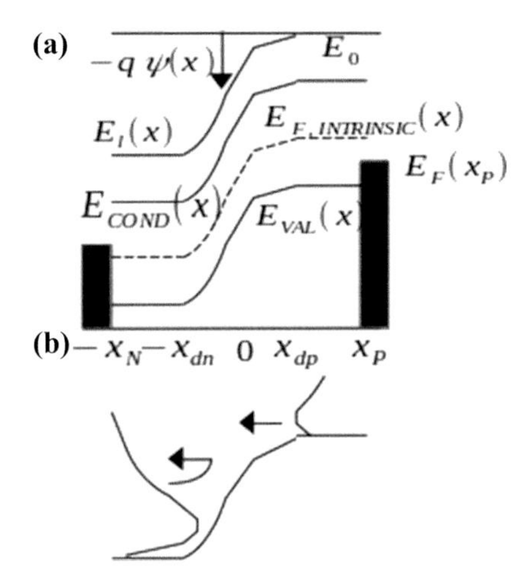

Fig. 3.3 **a**, **b** Reverse biased homogeneous np junction and effect of the reverse bias on electron movement across junction

If the barrier becomes so high that few electrons on the n side have enough energy to diffuse over the barrier, then the electron current is carried mainly by the equilibrium concentration of minority carriers on the p side drifting down the potential barrier. There are very, very few of these electrons, so the reverse n bias current is very small.

The space charge region has larger physical length in the reverse bias condition compared to forward bias condition. As the space charge region is practically free of mobile charges, it acts as a resistor. So the resistance value is much larger for the reverse bias case than the forward bias case. **So the simple np junction has two resistances associated with it—one for the forward bias case, and a larger value one for the reverse bias case. This is the basis for transconductance and so a very, very important property**.

3.2.2 Quasi Fermi Levels

In the quasi neutral regions for both equilibrium|non-equilibrium conditions, when large currents are present, the potential profile is flat, and the majority carrier concentration deviates slightly from its equilibrium value. So in the quasi neutral regions, carrier concentrations can be described by equilibrium carrier concentrations with perturbations. *Equilibrium Fermi levels for pure equilibrium conditions are replaced by Maxwell Boltzmann distributions and the corresponding energy levels are called quasi Fermi levels.*

$$n = n_i e^{\frac{E_{FERMIn} - E_{FERMIi}}{k_B T_{lattice}}} \qquad p = n_i e^{\frac{E_{FERMIi} - E_{FERMIp}}{k_B T_{lattice}}} \qquad (3.13a, b)$$

Here E_{FERMIn}, E_{FERMIp} are the electron|hole quasi Fermi levels respectively. **Quasi Fermi levels do not have any basis in thermodynamics, unlike pure Fermi levels—these describe carrier concentrations in non-equilibrium conditions**. The electron|hole concentrations in terms of the potential Ψ are:

$$n = n_0(x_P)e^{\frac{\psi}{V_{THERMAL}} + \frac{E_{FERMIn} - E_{FERMI}}{k_B T_{lattice}}} \quad p = p_0(x_P)e^{\frac{-\psi}{V_{THERMAL}} + \frac{E_{FERMI} - E_{FERMIp}}{k_B T_{lattice}}} \tag{3.14a, b}$$

Then the quasi Fermi potentials are:

$$-q\Phi_n(x) = E_{FERMIn} - E_{FERMI} \quad -q\Phi_p(x) = E_{FERMIp} - E_{FERMI} \tag{3.15a, b}$$

The electron|hole concentrations, expressed in terms of the quasi Fermi potentials are:

$$n = n_0(x_P)e^{\frac{\psi - \Phi_n}{V_{THERMAL}}} \quad p = p_0(x_P)e^{\frac{\Phi_p - \psi}{V_{THERMAL}}} \tag{3.16a, b}$$

The total non-equilibrium current, the sum of the diffusion and drift currents is:

$$\vec{J}_{TOTAL} = \vec{J}_{electron} + \vec{J}_{hole} = -q\left(n\mu_{electron}\nabla\Phi_n + p\mu_{hole}\nabla\Phi_p\right) \tag{3.17}$$

3.3 The Ideal Homogeneous np Junction (Diode) Equation

The current|current density (I|J) versus voltage equation for a **forward biased** np junction is based on following assumptions.

- Carrier temperature is zero—the diffusion-drift equations can be used.
- Negligible electric fields in the quasi neutral regions, allowing minority carrier transport in these regions to occur via diffusion only. The applied voltage is dropped entirely across the space charge region.
- Quasi neutral regions are so long that the few injected minority carriers recombine before reaching the end contacts.
- The number of injected carriers is low enough so that quasi Fermi levels are constant across the space charge layer.
- There is no generation|recombination in the space charge region so that the electron|hole currents at the boundaries of the space charge region can be added to calculate the total current.
- Thermal generation is the only process to generate charge carriers.
- Impurity doping is uniform in the n, p regions.

For electrons injected into the p side, the electron current density is:

$$J_{electron} = D_{electron}\left(\frac{\partial J_{electron}}{\partial x} - \frac{q(n - n_{0hole})}{\tau_{electron}}\right) \tag{3.18a}$$

The electron current density under steady state conditions becomes:

$$\frac{d^2n}{dx^2} - \frac{n - n_{0hole}}{D_{electron}\tau_{electron}} = \frac{n - n_{0hole}}{L_{electron}^2} = 0 \tag{3.18b}$$

where $L_{electron} = \sqrt{D_{electron}\tau_{electron}}$ is the minority carrier diffusion length. Using the boundary conditions:

$$n(x_p) \equiv n(\infty) = n_{0p} \quad n(x_{dp}) = n_{0p}e^{\frac{-V_{APP}}{V_{THERMAL}}} \tag{3.19a}$$

the general solution is:

$$n(x) - n_{0hole} = Ae^{\frac{-x}{L_{electron}}} + Be^{\frac{x}{L_{electron}}} \tag{3.19b}$$

Using the stated boundary conditions the solution is:

$$n(x) = n_{0hole}\left(1 + e^{\frac{-V_{APP}}{V_{THERMAL}}-1}e^{\frac{x_{dp}-x}{L_{electron}}}\right) \tag{3.19c}$$

Then the electron current density at the edge at the edge of the space charge region is:

$$J_{electron}(x_{dp}) = \frac{-qn_{0hole}D_{electron}e^{\left(\frac{-V_{APP}}{V_{THERMAL}}-1\right)}}{L_{electron}} \tag{3.20a}$$

Using identical reasoning, the hole current density at the edge of the space region is:

$$J_{electron}(x_{dp}) = \frac{-qn_{0electron}D_{hole}e^{\left(\frac{-V_{APP}}{V_{THERMAL}}-1\right)}}{L_{hole}} \tag{3.20b}$$

Combining these two results, the electron and hole current is:

$$J = -q\left(\frac{n_{0hole}D_{electron}}{L_{electron}} + \frac{n_{0electron}D_{hole}}{L_{hole}}\right)e^{\frac{-V_{APP}}{V_{THERMAL}}-1} \tag{3.20c}$$

This is the **ideal** diode equation.

A real world diode's properties differ noticeably from the ideal, the reasons being:

- Quasi neutral regions are often short, so not all injected minority carriers recombine before reaching the junction extremities. Appropriate boundary conditions must be used, and the saturation current density is different from the ideal value.
- For very high currents, the quasi Fermi levels in the space charge region deviate from constant values.
- High currents trigger increased injection of minority carriers, and a possible loss of charge neutrality in parts of the quasi neutral regions. This leads to local fields outside

of the depletion region, in which case not all of the applied voltage appears across the junction. This can be represented in the diode equation by replacing V_{APP} with $\frac{V_{APP}}{\gamma}$ $\gamma > 1$, where γ (often called η) is the diode ideality factor. At high bias, the diode current still increases exponentially, but the dependency on bias is weaker.

- At low bias including the recombination current of electrons and holes in the space charge region, leading to values of saturation current density greater than $\frac{n_{9hole} D_{ekectron}}{L_{electron}} + \frac{n_{0electron} D_{hole}}{L_{hole}}$, and $\gamma \approx 2$.

Electron hole pairs generated in the space charge region are often separated by the junction field, leading to a current in the reverse bias direction. At large reverse bias, the space charge region can be so wide that this generation current is much larger than the ideal diode saturation current. At even higher reverse bias, carriers traversing the space charge region can gain sufficient kinetic energy to impact ionize lattice atoms, leading to current multiplication and avalanche breakdown—high power transistors.

3.4 Heterogeneous Semiconductor Junctions Np

A heterogeneous semiconductor junction is created when the base material on the two sides of the junction are different e.g., GaAs (gallium arsenide) and InP (indium phosphide) since gallium and indium are two different elements on the periodic table. Very often the dopant concentration on the two sides of the junction are different. *Thus with the Np heterogeneous junction the n side has larger bandgap compared to the p side.*

The two key semiconductor properties of a heterogeneous junction that differ compared to a homogeneous junction are electron affinity and energy bandgap. Based on these two properties, heterogeneous semiconductor junctions are classified into Type I, Type II or Type III (broken heterogeneous junction) Fig. 3.4a, b. For a Np heterogeneous junction,, the wider bandgap straddles the narrower one. For a type II heterogeneous junction, the two bandgaps are staggered. With reference to Fig. 3.4a, b for both (Type I, II), the built-in potential is:

$$qV_{BI} = q(V_N + V_P) = \chi_P + qV_2 - \chi_N - qV_1 \tag{3.21a}$$

which gives after some manipulation (left as an exercise for the reader):

$$qV_{BI} = k_B T_{lattice} \ln\left(\frac{N_{CONDP} n_{0N}}{N_{CONDN} n_{0P}}\right) + \chi_P - \chi_N \tag{3.21b}$$

If the two semiconductors of the heterogeneous junction have similar density-of-states electron effective masses, then the built-in potential for Type I heterogeneous junction will be greater than that for the corresponding homogeneous junction and for Type II heterogeneous junction the built-in potential is lower than that for the corresponding homogeneous junction.

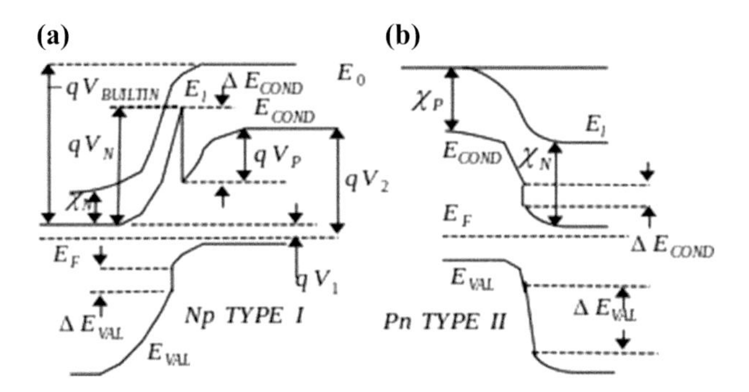

Fig. 3.4 **a**, **b** Energy band diagrams for heterogeneous Np and Pn semiconductor junctions

From Fig. 3.4a, b the local vacuum level $E_I(x)$ varies smoothly with position, as it tracks the electrostatic potential. **But electron affinities on the two sides differ and the band edges $E_{COND}(x)$, $E_{VAL}(x)$ show discontinuities at the interface (heterogeneous junction) between the two different semiconductors**. These discontinuities are called *band offsets*, and labelling them as ΔE_{COND}, ΔE_{VAL}, gives:

$$\Delta E_{COND} + \Delta E_{VAL} = |\chi_N - \chi_P| + \Delta E_{VAL} = E_{GAPN} - E_{GAPP} \tag{3.22}$$

The physical dimensions of the heterogeneous junction space charge region can be estimated by extending the above concepts. The potential across the space charge region at an Np heterogeneous junction is given by appropriate modification of the corresponding expression for a homogeneous junction, allowing for different permittivities of the two semiconductors. The ionic charge on both sides of the heterogeneous junction can be equated, provided there is no free charge density at the interface between the two semiconductors. Therefore the two dimensions of the space charge region, centered at the junction are:

$$x_{dN} = \sqrt{\frac{2\varepsilon_N \varepsilon_p N_{ACCEPTOR} V_{IUNC}}{q N_{DONOR}(\varepsilon_N N_{DONOR} + \varepsilon_P N_{ACCEPTOR})}}$$

$$x_{dP} = \sqrt{\frac{2\varepsilon_N \varepsilon_p N_{DONOR} V_{JUNC}}{q N_{ACCEPTOR}\left(\varepsilon_N N_{DONOR} + \varepsilon_p N_{ACCEPTOR}\right)}} \tag{3.23}$$

3.4.1 Heterogeneous Semiconductor Junction Quasi Fermi Level Splitting

For heterogeneous junctions with a conduction band potential 'spike-well' at the dissimilar semiconductor interface, e.g., Type-I Np junction shown in Fig. 3.5, the quasi Fermi level splits. In any junction (homo|heterogeneous), an applied forward bias reduces both $|qV_N|$, $|qV_p|$. **For a heterogeneous Np junction, this reduces the barrier to electron flow into the base, while increasing the barrier to electron flow from the base**. *So the net electron flow is not small compared to each of the counter directed flows. That means that unlike in a homogeneous junction, the flow of electrons is not a minor perturbation of the equilibrium condition.* **This significant departure from equilibrium results in discontinuous electron quasi Fermi levels at the dissimilar semiconductor interface**. The forward|backward electron current densities injected over the potential barrier, expressed as hemi-Maxwellian (*hemispherical Maxwellian*) fluxes are:

$$J_{electron,FORWARD} = -qn\left(0^{MINUS}\right)V_R \quad J_{electron,BACKWARD} = qn\left(0^{PLUS}\right)V_R \tag{3.24}$$

where the MINUS/PLUS indicate flow directions along the positive|negative x axis, about x = 0. After applying the Maxwell–Boltzmann distribution and some simplification $(n(0) = n\left(0^{PLUS}\right), V_{BI} = V_N + V_P n(-x_{dN}) = n(-x_{dn}))$, the current is:

$$\begin{aligned} J_{electron1} &= J_{electron,FORWARD} - J_{electron,BACKWARD} \\ &= -qV_R\left(n_0(x_{dP})e^{\frac{-V_{APP}}{V_{THERMAL}}} - n(x_{dP})\right)e^{\frac{-\Delta n}{k_B T_{lattice}}} \end{aligned} \tag{3.25a}$$

and the diffusion current is:

$$J_{electron2} = \frac{-qD_{electron}(n(x_{dP}) - n(x_2))}{W} \tag{3.25b}$$

where W is the width of the depletion region, and $x = x_2$ is the position of the end contact to region 2. The first version of the of the diffusion current equation is obtained by assuming that recombination in region 2 is negligible, and the second version implies that the contact at the end of region 2 is Ohmic. The negligible recombination assumption gives the boundary condition for $n(x_{dp})$:

$$n(x_{dP}) = n_{0P}\left(\frac{e^{\frac{-V_{APP}}{V_{THERMAL}}} + \frac{D_{electron}}{We^{\frac{-E_n}{k_B T_{lattice}}}}}{1 + \frac{D_{electron}}{We^{\frac{-E_n}{k_B T_{lattice}}}}}\right) \tag{3.26}$$

The new equation, via γ, *incorporates both the finite velocity* v_R *of carriers crossing the heterogeneous junction, and the presence of an energy difference at the interface.*

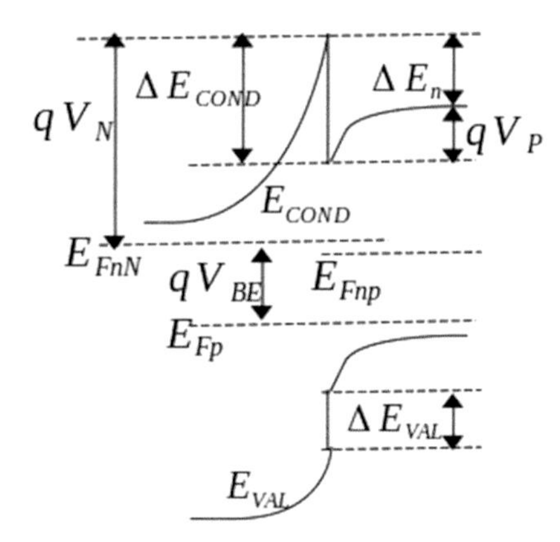

Fig. 3.5 Forward biased heterogeneous Type I Np junction, with quasi Fermi level splitting

Alternatively, recognizing that the impeding effect of the junction itself is via quasi Fermi level splitting, gives:

$$n(x_{dP}) = N_{CONDP}e^{\frac{E_{FnP}-E_{COND}(x_{dP})}{k_B T_{lattice}}} \quad n_0(x_{dP}) = N_{CONDP}e^{\frac{E_{FP}-E_{COND}(x_{dP})}{k_B T_{lattice}}} \tag{3.27}$$

Then the actual splitting of the electron quasi Fermi level at the heterogeneous junction interface is:

$$E_{FERMIFn} = E_{FERMInN} - E_{FERMInP} = -\left(qV_{APP} + k_B T_{lattice}\ln\left(\frac{n(x_{dP})}{n_{0p}}\right)\right) \tag{3.28}$$

Using (3.28), the diffusion equation becomes:

$$J_{electron2} = \frac{-qD_{electron}n_{0P}\left(e^{\frac{-(qV_{APP}+\Delta E_{Fn})}{k_B T_{lattice}}} - 1\right)}{W} \tag{3.29}$$

A key property of a heterogeneous Type I Np diode is emitter injection efficiency: the ratio of the injected electron emitter current to the total current. It is used in a heterogeneous bipolar|bijunction transistor(HBT) analysis. The electron current density is given by (3.29) and the hole current is:

$$J_{hole}(-x_{dN}) = \frac{-qD_{hole}p_{0N}\left(e^{\frac{-(qV_{APP}+\Delta E_{Fn})}{k_B T_{lattice}}} - 1\right)}{L_{hole}} \tag{3.30a}$$

Assuming that the effective densities of states are the same in both materials, and in forward bias the '−1' terms in the expressions for the current components can be dropped, the emitter injection efficiency is:

$$\eta_{EMITTER} = \frac{1}{1 + \frac{D_{hole} N_{BASE} W_{BASE} e^{\frac{-(\Delta E_{GAP} - \Delta E_{Fn})}{k_B T_{lattice}}}}{D_{EMITTER} L_{hole} N_{EMITTER}}} \quad (3.30b)$$

where 'BASE, EMITTER' refer to the conventional bipolar transistor device regions.

References

1. Streetman, B. G., & Banerjee, S. (2000). *Solid state electronic devices*. Prentice Hall.
2. Sze, S. M. (1969). *Physics of semiconductor devices*. Wiley and Sons. ISBN 0-471-84290-7; 2nd ed., 1981, ISBN 0-471-05661-8; 3rd ed., with Kwok K. Ng, 2006.
3. Neaman, S. A. *Semiconductor physics and devices basic principles* (4th ed.). https://www.amazon.in/Semiconductor-Physics-Devices-Basic-Principles/dp/935460112X/ref=asc_df_935460112X/?tag=googleshopdes-21&linkCode=df0&hvadid=545233384715&hvpos=&hvnetw=g&hvrand=3864533311912858920&hvpone=&hvptwo=&hvqmt=&hvdev=c&hvdvcmdl=&hvlocint=&hvlocphy=1007828&hvtargid=pla-1435439439101&psc=1
4. Fiore, J. M. (2018). *Semiconductor devices: Theory and applications*. ISBN 13: 9781796543537. https://open.umn.edu/opentextbooks/textbooks/573
5. Hess, K. Advanced theory of semiconductor devices. https://ieeexplore.ieee.org/book/5265897. https://www.amazon.com/Advanced-Theory-Semiconductor-Devices-Karl/dp/0780334795
6. Nair, B. S., & Nair, R. S. *Solid stare devices*. https://www.amazon.in/Books-B-Somanathan-Nair/s?rh=n%3A976389031%2Cp_27%3AB.+Somanathan+Nair
7. Pulfrey, D. L., & Tarr, N. G. (1989). *Introduction to microelectronic devices*. Prentice Hall. ISBN 10 0134881079. ISBN 13 9780134881072.

Basic Heterogeneous Bipolar|Bijunction Transistor (HBT) Properties

4

4.1 Types of Heterogeneous Bipolar Transistor and Characteristics [1–12]

A bipolar transistor has two semiconductor junctions np inside it, hence *bipolar|bijunction* transistor. For a homogeneous [1–12] bipolar transistor, the base material for each side of both the junctions is the same—silicon. A heterogeneous transistor can be either single heterogeneous junction, or double heterogeneous [1–12] junction. A single heterogeneous junction transistor is a combination of a homogeneous junction and a heterogeneous junction. A double heterogeneous junction bipolar transistor has a different base material for each of the base, collector and emitter regions. *To design|fabricate a Npn(single heterogeneous junction bipolar transistor), the starting point is:*

- **Select two materials that give a barrier height for holes that is much higher than that for electrons, allowing a much larger doping density in the base, compared to homogeneous junction devices. The base resistance is minimized and correspondingly the base-emitter hole current is very low.** *This is bandgap engineering.*

This type of transistor is constructed using molecular beam epitaxial semiconductor layers deposited sequentially. ***The key property of epitaxial layers is that the substrate layer's lattice constant must be as close as possible to that of the deposited layer.*** *This key property eliminates interface defects responsible for unwanted recombination generation centers, and charge carrier scattering.* For gallium arsenide(GaAs) substrates, the AlAs-GaAs(aluminum arsenide-gallium arsenide) materials, and GaP-InP(gallium phosphide indium phosphide) are ideal candidates for HBTs.

Silicon does not naturally have these advantages. *Defect free junctions are constructed between silicon and a dilute alloy of* Si_{1-x} *and germanium(Ge) with the germanium mole*

A. Banerjee, *Semiconductor Devices*, Synthesis Lectures on Engineering, Science, and Technology, https://doi.org/10.1007/978-3-031-45750-0_4

fraction x not exceeding 15%. Increasing Ge mole fraction through the base forces the bandgap decrease, and a field is created in the base. This field enhances electron transport from the base to the collector, in turn improving the high frequency performance of the transistor. All HBTs are bandgap engineered devices.

Figure 4.1a shows the basic physical structure of a HBT using semiconductor materials of the periodic table group III-V elements, for which various semiconducting layers are deposited epitaxially. *The pairs of contacts for both the base and collector, reduce both the base spreading resistance and the series resistance and access regions to the main part of the transistor*. The energy band diagram for a single heterogeneous HBT (Fig. 4.2) is constructed using the sequential procedure detailed in the previous chapter.

This HBT consists of: emitter $Ind_{0.49}Ga_{0.51}P$, a base of p + gallium arsenide(GaAs) and a collector of n doped GaAs. The particular mole fractions for In and Ga in the indium gallium phosphide(InGaP) layer are chosen to give a good lattice match to GaAs. The bandgap for this material is 1.86 eV. The electron affinities of GaAs and $Ind_{0.49}Ga_{0.51}P$

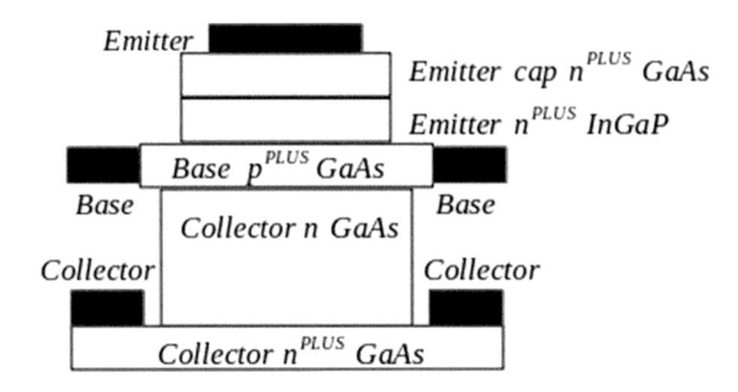

Fig. 4.1 **a** Basic layered physical structure of a HBT (Heterogeneous Junction Bipolar Transistor)

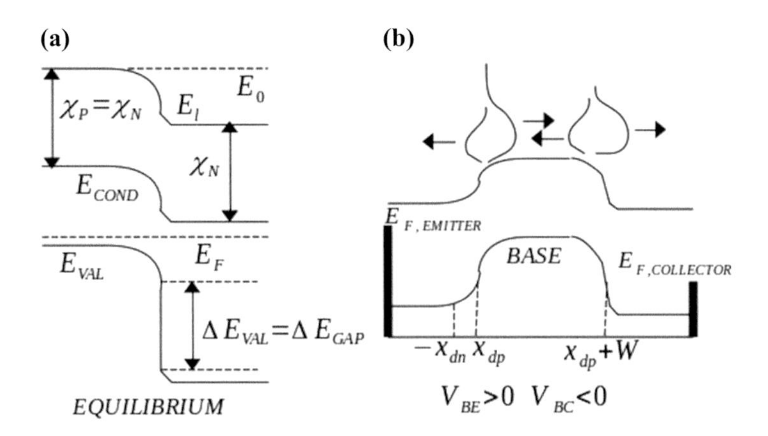

Fig. 4.2 **a, b** Energy band diagram corresponding to Fig. 4.1, in equilibrium and active mode

are the same, so the difference in bandgaps (0.4 eV) is entirely due to the band edge offset in the valence band. **This clever bandgap engineering trick results in different energy barriers for electrons and holes at the emitter base junction. Consequently, performance characteristics of heterogeneous bijunction transistors is vastly superior to that of homogeneous junction transistors.** In all subsequent discussions, regions of higher bandgap is denoted with uppercase symbol.

A higher hole barrier does not interfere with the doping density in the base which can be increased without compromising the current gain. Also, the high base doping density allows a very narrow base to be used without boosting the base access resistance. The advantages of a narrow base are:

- The minority carrier concentration has a steep profile leading to a high collector current resulting in less minority carrier storage and consequently reduced base storage capacitance.

These properties in combination result in a ultra high frequency transistor have very high values for two key performance metrics unity power gain frequency, unity current gain frequency f_{max}, f_T.

$$qV_{BI} = k_B T_{lattice} \ln\left(\frac{N_{CONDp} n_{0N}}{N_{CONDP} n_{0P}}\right) \tag{4.1a}$$

An extremely important distinction between an ideal transistor and a real world HBT is the width of the base region—important for accurate current calculation. *An ideal diode has infinite base width, so any carriers injected from the emitter can collide with other carriers* etc., *so that their distribution at the edge of the emitter–base space charge region is of a near equilibrium form.*

In a modern HBT, the base width can be≈30 nm. To determine the approximate mean free path for the example HBT mentioned previously, the required parameters are the electron mobility in GaAs, and the mean unidirectional velocity for injection from an emitter. Using available data for these parameters from standard references [1–12], the mean free path for the carriers is approximately 10 nm. This is insufficient distance for all the carriers injected into the base to get 'thermalized' by collisions. *A key assumption is that the electrons injected from the emitter maintain a hemi-Maxwellian('hemisperical Maxwellian') distribution through the emitter–base space charge region.* Denoting the electron distribution moving towards the collector that overcomes the barrier as $\frac{n^P_{EMITTER}}{2}$ and assuming quasi equilibrium conditions gives:

$$n^P_{EMITTER} = n_{0EMITTER} e^{\frac{-(V_{BI}-V_{BE})}{V_{THERMAL}}} = n_{BASE} e^{\frac{V_{BE}}{V_{THERMAL}}} \tag{4.1b}$$

4.2 The Collector and Base Current of a HBT

Just like a homogeneous bipolar|bijunction transistor, an expression for the collector current of a heterogeneous bipolar transistor must be evaluated, with an applied bias. The transistor's controlling voltage is $V_{APP,JUNCTION} = V_{BE}$, and the collector current is estimated by first evaluating the electron flow in the transistor's base region. The electric field free portion of the base between the junction space charge regions is the quasi neutral base. Minority diffusion carrier currents this region are controlled by carrier concentration gradients. So the electron profile in the quasi neutral base is estimated first, by combining equations for electron continuity and transport.

Under steady state conditions the drift–diffusion equations give:

$$\frac{d^2n}{dx^2} - \frac{n - n_{0BASE}}{L^2_{electron}} = 0 \tag{4.2}$$

where $L_{electron}$ is the electron minority-carrier diffusion length. This equation is the same as in the case of homogeneous junction ideal diode. For the heterogeneous junction device, the solution is different different because of different boundary conditions. *The minority-carrier diffusion length is a measure of the distance a minority carrier diffuses before it recombines with a majority carrier.* To eliminate recombination in the base region, the quasi neutral base width in a Npn HBT must be less than electron minority-carrier diffusion length: $W_{BASE} < L_{electron}$, and the electron diffusion current is a constant. *A short base guarantees a high collector current—essential for ultra high frequency operation of the HBT.*

With reference to Fig. 4.2b, the forward bias lowers the energy barrier at the base-emitter junction, boosting electron injection from the emitter into the base. These electrons are drawn from the positive-going hemi-Maxwellian at $x = -x_{dn}$. The total width of the depletion region is $W = x_{dp} + x_{dn}, \quad W < L_{electron}$. Then the right going part of the electron distribution at $x = x_{dp}$, upon entering the base will scatter, and some will be re-directed towards the emitter, thus contributing to the left going concentration at $x = x_{dp}$. Another small contribution (due to reverse biased base collector junction) to the left going distribution is due to electrons injected into the base from the collector. Considering scattering of these electrons from the collector:

$$n(x_{dP}) = \frac{n^P_{EMITTER}}{2} + n_{LEFTGOING} \tag{4.3a}$$

where $n_{LEFTGOING}$ are the left going electrons.

Scattering reduces the left going electron distribution to a hemi-Maxwellian, so that the electron current density at $x = x_{dp}$ is:

$$J_{electron}(x_{dP}) = 2qv_R\left(n_{LEFTGOING} - \frac{n^P_{EMITTER}}{2}\right) \tag{4.3b}$$

Then the boundary condition is:

$$n(x_{dP}) = \frac{J_{electron}(x_{dP})}{2qv_R} + \frac{n^P_{electron}}{2} \tag{4.3c}$$

Under quasi neutral conditions the electron concentration at the extremity of the base region is:

$$n(x_{dP} + W_{BASE}) = n^P_{COLLECTOR} - \frac{J_{electron}(x_{dP} + W_{BASE})}{2qv_R} \tag{4.3d}$$

where $\frac{n^P_{COLLECTOR}}{2}$ is the concentration of electrons in the collector with enough left directed kinetic energy to cross over the collector base barrier:

$$n^P_{COLLECTOR} = n_{BASE}e^{\frac{V_{BC}}{V_{THERMAL}}} \tag{4.3e}$$

So the current density flowing through the base region is the difference of the current densities at the two boundaries of the same region:

$$J_{electron} = \frac{n(x_{dP} + W_{BASE}) - n(W_{BASE})}{W_{BASE}} \tag{4.3f}$$

Substituting the charge carrier concentrations in 4.3f and some rearrangement gives:

$$J_{electron} = \frac{-qn_{0BASE}\left(\mathrm{e}^{\frac{V_{BE}}{V_{THERMAL}}} - \mathrm{e}^{\frac{V_{BC}}{V_{THERML}}}\right)}{\frac{W_{BASE}}{D_{electron}} + \frac{1}{v_R}} \tag{4.3g}$$

Clearly the charge and velocity terms influence the current density. The reciprocal of the velocity is the sum of two reciprocal velocities. *If the 'diffusion velocity'* $\frac{D_{electron}}{W_{BASE}}$ *is much less than the 'injection' velocity* v_R, *then diffusion limits the current, strictly for real world bipolar transistors.* State-of-art fabrication techniques routinely allow base widths < 50 nm, so that the $\frac{1}{v_R}$ term also must be taken into account, without which the current would be overestimated, non-asymptotic to its ballistic limit. The electron current exiting the collector is viewed as a positive charge flow into the collector from the external circuit: Adhering to IEEE (Institute of Electrical and Electronics Engineers) convention this is a positive current:

$$I_{COLLECTOR} = -J_{electron}A = I_S\left(\mathrm{e}^{\frac{V_{BE}}{V_{THERMAL}}} - \mathrm{e}^{\frac{V_{BC}}{V_{THERMAL}}}\right) \tag{4.3h}$$

where A is the area of cross section, and I_S collects all the non exponential terms from the equations that are combined in this equation. Each of these parameters are obtained from experimental data of HBT performance characteristics, using sophisticated numerical techniques—*parameter extraction.*

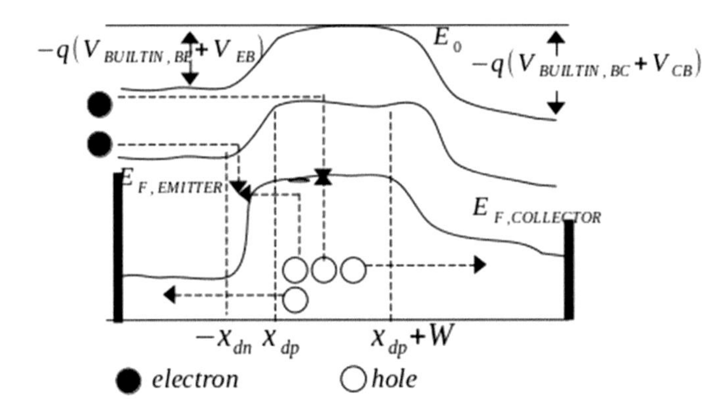

Fig. 4.3 Active mode operation, forward biased electron–hole flow from|to base—recombination currents

In a bipolar junction transistor there is a DC current at the controlling base electrode. Holes flow into the base to (Fig. 4.3):

- Replace holes lost to recombination with electrons in the base.
- Replace holes lost to recombination with electrons in the base emitter depletion region.
- Supply hole current across the reverse biased base–collector junction.
- Supply the hole current due to injection of holes into the emitter across the forward biased base-emitter junction.

This analysis for the collector current ignores the recombination current in the HBT base, i.e., *any recombination current in the base is small compared to the collector current*. In real world transistors, the recombination current may|may not be small compared to the base current, and so cannot be ignored in the base current calculation process.

The large number of electrons and holes in the forward biased base-emitter depletion region results in recombination. Recombination current will manifest itself in the emitter and base leads at low base-emitter voltages, when the base-emitter barrier is high and there is little net hole injection into the quasi neutral emitter. At moderate forward bias, this injection current becomes large and dominates over the space charge region recombination current.

The expressions for the base recombination current are obtained by starting with the continuity equation:

$$J_{electron,RECOMBINATION} = \frac{\int n_{BASE}(x) - n_{0BASE}}{\tau_{electron}} \quad x_{dP} \leq x \leq x_{dP} + W_{BASE} \tag{4.4}$$

where $n_{BASE}(x)$ is the coordinate dependent electron concentration in the HBT base region, expressed as:

$$n_{BASE}(x) = \left(n^{P}_{EMITTER} + \frac{J_{electron}(x_{dP})}{2qv_R}\right)\left(1 - \frac{x}{W_{BASE}}\right) + \frac{x}{W_{BASE}}\left(n^{P}_{COLLECTOR} - \frac{J_{electron}(x + W_{BASE})}{2qv_R}\right) \quad (4.5a)$$

After substituting (4.5a) into (4.4) and integrating the current density over the area of cross section('A' in expression below) of the base region, the base current, is:

$$I_{BASE,\,RECOMBINATION} = Aqn_{0BASE}\left(\left(e^{\frac{V_{BE}}{V_{THERMAL}}} - 1\right) + \left(e^{\frac{V_{Bc}}{V_{THERMAL}}} - 1\right)\right)\left(\frac{1}{\frac{2T_{electron}}{W_{BASE}} + \frac{1}{V_R}}\right) \quad (4.5b)$$

4.2.1 Emitter Hole Current in a HBT

The hole current flowing through the quasi neutral emitter region $-(W_{EMITTER} + x_{dn}) \leqslant x \leqslant -x_{dn}$ due to hole injection through the base-emitter barrier is evaluated using the drift–diffusion equations. The emitter is much wider than the base so recombination of the injected holes cannot be ignored. Thus for the hole diffusion current and the hole charge continuity in the drift diffusion approximation gives:

$$\frac{d^2p}{dx^2} - \frac{p - p_{0,EMITTER}}{L^2_{electron}} = 0 \quad (4.6a)$$

where $p_{0EMITTER}$, L_{hole} are respectively the equilibrium hole concentration in the emitter and the hole minority carrier diffusion length in the emitter region. The general solution to (4.6a) is:

$$p(x) = p_{0EMITTER} + B\cosh\left(\frac{x}{L_{hole}}\right) + C\sinh\left(\frac{x}{L_{hole}}\right) \quad (4.6b)$$

where B, C are constants to be estimated from boundary conditions.

As the emitter length is greater than the hole minority carrier diffusion length, recombination of the injected holes in the quasi neutral emitter cannot be ignored. The relatively long length of the emitter means that the carrier velocity can have any permissible value. Thus, the boundary conditions for the hole concentration at the emitter edge of the depletion region and the other end are:

$$p(-x_{dN}) = p_{0EMITTER}e^{\frac{V_{BE}}{V_{THERMAL}}} \quad p(-x_{dN} - W_{EMITTER}) = p_{0EMITTER} \quad (4.6c)$$

Using a simple change of variables $x^P = x + x_{dN}$ the boundary conditions give the following values for the constants B,C:

$$B = p_{0EMITTER}\left(e^{\frac{V_{BE}}{V_{THERMAL}}} - 1\right) \quad C = -p_{0EMITTER}\left(e^{\frac{V_{BE}}{V_{THERMAL}}} - 1\right)\coth\left(\frac{-W_{EMITTER}}{L_{hole}}\right) \tag{4.6d}$$

Combining Eqs. 4.6a–4.6d the emitter region hole current density is:

$$J_{hole}\left(x^P\right) = \frac{-qD_{hole}dp}{dx} = \frac{-qD_{hole}}{L_{hole}}\left(B\sinh\left(\frac{x_{hole}}{D_{hole}}\right) + C\cosh\left(\frac{x_{hole}}{D_{hole}}\right)\right) \tag{4.6e}$$

Now substituting for the values for B, C gives the hole current at a particular location, as:

$$J_{hole}(-x_{dN}) = \frac{-qD_{hole}p_{0EMITTER}}{L_{hole}}\left(e^{\frac{V_{BE}}{V_{THERMAL}}} - 1\right)\coth\left(\frac{W_{EMITTER}}{L_{hole}}\right) \tag{4.6f}$$

4.3 Heterogeneous Junction Transistor (HBT) Simple DC Equivalent Circuit Model

A simple DC equivalent circuit model of the HBT (Fig. 4.4) combines the base, collector and emitter currents. This equivalent circuit includes the current due to holes injected from the base into the collector. In the equivalent circuit, the base recombination term has been separated into two diode like terms. Each of the two terms are associated with each junction. This way, the origin of the electrons going into the base is identified. Base, collector and emitter resistors represent the resistances of the various quasi neutral and access regions.

A key performance metric for a bipolar transistor is the forward DC common emitter current gain β. In this configuration, the emitter is common to both the input and output, so that the controlling parameter is the base current.

$$\beta = \frac{I_{COLLECTOR}}{I_{BASE}}$$

In the active mode of operation the upper two diodes have negligible contributiom to the output, and so these two in the equivalent circuit can be ignored, and the current gain is expressed in a constant, bias independent form Thus, if the parameter in Fig. 4.4 were I_B rather than V_{BE}, then a family of curves can be generated, which have equal base steps and equally spaced collector currents in the active region. In the common base connection the input current is the emitter current and the DC common base current gain is:

$$\alpha = \frac{I_{COLLECTOR}}{|I_{EMITTER}|}$$

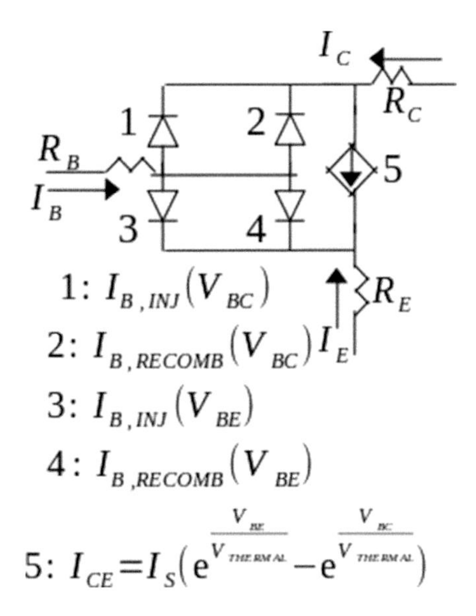

Fig. 4.4 Simple DC equivalent circuit model for the Npn heterogeneous bipolar|bijunction transistor

References

1. Pulfrey, D. L. (1999). Heterojunction bipolar transistor, wiley encyclopedia of electrical and electronics engineering, Webster, J. G., Ed., John Wiley & Sons, Inc., vol. 8, 690–706.
2. Pulfrey, D. L., & Tarr, N. G. (1989). Introduction to microelectronic devices, Prentice-Hall p. 348.
3. Anholt, R. (1995). *Electrical and thermal characterization of MESFET's HEMTs, and HBTs.* Boston: Artech House.
4. McAndrew, C., et al. (1996). VBIC95. *The Vertical Bipolar Inter Company Model, IEEE Journal of Solid State Circuits, 31*(10), 1475–1483.
5. Scott, J., Nonlinear III-V HBT Compact Models: Do we have what we need? *2001 IEEE Transactions on Microwave Technology and Techniques Symposium Digest*, pp. 663–666.
6. Pehlke, D., & Pavlidis, D. (1992). Evaluation of the factors determining HBT high frequency performance by direct analysis of S-parameter data. *IEEE Transactions on Microwave Theory and Techniques, 40*(12), 2367–2373.
7. Teeter, W., & Curtice, W. Comparison of hybrid Pi and Tee HBT circuit topologies and their relationship to large signal modelling, *1997 Microwave Theory and Techniques Symposium Digest*, pp.375–378.
8. Rudolph, M., Lenk, F., Doerner, R., & Haymann, P. On the implementation of transit time effects in compact HBT Large-Signal models, *2002 Microwave Theory and Techniques Symposium Digest*, pp.997–1000.
9. Linder, M., Ingvarson, F., Jeppson, K., & Grahn, J. (2000) Extraction of emitter and base series resistances of bipolar transistors from a single DC measurement, *IEEE Transactions On Semiconductor Manufacturing, 13*(2), pp.119–125.
10. Angelov, I., Choumei, K., & Inoue. An empirical HBT large signal model for CAD, *2002 Microwave Theory and Techniques Symposium Digest,* pp. 2137.

11. Dawson, D., & Gupta, A. (1992). CW measurement of HBT thermal resistance. *IEEE Transactions on Electron Devices, 39*(10), 22–35.
12. Maas, S., & Tait, D. (1992). Parameter-Extraction method for heterojunction bipolar transistors. *IEEE Microwave and Guided Wave Letters, 2*(12), 5.

5 Advanced VBIC and Angelov-Chalmers Models for Heterogeneous|Homogeneous Bipolar|Bijunction Transistor

5.1 The VBIC Specification

The VBIC (Vertical Bipolar Inter Company) [1] model enhances the original Gummel Poon model of the bipolar transistor by including unavoidable parasitic capacitances and inductances, missing in the original Gummel Poon model. *The original Gummel Poon bipolar transistor model included in the gold standard electrical\electronic circuit simulation and performance evaluation tool SPICE (Simulation Program with Integrated Circuit Emphasis)* [10–14]*, is enhanced by including the VBIC model. The 'Vertical Bipolar' arises from the fact that components (base, collector, emitter) of a modern BJT are stacked vertically during fabrication.*

For decades the SPICE [10–14] Gummel Poon [2] (SGP) model Nagel [3] has been the semiconductor industry standard for circuit simulation for bipolar junction transistors. Despite its sound physical basis, the SGP model has deficiencies.

- SGP model ignores collector resistance modulation (quasi saturation).
- SGP model does not account for unavoidable parasitic substrate transistor action.
- Early effect is not accurately modelled and output conductance not accounted for.

The approximations that underlie the SGP Early effect model [3], McAndrew [4] are sufficient for modelling wide base BJTs (Fig. 5.1a), but generate inaccurate results when applied to modelling narrow base BJTs (Fig. 5.1b). For narrow base BJT, a significant portion of the capacitances are from bias independent dielectric capacitances, not included in the original SPICE BJT model.

Although improved BJT models have been presented [5–9], none have become an industry standard to replace the SGP model. VBIC (Vertical Bipolar Inter-Company model) was defined by a group of representatives from the semiconductor and CAD

A. Banerjee, *Semiconductor Devices*, Synthesis Lectures on Engineering, Science, and Technology, https://doi.org/10.1007/978-3-031-45750-0_5

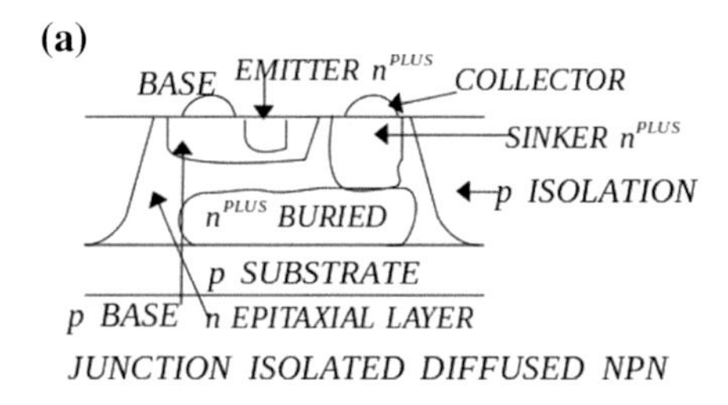

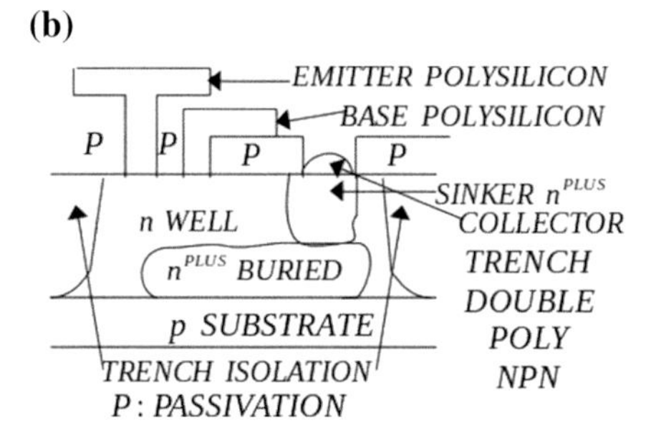

Fig. 5.1 **a** Junction isolated diffused NPN. **b** Trench isolated double polysilicon NPN

industries to try to rectify this situation. VBIC is public domain, and complete source code is publicly available. *VBIC is as similar as possible to the SGP model, to leverage the existing knowledge and training of characterization and semiconductor design engineers.* The main modelling enhancements of VBIC that overcome the deficiencies of the original SGP BJT model are:

- Improved Early effect modelling
- Quasi-saturation modelling
- Parasitic substrate transistor modelling
- Parasitic fixed (oxide) capacitance modelling
- Avalanche multiplication modelling
- Improved temperature dependence modelling
- Decoupling of base and collector currents
- Electrothermal (self heating) modelling
- C ∞ continuous (smooth) modelling
- Improved heterogeneous junction bipolar transistor (HBT) modelling.

Although VBIC offers many advantages over SGP, it was designed to default to being as close to SGP as possible, *so that existing SPICE* [10–14] *simulators could be used to estimate performance characteristics of electronic circuits that include VBIC based bipolar transistor device models.* The Early effect model used in VBIC is vastly improved version of the rudimentary model used in the original SGP. The other features of VBIC are additions that, with the default parameters, are not active. To include the enhancements offered by the VBIC formulation in a circuit simulation with SPICE [10–14], the VBIC

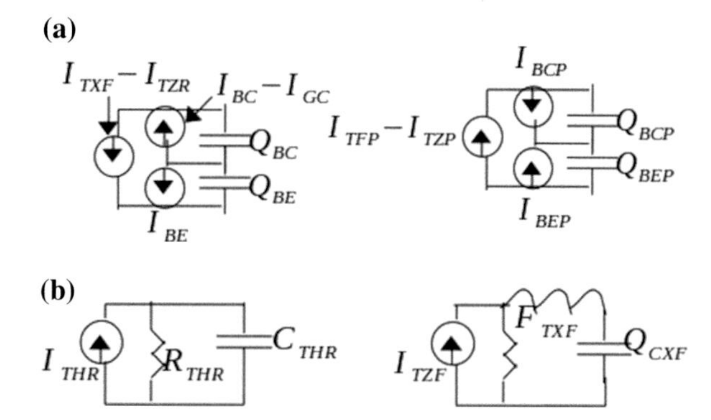

Fig. 5.2 **a**, **b** Intrinsic NPN and parasitic PNP transistors. **c**, **d** Thermal and extra phase circuits

enhancements are incrementally included to refine and fine tune the SPICE simulation results. The mapping between VBIC and pure SGP parameters is provided in [1].

5.1.1 VBIC Homogeneous Junction Bipolar Transistors [1]

The equivalent electrical circuit network [1] of VBIC NPN transistor includes:

- Intrinsic NPN transistor
- Parasitic PNP transistor (Fig. 5.2a)
- Parasitic resistances and capacitances (Fig. 5.2b)
- A local thermal network (Fig. 5.2c) used only with the electrothermal version of the model). A circuit that implements excess phase for the forward transport current I_{tzf} (Fig. 5.2d).

The electrothermal version of VBIC, branch currents and charges in the electrical part of the model depend on the local temperature rise. The thermal equivalent circuit includes two external nodes that allow the local heating and dissipation to be connected to a thermal network that models the thermal properties of the BJT substrate material Although the equivalent network is often shown with fixed value resistances, bias dependent capacitances, and current sources, in reality the elements are voltage controlled current sources $I(V_1, V_2, V_3, \ldots)$ and voltage controlled charge sources $Q(V_1, V_2, V_3, \ldots)$. Simple resistors and capacitors are current and charge sources, controlled only by the voltage across them. If a branch current andlor charge is controlled by more than one branch voltage, can include transconductance and transcapacitance elements when they are linearized, as is required for DC solution and for AC, noise and transient simulations.

Elaborate explanation of how the VBIC model addresses the deficiencies of the original SPICE [10–16] SGP transistor model (e.g., accurate treatment of Early effect etc.,) is in [1].

5.2 VBIC Based Group 4 Silicon Germanium (SiGe) Heterogeneous Bijunction Transistor (HBT) Model

The main features of the SiGe HBT in terms of the VBIC model are enumerated here, as compared to the original SPICE Gummel Poon (SGP) model [15].

- The VBIC HBT is a four terminal device—base, collector, emitter and substrate. The SGP bipolar transistor is a three terminal device.
- The VBIC HBT has self heating properties, absent in the Gummel Poon based SGP model.
- Parasitic capacitances C_{BE0}, C_{BC0} are included by default. SGP requires manual insertion of these parasitic capacitances.
- Detailed substrate model—absent in SGP.
- Excess phase during signal transit through device is modelled with sub-circuit and delay time—SGP has forward delay time.
- Temperature mapping Is done through 21 parameters for HBT, compared to 4 for SGP.
- Enhanced Flicker noise model for HBT, compared to very simple one for SGP.

The HBT collector current equation is:

$$I_{CE} = \frac{I_S}{q_b}\left(\mathrm{e}^{\frac{V_{BEI}}{NFV_{THERMAL}}} - \mathrm{e}^{\frac{V_{BCI}}{NRV_{THERMAL}}}\right) \tag{5.1a}$$

The related Early effect equations, included into the VBIC model without any simplifications are:

$$q_b = \frac{q_1}{2} + \sqrt{\frac{q_1^2}{4} + q_2} \quad q_1 = 1 + \frac{q_{JE}}{V_{ER}} + \frac{q_{JC}}{V_{EF}} \tag{5.1b}$$

$$q_2 = \frac{I_S}{I_{KF}}\left(\mathrm{e}^{\frac{V_{BEI}}{V_{THERMAL}}} - 1\right) + \frac{I_S}{I_{KR}}\left(\mathrm{e}^{\frac{V_{BCI}}{NRV_{THERMAL}}} - 1\right) \tag{5.1c}$$

The SGP model includes a rudimentary Early effect model. The independent and distributed forward base current for a SiGe HBT is:

$$I_{BE} = I_{BEI} + I_{BEN} = W_{BE} I_{B,GES} \quad I_{BEX} = I_{BEXI} + I_{BEXN} = (1 - W_{BE}) I_{B,GES} \tag{5.2a,b}$$

$$I_{BEN} + I_{BEI} = I_{BCI}\left(e^{\frac{V_{BCI}}{NCIV_{THERMAL}}} - 1\right) + I_{BCN}\left(e^{\frac{V_{BCI}}{NCNV_{THERMAL}}} - 1\right) \tag{5.1c}$$

The forward base current in SGP is coupled to the collector current. The reverse base current is decoupled from the emitter current, in the VBIC model, unlike that in the SGP model:

$$I_{BEN} + I_{BEI} = I_{BCI}\left(e^{\frac{V_{BCI}}{NCIV_{THERMAL}}} - 1\right) + I_{BCN}\left(e^{\frac{V_{BCI}}{NCNV_{THERMAL}}} - 1\right) \quad (5.3a)$$

The VBIC model includes a weak base collector avalanche current, absent in the original SGP formalism.

$$I_{GC} = A_{VC1}(I_{CC} - I_{BC})(PC - V_{BCI})e^{-A_{VC1}(PC - V_{BCI})^{(MC-1)}} \quad (5.3b)$$

For both the SGP and VBIC models, the emitter resistance is constant. The VBIC base resistance is modulated by the conductivity; empirically obtained in the original SGP formalism.

$$R_B = \frac{R_{BI}}{q_B} + R_{BX} \quad (5.3d)$$

The VBIC collector resistance consists of both a constant and variable parts to model quasi saturation.

$$R_C = R_{CX} + R_C(V_{BCX}, V_{BCI}) \quad (5.4a)$$

$$I_{RC} = \frac{I_{EPI0}}{\sqrt{1 + \left(\frac{I_{EPI0}R_{CI}}{V_0(1+AA)}\right)^2}} \qquad AA = \frac{0.5\sqrt{0.01 + V_{RC1}^2}}{V_0 HRCF} \quad (5.4b)$$

$$I_{EPI0} = \frac{\left(V_{RCI} + V_{THERMAL}\left(K_{BCI} - K_{BCX} - \ln\left(\frac{1+K_{BCI}}{1+K_{BCX}}\right)\right)\right)}{R_{CI}} \quad (5.4c)$$

$$K_{BCI} = \sqrt{1 + GAMMe^{\frac{V_{BCI}}{V_{THERMAL}}}} \qquad K_{BCX} = \sqrt{1 + GAMMe^{\frac{V_{BCX}}{V_{THERMAL}}}} \quad (5.4d)$$

The SGP model excludes substrate effects, while the VBIC model allows a constant substrate resistance. The VBIC space charge capacitance model is the same as that for the SGP model under certain conditions:

$$C_{JC,E,S} = \frac{C_{JC,E,S0}}{\left(1 - \frac{V_{BASEC,E,S}}{P_{C,E,S}}\right)^{M_{C,E,S}}} \qquad V_{BASEC,E,S} < FC \cdot P_{C,E,S} \quad (5.5)$$

In the VBIC model, the SGP diffusion capacitance model is enhanced by QTF. The interested reader is requested to check the details [1]. Both germanium and silicon (Ge, Si) belong to group 4 of the periodic table. The next section examines in detail the case of group 3–5 element based HBT, without using any VBIC model.

5.3 Non-VBIC Group 3–5 and 4 Heterogeneous Junction Transistor (HBT) Large Signal Model (Angelov Chalmers) [16, 17]

Formulating a large signal model for high frequency and power transistors is complicated because:

For high power devices, the operating power densities and currents are very broad.

- Poor thermal conductivity of Group 3–5 elements used in some of these devices leads to self heating problems at high operating frequencies and power densities.
- The forward DC current gain β is not constant. It rises to a maximum value and then drops, as in a Lorentzian curve.
- Both the base and collector currents show exponential and logarithmic behaviour, depending on the applied bias (e.g., base-emitter).
- The complicated underlying physics makes it difficult to include the corresponding large signal model in a CAD (Computer Aided Design) tool. A CAD tool that generates inaccurate results is meaningless.

These issues have been addressed by [16, 17] *which presents a general large signal model applicable to transistors based on group 3, 4 and 5 elements (AlGaAs/GaAs, InGaP and Si).* The data obtained with steady state DC sweeps and AC (small signal) S parameters are carefully analysed numerically to extract the characteristic device parameters—parameter extraction.

The equivalent circuit model of a HBT is in Fig. 5.3, used to formulate the base and collector current expressions. Although the device physics dictate use of an exponential function to describe the junction current, the reference point for measuring I_B, I_C is changed to operating currents and voltages e.g., close to currents (voltages) at which β is maximum. The junction current equation has a power series as its argument. These modifications allow a flexible device model that can be used with a number of related but different devices. The base current equations are:

$$I_{BE} = I_{BEJ}\left(e^{P_{BE}} - e^{P_{BE0}}\right) \quad I_{BEJ} = \frac{I_{PKC}}{\beta_{MAX}} \quad P_{BE1} = \frac{q}{K_B T_{lattice} N_{b1}} \tag{5.7a,b,c}$$

$$P_{BE} = \frac{P_{b1}}{2}\tanh\left(2\left((V_{BE} - V_J) + N_{b2}(V_{BE} - V_J)^2 + N_{b3}(V_{BE} - V_J)^3 + \cdots\right)\right) \tag{5.7d}$$

$$P_{BE0} = \frac{P_{b1}}{2}\tanh\left(2\left(-V_J + N_{b2}V_J^2 - N_{b3}V_J^3 + \cdots\right)\right) \quad P_{b1} = \frac{1}{N_{b1}V_{THERMAL}} \tag{5.7e,f}$$

When $V_{BE} = V_{JE}$ base current $I_{BE} = I_{JBE}$ $V_{BE} = 0$ $I_{BE} = 0$. The P_{BE0} defines the junction current equals 0 when the junction voltage is 0, and its equation is

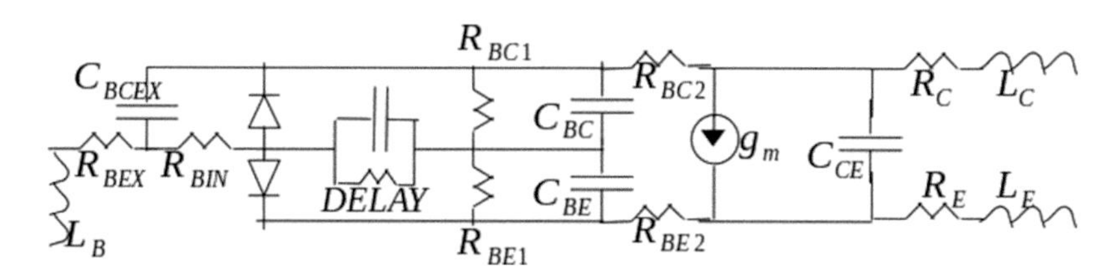

Fig. 5.3 Heterogeneous junction bipolar transistor (HBT) equivalent electrical circuit corresponding to Angelov Chalmers scheme

$P_{BE}\ for\ V_{BE} = 0$. Usually, one term in the power series is sufficient for better than 5% accuracy in 5 to 6 decades of the diode current. For higher accuracy or to maintain accuracy in a range of currents more then 6 decades, more terms can be used.

Similarly, the collector current equations are obtained with similar reasoning.

$$I_C = I_{CF}\tanh(\alpha V_{CE})(1 + \lambda V_{CB} + \lambda_{MAX}) \quad I_{CF} = \frac{I_{PKC}\left(e^{P_{CF}} - e^{P_{CF0}}\right)}{\cosh(B(V_{BE} - V_{BEPM}))} \tag{5.8a,b}$$

$$P_{CF} = \frac{19.347}{N_{C1}}\tanh\left(2\left((V_{BE} - V_{BEP}) + N_{C2}(V_{BE} - V_{BEP})^2 + N_{C3}(V_{BE} - V_{BEP})^3 + \cdots\right)\right) \tag{5.8c}$$

$$P_{CF0} = \frac{19.347}{N_{C1}}\tanh\left(2\left(-V_{BEP} + N_{C2}V_{BEP}^2 - N_{C3}V_{BEP}^3 + \cdots\right)\right)$$

$$\alpha = \alpha_I + \alpha_S\left(e^{\frac{38.695V_{CE}}{N_{C1}}} - 1\right) \tag{5.8d}$$

$$V_{BEPM} = V_{BEP} + \Delta V_{BEP}(1 + \tanh(P_{CF1}V_{CE}) - V_{SB2}(V_{CB} - V_{TR})) \tag{5.8e}$$

$$\lambda_{MAX} = \lambda_2 V_{CB}\left(1 + \tanh\left(\frac{38.685(V_{BE} - V_{BEP})}{N_{CF}}\right)\right) + \frac{\lambda_2}{1 + V_{CE}^2} \tag{5.8f}$$

With 3 terms in the power series, the number of model parameters is 15 and 11 in the single term case for the collector and base currents. These are:

$$\alpha_I, \alpha_S, \beta_{MAX}, I_{PKC}, V_{BEP}, V_{BE}, \lambda, N_{B1}, N_{C1} \quad V_{JE} = V_{BEP}$$

The model flexible and more parameters can be added to model more complicated effects in HBTs and BJT.

The HBT capacitances consist of both diffusion and depletion capacitances. The diffusion capacitances are:

$$C_{BE,DIFFUSION} = C_{BEP} + C_{BE0}(1 + \tanh(C_{BE10} + C_{BE11}V_{BE}))$$

$$C_{BC,DIFFUSION} = C_{BCP} + C_{BC0}(1 + \tanh(C_{BC20} + C_{BC21}V_{BE})) \tag{5.10a,b}$$

The depletion capacitance is:

$$C_{DEPLETION} = C_{DEPP} + \frac{C_{DEP0}\left(m - (2n-1)\left(\frac{V_{BE}}{V_{BEI}}\right)^2\right)}{\left(m + \left(\frac{V_{BE}}{V_{BEI}}\right)^2\right)^{n+1}} \tag{5.11}$$

where n is the grading coefficient. The parameter m denotes the maximum to minimum capacitance ratio. C_{DEPP} represents undesirable parasitic elements.

HBTs suffer from self heating issues, because of the temperature dependencies of collector-emitter, base emitter currents and the forward DC current gain. The changes of the model parameters with changes in I_{CE}, I_{BE}, β are quasi linear with respect to the temperature. This can be used to model self-heating in a simple way; but biasing the device with a constant DC voltage source at the base results in a thermal runaway. However, as the temperature dependency of the model parameters on temperature is of the order of 1 in 1000, a linear model can be used to model self heating in small temperature ranges. Details may be found in [16, 17].

An accurate model of the signal delay through a HBT is very important to analyse the performance characteristics of the circuit in which it is embedded. The delay increases with the collector current and saturates and decreases as the collector voltage is increased. In simulators the bias dependency of the delay is defined via the collector emitter current. It can be defined using the intrinsic controlling voltages and expressed empirically as:

$$T_{FF} = T_F\left(1 + \frac{X_{TF}}{2}\left(1 + \tanh\left(\frac{V_{BE} - V_{JE}}{N}\right)\right)\right)e^{\frac{V_{BE}}{1.44V_{TF}}} \tag{5.13}$$

There are two DC transfer characteristics of a bipolar|bijunction transistor:

- Collector emitter current with linearly or pulsed varied base emitter voltage and constant collector emitter voltage.
- Collector emitter current with linearly or pulse varied collector emitter voltage and constant base emitter voltage.

  Both methods create a family of curves.

References

1. https://designers-guide.org/vbic/documents/VbicText.pdf
2. Gummel, H. K., & Poon, H. C. (1970). An integral charge control model of bipolar transistors. *Bell System Technology Journal, 49*, 827–852.

3. Nagel, L. W. (1975) *SPICE2: A computer program to simulate semiconductor circuits. Memo. no. ERL-520.* Berkeley: Electronics Research Laboratory, University of California.
4. McAndrew, C. C., & Nagel, L. W. (1996). Early effect modeling in SPICE. *IEEE Journal of Solid State Circuits, 31*, 136–138.
5. Turgeon, L. J., & Mathews, J. R. (1980) A bipolar transistor model of quasi-saturation for use in Computer-Aided Design (CAD). In *Proceeding of IEEE International Electron Devices Meeting* (pp. 394–397).
6. Kull, G. M., Nagel, L. W., Lee, S.-W., Lloyd, P., Prendergast, E. J., & Dirks, H. K. (1985). A unified circuit model for bipolar transistors including quasi-saturation effects. *IEEE Transactions on Electron Devices, 32*, 1103–1113.
7. de Graaff, H. C., & Kloosterman, W. J. (1985). New formulation of the current and charge relations in bipolar transistors for modeling for CACD purposes. *IEEE Trasactions Electron Devices, 32*, 2415–2419.
8. Stubing, H., & Rein, H.-M. (1987). A compact physical large-signal model for high-speed bipolar transistors at high current densities—part i: one-dimensional model. *IEEE Transactions on Electron Devices, 34*, 1741–1751.
9. Jeong, H., & Fossum, J. G. (1989). A charge-based large-signal\bipolar transistor model for device and circuit simulation. *IEEE Transactions on Electron Devices, 36*, 124–131.
10. https://ngspice.sourceforge.io/docs/ngspice-34-manual.pdf
11. https://resources.pcb.cadence.com/i/1180526-pspice-user-guide/25?
12. https://www.ti.com/tool/TINA-TI
13. https://www.analog.com/en/design-center/design-tools-and-calculators/ltspice-simulator.html
14. https://www.synopsys.com/implementation-and-signoff/ams-simulation/primesim-hspice.html
15. https://citeseerx.ist.psu.edu/viewdoc/download?doi=10.1.1.200.1131&rep=rep1&type=pdf
16. https://ep.liu.se/ecp/008/posters/034/ecp00834p.pdf
17. https://www.researchgate.net/profile/Iltcho-Angelov/publication/292138930_10AngelovHBT article/links/56a9c81308aeaeb4cef96107/10AngelovHBTarticle.pdf

6 Heterogeneous Junction Field Effect Devices—Schottky Diode, Metal Semiconductor Field Effect Transistor (MESFET), High Electron Mobility Transistor (HEMT)

6.1 Heterogeneous Junction Properties

The heterogeneous junction semiconductor [1–6] has dissimilar base materials on the two sides of the junction. The dissimilarity in the bandgaps of the two materials create unique electrical properties (when the materials are brought in contact) that make a transistor with one|more heterogeneous junctions ideal for ultra high frequency(100 s of MHz–10 s of GHz) and high power(100's of Amperes and Volts) switching applications.

The very high input impedance of a MOSFET (teraOhm range due to its gate insulation) differentiates it from the BJT. *So heterogeneous junction field effect transistors exploit the advantages of non-silicon semiconductor materials (e.g., vastly superior electron mobility) and the insulating properties of silicon dioxide. The electric field is created by using a metal–semiconductor junction, that enables a vertical electric field to control the charge in the channel.* The two main devices that exploit this property are the MESFET (metal–semiconductor FET), and HEMT (high-electron-mobility transistor). These transistors are HJFETs (heterogeneous junction field effect transistors) because of the presence of a metal/semiconductor heterogeneous junction in their structure.

The high electron mobility semiconductors in MESFETs and HEMTs ensure high transconductance, which is a prerequisite for ultra high frequency(100 s of MHz–10 s of GHz) and high power applications. Electron mobility is enhanced in a HEMT is boosted using a variety of device fabrication tricks—e.g., by reducing the doping density in the barrier layer neighbouring the channel, exploiting spontaneous and piezoelectric polarization. *The key feature of a heterogeneous junction field effect transistor is the two dimensional electron gas (2DEG) which behaves like free electrons in a metal. The heterogeneous junction field effect transistor's superb performance characteristics result from carefully controlling the two dimensional electron gas.*

A. Banerjee, *Semiconductor Devices*, Synthesis Lectures on Engineering, Science, and Technology, https://doi.org/10.1007/978-3-031-45750-0_6

6.2 Metal Semiconductor Junction—Schottky Barrier Diode—Ohmic, Rectifying

A metal–semiconductor junction that normally acts as a rectifier are called Schottky barrier. The simplest case is when the electrostatics of the junction is determined by the work functions Φ of the two components, and by the electron affinity X_S of the semiconductor. Figure 6.1a–d show the energy bands for the following cases of this simple metal–semiconductor junction.

- The metal and the semiconductor are not in contact.
- The metal and semiconductor are in contact and the junction is in equilibrium.
- The metal semiconductor junction is forward biased.
- The metal semiconductor junction is reverse biased.

When brought in contact, electrons flow from the metal to the semiconductor, resulting in a depletion region. The width of this depletion zone can be controlled (modulated) by applying a bias (forward|reverse).

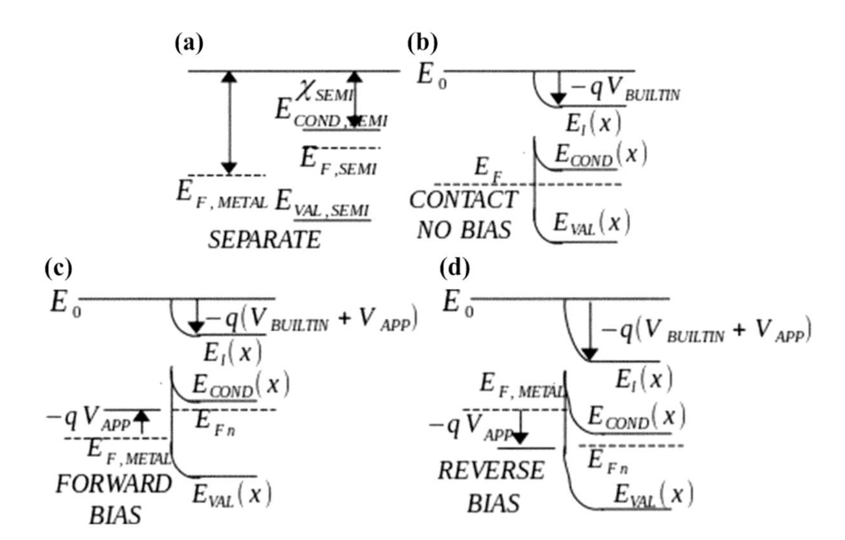

Fig. 6.1 **a**, **b** metal and semiconductor junction before contact and after contact. **c**, **d** forward and reverse biased metal semiconductor junction

The applied bias also changes the the barrier height for electrons crossing the junction from the semiconductor to the metal—just like that of an homogeneous np junction. *A key difference between a heterogeneous and homogeneous semiconductor junction is that for a homogeneous np junction, the net flow across the junction is determined to by the subsequent diffusion of the injected electrons in the p-region.* This impedes charge transport, and perturbs the junction from the equilibrium state. Thus, the quasi Fermi levels are almost constant across the junction.

For a metal semiconductor junction there is an abrupt change in the Fermi levels at the interface—equilibrium conditions are irrelevant. *This arises because the electrons injected into the metal are not controlled by diffusion—they simply join other electrons already in the conduction band of the metal, and a large current can be maintained by an infinitesimal field.* Then charge transport across the junction is controlled by itself and the current across the junction has to be driven by a change in the Fermi level. An identical situation arises in an Np heterogeneous junction with a very short base. From the current density due to electrons injected into the metal is:

$$J_{electron} = q n_0 v_R \mathrm{e}^{\frac{-V_{APP}}{V_{THERMAL}}} = q N_{COND} v_R \mathrm{e}^{\frac{\chi_{SEMI} - \Phi_{METAL} - V_{APP}}{V_{THERMAL}}} \tag{6.0}$$

where $\Phi_{METAL} - \mathrm{X}_{SEMI}$ is the Schottky barrier height. *At zero-bias, electron current density from the semiconductor to the metal must be matched by the electron current density from the metal to the semiconductor.* This flow from the metal is not affected by the applied bias. The full expression for the current is:

$$J_{electron} = q n_0 v_R \mathrm{e}^{\frac{-V_{APP}}{V_{THERMAL}}} = q N_{COND} v_R \left(\mathrm{e}^{\frac{\chi_{SEMI} - \Phi_{METAL} - V_{APP}}{V_{THERMAL}}} - \mathrm{e}^{\frac{\chi_{SEMI} - \Phi_{METAL}}{V_{THERMAL}}} \right) \tag{6.1}$$

The two main modes of charge transport for a metal semiconductor junction are thermionic emission and tunnelling. Although there is an abrupt discontinuity in the quasi Fermi level at the junction, quasilweak equilibrium conditions can still be applied—Fig. 6.2a, b. *The current due to a small number of highly energetic electrons, from a much larger pool of electrons (that remain in an equilibrium distribution), lead to thermionic emission. The number of electrons contributing to thermionic emission is small if the depletion-region potential drops by at least $k_B T$ eV within one mean free path length l of the junction. If the doping density is very large, then the bands bend very steeply in the semiconductor—if the barrier becomes less than ≈5 nm, then tunnelling of electrons can take place. Electrons can pass through such a barrier in either direction. The junction loses its rectifying property. The I-V characteristic becomes linear about the origin, and the contact is said to be* ***ohmic***—essential when the metal is required merely to connect the semiconductor to external circuitry with minimal voltage drop. Both ohmic and rectifying contacts are used in heterogeneous junction field effect transistors.

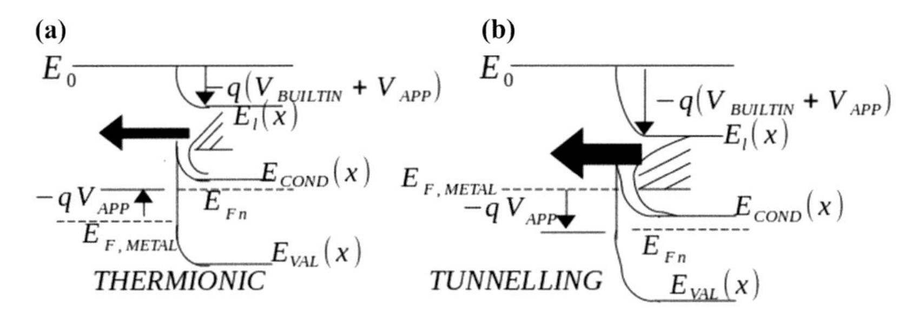

Fig. 6.2 **a**, **b** Thermionic and tunnelling across metal semiconductor junction

6.3 Metal Semiconductor Field Effect Transistor (MESFET)

The physical structure of a metal semiconductor field effect transistor (MESFET) is in Fig. 6.3a. The n^{PLUS} 'cap' layer provides ohmic contacts for the source and drain, while the gate metallization is chosen to make a rectifying contact to the less heavily doped n type active layer. The n regions are grown epitaxially on a weakly doped p region or on a semi-insulating substrate.

Figure 6.3b shows the space-charge region underneath the gate in a MESFET with $V_{DS} > 0$. This bias makes the semiconductor potential at the drain more positive than the source. So the drain side of gate-semiconductor Schottky barrier is reverse biased with reference to the gate—the space-charge region is wider in the vicinity of this junction. The band bending in this layer, and at the n semiconductor insulating junction creates a channel through which charge must pass while traversing to the drain from the source, Fig. 6.3c. The space-charge region width *W(x)* is controlled by the bias V_{GS} applied to the Schottky diode. If, at zero bias, W is less than the thickness of the active layer, then the channel is 'open'. To 'close' it the Schottky diode must be reverse biased sufficiently for the depletion-region edge to reach the interface between the active and semi-insulating

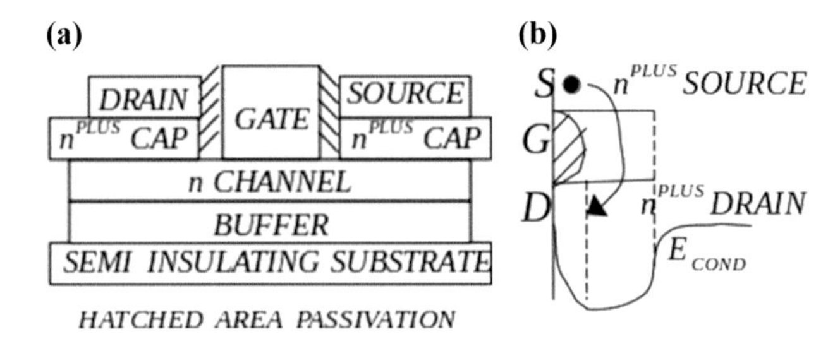

Fig. 6.3 **a**, **b** physical structure and channel formation in MESFET

regions. The gate source voltage required to achieve this is the threshold voltage V_T—negative for n-type semiconductors. FETs which are conducting when $V_{GS} = 0$ are called depletion mode FETs.

A closed channel for a MESFET is equivalent to a MOSFET's *'pinch-off'* condition. *It denotes drain current saturation in 'Level 1' model FET. The channel charge goes to zero at the drain end of the channel, though in reality the charge is small, but finite, and the current is sustained at a high value because the electrons are moving very quickly in this part of the channel.* No channel charge condition occurs if the semiconductor active layer is so thin that the depletion region at zero bias reaches right through the semiconductor. The channel is opened up by forward biasing the Schottky junction—then current grows exponentially with forward bias. So the upper limit on the forward bias is set by the *leakage current* to the gate that can be tolerated, and is only a few tenths of a volt. A MESFET working in this mode has a positive threshold voltage and is an enhancement-mode device.

A depletion mode Level 1 MESFET has drift current density:

$$\overrightarrow{J_{electron,DRIFT}} = -nq\mu_{electron}\frac{\partial V_{CHANNEL}}{\partial x} \tag{6.2}$$

The drain current is by definition:

$$I_{DRAIN} = -J_{electron}w_{transistor}(Channel_{thickness} - W(x))$$

$$W(x) = \sqrt{\frac{2\varepsilon_S}{N_{DONOR}}(V_{BI} + V_{CS}(x) - V_{GS})} \tag{6.2a}$$

where ϵ_S, N_D are the semiconductor material permittivity and doping density of the active layer. Combining the Eq. 6.2a–d gives the general expression for the drain current:

$$I_{DRAIN} = \frac{Aq\mu_{electron}N_{DONOR}W_{transistor}}{L}\left(V_{DS} - \frac{2}{3\sqrt{V_{BI} - V_{THRSH}}}\left((V_{DS} + V_{BI} - V_{GS})^{1.5} - (V_{BI} - V_{GS})^{1.5}\right)\right) \tag{6.3a}$$

After some manipulation, the the saturation drain current is:

$$I_{DRAIN,SAT} = \frac{Aq\mu_{electron}N_{DONOR}W_{transistor}}{L}\left(V_{GS} - V_{THRSH} - \frac{2}{3\sqrt{V_{BI} - V_{THRSH}}}\left((V_{BI} - V_{THRSH})^{1.5} - (V_{BI} - V_{GS})^{1.5}\right)\right) \tag{6.3b}$$

6.4 High Electron Mobility Transistor (HEMT)

In a high electron mobility transistor (HEMT) unlike a MESFET, the channel exists in an underlying, undoped, semiconducting layer Fig. 6.4. Absence of impurity dopant atoms translates to elimination of ionized impurity scattering. *Consequently, electrons move inside this transistor just like free electrons in a metal.* The barrier semiconductor forms a heterogeneous junction with the gate metal, and a heterogeneous junction with the channel layer—Fig. 6.5. *For the AlGaAs/GaAs case examined here, the electron affinities of the two semiconductors are not equal, so there is a discontinuity in the conduction band edge.* This discontinuity defines one side of the channel, containing confined the electrons induced into the undoped semiconductor during charge transfer at the time the system equilibrates. The large band-bending in the undoped semiconductor forms the other side of the confining potential well. *The confinement in the x, y directions is such that the electrons are free to move only in the two, mutually orthogonal directions (x, y). These highly mobile electrons form a two dimensional electron layer (two dimensional electron gas—2DEG). Chapter 15 is dedicated to detailed discussion on state-of-art gallium nitride (GaN) HEMTs.*

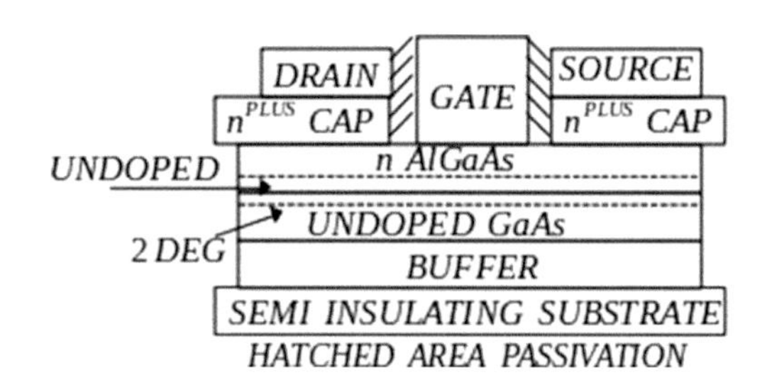

Fig. 6.4 Physical planar layered structure of a AlGaAs-GaAs HEMT

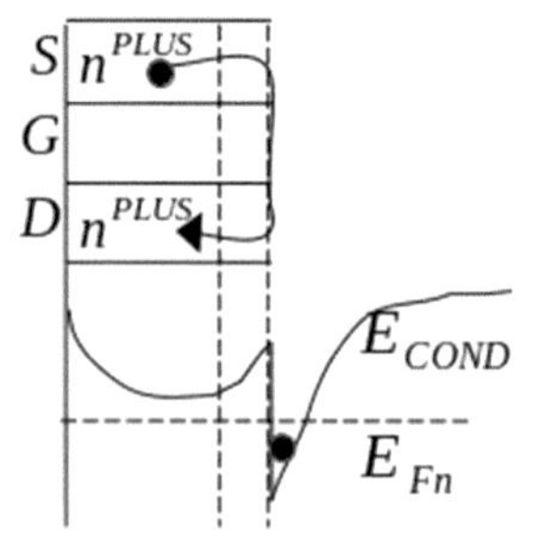

Fig. 6.5 Heterogeneous junction structure and formation

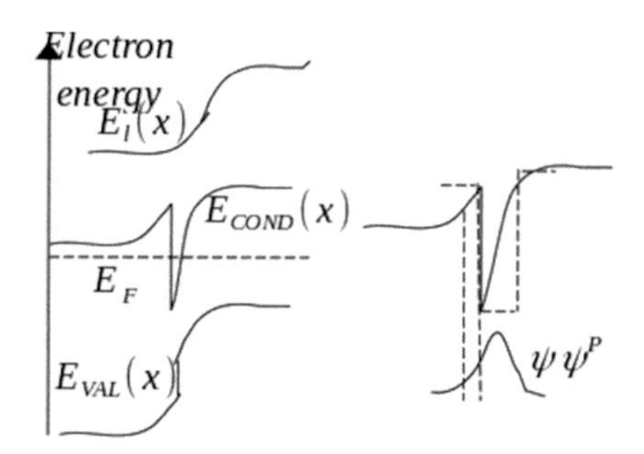

Fig. 6.6 Triangular potential well and rectangular well approximation for heterogeneous junction of HEMT

6.4.1 The Two Dimensional Electron Gas (2 DEG) [1–6]

The two dimensional electron gas is so unique that it requires careful analysis, with two simplifying assumptions.

- The triangular potential well at the AlGaAs/GaAs interface is approximated by a rectangular potential well Fig. 6.6.
- This asymmetrical finite barrier is approximated by a symmetrical barrier stretching to infinite energy.

In this potential well, Schrodinger's wave equation becomes:

$$\frac{d^2\psi(x)}{dx^2} + k_x\psi(x) = 0 \quad k_x = \frac{\sqrt{2m^P E}}{\hbar} \tag{6.4a}$$

The boundary conditions for an infinitely high potential barrier, the corresponding simplest solution, the wave vectors and associated energy levels are:

$$\psi_x(0) = \psi_x(a) = 0 \quad \psi_x(x) = A_0 \sin(k_x x) \quad k_x = \frac{n\pi}{a} \quad n = 1, 2, 3, \ldots \quad E_{x,n} = \frac{\hbar^2 n^2 \pi^2}{2a^2 m^P} \tag{6.4b}$$

The allowed energy levels are quantized. Each allowed energy level depends on k, which is quantized—integer multiples of the reciprocal of the length of the region of interest. The length a is small, so the allowed energy levels are widely separated, and cannot be viewed as a continuum. However there are bands in the orthogonal directions to that of the channel thickness. Each of the allowed energy levels are allowed bands of energy in the unconstrained perpendicular directions y and z. Assuming that electrons are confined near the bottom of these bands, then a parabolic E-k dispersion relationship is used to characterize these bands. Assuming an isotropic effective mass, the total energy is:

$$E = E_{x,n} + \frac{\hbar^2 k_{PERP}^2}{2m^P} \tag{6.4c}$$

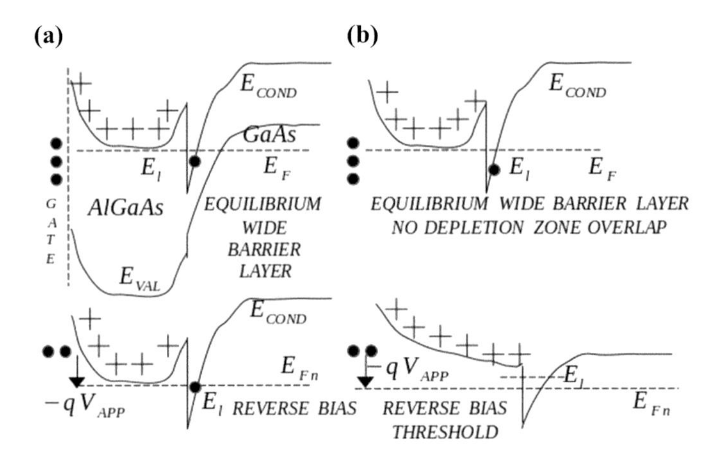

Fig. 6.7 **a–d** Energy band diagrams for AlGaAs-GaAs heterogeneous junction interface under various HEMT operating conditions. Dark dots are electrons, ' + ' are holes

Electrons are trapped in an infinite potential well. *This is ensured by setting $\psi = 0$ at $y = 0$, in deriving the discrete energy level expressions.* The wavefunctions are sinusoidal and the probability density function, for the first energy level, e.g., indicates the greatest probability of finding the electrons at the centre of the well. *This is the ideal condition for correct FET operation since it keeps the electrons away from the vicinity of the junction—crystalline imperfections at the interface can trigger scattering and a reduction in mobility in real world devices*—Fig. 6.7a–c. The well is not finite, so electrons can escape through the sides of the potential well—i.e., the probability density function for channel electrons is not zero in the barrier.

Although penetration into the barrier cannot be prevented, its effect on the mobility can be minimized by inserting a thin, undoped region of AlGaAs barrier material spacer layer, next to the interface. This minimizes ionized impurity scattering.

The concentration of electrons in the potential well region can be estimated easily, by estimating the number of states in a two dimensional region of area πk_{PERP}^2. The annular surface of this circle has an area of $2\pi k_{PERP} \partial k_{PERP}$. In one dimension a state occupies $2\pi/L$ of k-space, where L is the real space length in that direction. The number of states in the annulus, is:

$$N_{2DIM} = \frac{4\pi k_{PERP} \partial k_{PERP}}{\frac{4\pi^2}{L_y L_z}} = \frac{k_{PERP} \partial k_{PERP} L_y L_z}{\pi} \tag{6.5a}$$

where L_y, L_z are the physical dimensions of this region. *After dividing the above expression by the real space area $L_y L_z$, and after converting ∂k_{PERP} to ∂E and finally dividing by ∂E, the two dimensional density of states per unit area and per unit energy, for each sub-band, is*

$$g_{2DIM} = \frac{m^P}{\pi \hbar^2} \tag{6.5b}$$

This density of states expression for each sub band is independent of energy. Estimating electron concentration in each sub-band is straightforward:

$$n_{2DIM,n} = \int \left(\frac{m^P f(E_x, n) dE_{x,n}}{\pi \hbar^2} \right) \tag{6.5c}$$

Each sub band starts at an energy $E_{x,n}$, and assuming that the top of each sub band is at infinity, then using Fermi–Dirac distribution function and integrating gives:

$$n_{2DIM,n} = \frac{m^P k_B T_{lattice} \left(1 + \mathrm{e}^{\frac{E_{FERMI} - E_{x,n}}{k_B T_{lattice}}} \right)}{\pi \hbar^2} \tag{6.5d}$$

Finally, the total concentration of electrons in the channel is a summation of the number of electrons for each sub band:

$$n_{TOTAL} = \sum n_{2DIM,n} \quad 1 \leq n \leq \infty \tag{6.5e}$$

The underlying physical processes of the heterogeneous junction inside a HEMT can be understood with reference to Fig. 6.7a–c. At start, to equilibrate the transistor, the donors in the AlGaAs barrier layer supply electrons to both the gate metal and to the 2-DEG in the channel, creating depletion regions at both ends of the barrier layer, *provided that the barrier layer is sufficiently thick so that the two depletion regions do not overlap.* So, n_S will reach its highest value n_{S0}. Figure 6.7b shows two depletion regions just meeting at equilibrium without any overlap. This is the ideal equilibrium for HEMT operation.

Now a reverse bias applied to the gate barrier Schottky diode forces electrons from to the channel region to move into the gate region—n_S decreases—Fig. 6.7c. This shows the electron quasi Fermi level E_{Fm} dropping towards the first allowed energy level. With gradual increase of the reverse bias, a situation will be reached where all the donors donate electrons to the gate. In this case n_S tends to 0, and the applied voltage $V_{APP} = V_T$.

References

1. Sze, S. M. (2002). *Semiconductor devices, physics and technology* (2nd ed.). Wiley.
2. Berz, F. (1985). The bethe condition for thermionic emission near an absorbing boundary. *Solid State Electronics, 28*, 1007–1013.
3. Kabiraj, D., Grötzschel, R., & Ghosh, S. (2008). Modification of charge compensation in semi-insulating semiconductors by high energy light ion irradiation. *Journal of Applied Physics., 103*, 053703.
4. Roblin, P., & Rohdin, H. (2002). *High-speed heterostructure devices*. Cambridge University Press.

5. Griffiths, D. W. (1995). *Introduction to quantum mechanics*. Prentice-Hall.
6. Pulfrey, D. L. (2010). *Understanding modern transistors and diodes*. Cambridge University Press. ISBN 13 978-0-521-51460-6.

7 AlGaAs-GaAs, AlGaN-GaN, SiC, HEMT Large Signal Equivalent Electrical Circuits (Angelov Chalmers Model)—Normally On|Off HEMT, pHEMT, mHEMT and MODFET

7.1 Large and Small Signal Models of HEMTs and MESFETs

The large signal model and the corresponding equivalent electrical circuit of a HEMT represent the model of the transistor that can be used to analyze and understand the behaviour and performance of the transistor under steady state operating conditions. **As both the HEMT and the MESFET are unipolar, field effect transistors, the large signal model represents the drain source current under steady state conditions**. *In contrast, the large signal model for a HBT (heterogeneous bipolar|bijunction transistor) is more complicated as the large signal model must represent both the collector-emitter and base emitter currents under steady state operating conditions.*

The small signal model representation of the same HEMT|MESFET is essential for all signal frequency based analysis. The small signal analysis is used to extract the scattering parameters (S-parameters). The measured S-parameters are used to estimate essential frequency dependent device properties as input|output impedances, device internal capacitances (parasitic|non-parasitic), parasitic impedances etc., Of large number of available large|small signal device models, widely used ones are the Angelov [1–12], Curtice, CMC (Compact Modelling Consortium), MVSG (MIT Virtual Source GaN field effect transistor) have been included in available commercial CAD tools.

7.2 Angelov-Chalmers Large Signal Model for HEMTs and MESFETs

The equivalent electrical circuit for the Angelov [1–12] large signal model for a HEMT Fig. 7.1. The drain source current is nonlinear:

A. Banerjee, *Semiconductor Devices*, Synthesis Lectures on Engineering, Science, and Technology, https://doi.org/10.1007/978-3-031-45750-0_7

$$I_{DS} = \frac{I_{DSP} - I_{DSN}}{2} \tag{7.1a}$$

$$I_{DSP} = I_{PK0}\Big(1 + \tanh(\psi_P)\Big(1 + \tanh(\alpha_P V_{DS})\Big(1 + \lambda_P V_{DS} + L_{SB0} e^{V_{DG} - V_{TR}}\Big)\Big)\Big) \tag{7.1b}$$

$$I_{DSN} = I_{PK0}(1 + \tanh(\psi_N)(1 - \tanh(\alpha_N V_{DS})(1 - \lambda_N V_{DS}))) \tag{7.1c}$$

$$\psi_P = P_{1M}(V_{GS} - V_{PK}) + P_{2M}(V_{GS} - V + PK)^2 + P_{3M}(V_{GS} - V_{PK})^3 \tag{7.1d}$$

$$\psi_N = P_{1M}(V_{GD} - V_{PK}) + P_{2M}(V_{GD} - V + PK)^2 + P_{3M}(V_{GD} - V_{PK})^3 \tag{7.1e}$$

$$V_{PK}(V_{DS}) = V_{PKS} - \Delta V_{PKS} + \Delta P_{PKS} \tanh(\alpha_S V_{DS} + K_{BG} V_{BG}) \tag{7.1f}$$

$$P_{1M} = P_1(f(T))((1 + \Delta P_1)(1 + \tanh(\alpha_S V_{DS}))) \tag{7.1g}$$

$$P_{2M} = (1 + \Delta P_2)(1 + \tanh(\alpha_S V_{DS})) \quad P_{3M} = (1 + \Delta P_3)(1 + \tanh(\alpha_S V_{DS})) \tag{7.1h}$$

$$\alpha_P = \alpha_R + \alpha_S(1 + \tanh(\psi_P)) \quad \alpha_N = \alpha_R + \alpha_S(1 + \tanh(\psi_N)) \quad P_1 = \frac{g_{m,MAX}}{I_{PK0}} \tag{7.1i}$$

Here Ψ_P is a power series centred around V_{PK}. K_{BG} controls the dispersion through the intrinsic gate voltage at RF frequencies. ΔP_1, ΔP_2 control the second and third harmonic dependencies through V_{DS}.

The temperature dependency of P_1 is embodied in f(T) which represents the temperature dependencies of the carrier velocities. The temperature dependency of carrier concentrations is embodied in the temperature dependency of I_{PK0}. The other parameters are:

- I_{PK}, V_{PK} are the drain current and gate voltage for $g_{m,MAX}$

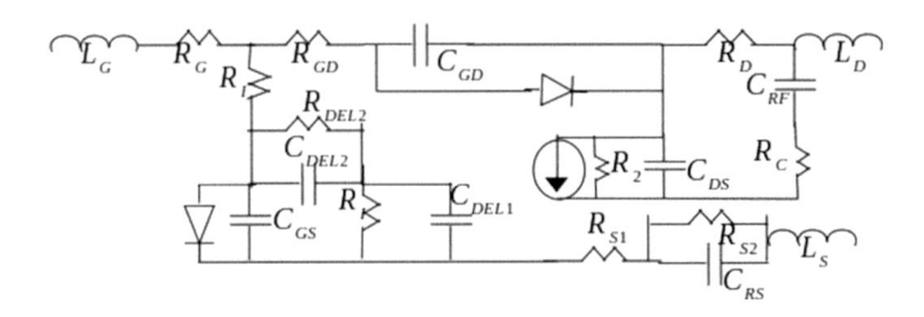

Fig. 7.1 Basic Angelov Chalmers HEMT equivalent circuit model

- α_n, α_p are the saturation parameters
- λ_n, λ_P are channel length modulation parameters

The total number of parameters is 11. The numerical values of these parameters for a given transistor are estimated from a combination of DC sweep (large signal) and small signal AC frequency dependent experimental data.

The large and small signal models do not include any transcapacitances, as the time derivatives of the charge depend only on their own terminal voltage. So the small signal equivalent circuit consists of these capacitances evaluated at the corresponding DC voltage. C_{GS}, C_{GD} are continuous functions of voltages with well defined derivatives—this guarantee that they converge in harmonic balance simulations.

To measure the frequency dispersion and signal delay, a back-gate scheme is used. In this case the RF feedback voltage, V_{BGATE}, controls the RF voltage at the gate and provides both a large and small signal description.

With reference to Fig. 7.1 the delay network (C_{DEL1}, C_{DEL2}, R_{DEL}), connected at the input accounts for high frequency delay effects. At high frequency, the capacitor C_{DEL1} shunts the input and decreases the amplitude of the control voltage V_{GSC}—this is experimentally observed delay. The value of the delay capacitance is estimated to be in the femtoFarad range [1–12]. The time constant $C_{DEL}R_{DEL}$ determines the frequency at which high frequency|power limitations. The frequency dependency of the output power can be tuned using the capacitance C_{DEL2}. Both delay capacitors are similar, often assumed to be equal.

Silicon carbide (SiC) devices exhibit effects found in some GaAs FETs,—an increase of magnitude of S21 versus frequency. These are due to channel on the spreading resistance Rs. This basic model has been extended [12]. Parameter extraction measurement is performed with both pure DC sweep and pulsed DC sweeps, resulting in a more robust large signal model.

7.3 Gallium Nitride (GaN) Properties Normally On|Off HEMTs

Gallium nitride has become the material of choice for constructing HEMTs, because of its unique properties [13]. **State-of-art GaN HEMTs derive their outstanding performance characteristics by exploiting both spontaneous and piezoelectric polarization induced inside the device by virtue of its physical layered structure. Basically, the two degree electron gas (2DEG) is generated without doping**. *Chapter 15 is devoted to gallium nitride (GaN) properties and transistors.* It (GaN) is a wide band gap semiconductor with both face centered cubic (FCC) and hexagonal close packed (HCP) wurtzite crystalline structure, with HCP being the preference for HEMT fabrication. One very attractive property is the wide band gap (3.4 eV) that gives an intrinsic carrier concentration $n_{INTRINSIC}$ several orders of magnitude lower than in Si—*the advantages being reduced leakage current* *and high operating high temperatures.* The high critical electric field and the maximum reachable breakdown of GaN are very important for fabricating high current|voltage (10 s of Amperes and 100 s of Volts) devices. **The breakdown voltage can be engineered to a target value, and in combination with an equally engineered thin drift layers vastly reduces the specific on-resistance compared with Si based devices**. Consequently, very compact devices are fabricated, with minimum static and dynamic losses. GaN's high electron saturation velocity allows ultra high frequency switching.

High frequency switching is possible mainly due to the two dimensional electron gas (2DEG) in typical heterogeneous junctions (e.g., AlGaN/GaN) for which the electron mobility values exceed $\frac{1000\,\text{cm}^2}{V\,s}$.

The only disadvantage of the 2DEG is that the GaN HEMT is *normally on* (**depletion mode**) device. Such a GaN HEMT can achieve switching action if a negative voltage is applied to the gate—*channel interruption.*

However, normally-off devices are preferred in both ultra high frequency RF|microwave circuits and high power electronics—creating negative gate voltage pulses is difficult.

To push the threshold voltage above zero for a normally-off HEMT, the region near the gate must be appropriately modified, e.g., by near surface processing, or bandgap engineering techniques.

- The "recessed gate" approach to creating a "normally-off" HEMT, involves reduction of the AlGaN barrier layer thickness under the gate. Below a certain AlGaN thickness, the Fermi level at the interface will lie below the AlGaN conduction band minimum—effectively depleting the 2DEG below the gate, and thus a positive threshold voltage.
- Another approach is the "fluorine gate" HEMT. Negatively charged fluorine ions are added below the gate electrode, either by plasma or ion implantation. The freshly added negative charge produces a positive shift of the threshold voltage and depletes the 2DEG. Also the negative fixed charges will force an upward conduction band bending of the AlGaN, increasing the metal/AlGaN barrier height and reduce the gate leakage current.

These and some related solutions are difficult to implement in large scale real-world HEMT fabrication processes. Currently, three solutions for fabricating normally-off GaN HEMTs are the "cascode" configuration, the "p-GaN gate" and the "recessed gate hybrid MISHEMT".

7.3.1 P-GaN Gate P-GaN-AlGaN-GaN Normally-Off HEMT [1–19]

Figure 7.2a–c show respectively the physical layered structure and energy band diagram of a p-GaN gate AlGaN-GaN HEMT [14]. **The p-GaN cap layer forces the the Fermi level**

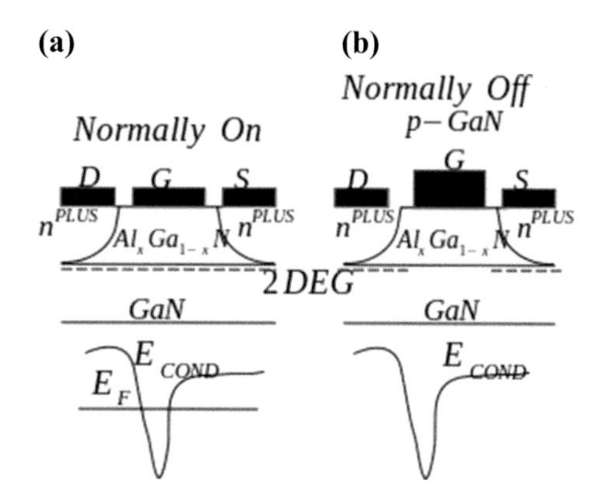

Fig. 7.2 **a–c** p-Gate scheme to convert normally on AlGaN-GaN HEMT **a** to normally off HEMT **b** and corresponding conduction energy bands

of the AlGaN to move below its conduction minimum, thereby disrupting the 2DEG. *So a normally on AlGaN-GaN HEMT becomes a normally-off HEMT. In a normally-on AlGaN-GaN HEMT, the Fermi level is in the middle of the potential well at the junction interface, allowing electrons to collect and form the 2DEG.*

To achieve an efficient 2DEG depletion and push the HEMT threshold voltage to above zero, the properties of the AlGaN-GaN heterostructure such as thickness of the AlGaN barrier and Al-concentration must be accurately monitored. Typically, in a normally-off p-GaN-AlGaN-GaN heterostructure the AlGaN barrier layer thickness is approximately 10–15 nm, with Al concentration of about of 15–20%. A high doping level ($>\frac{10^{18}}{\text{cm}^3}$) of the p-GaN layer is necessary. To improve the threshold voltage for a fixed Mg concentration of the p-GaN layer the Mg electrical activation is boosted by carefully controlling p-GaN layer growth parameters and annealing conditions. This is an active topic of research and development [13].

7.3.2 Recessed Gate Hybrid MISHEMT (Metal–Insulator–Semiconductor HEMT)

In this novel HEMT structure (Fig. 7.3), shows the recessed gate hybrid GaN MISHEMT. **In the gate region the AlGaN barrier layer is removed by plasma etch and the recessed GaN region is passivated by an insulator. This device is a hybrid transistor connecting in series the recessed MIS channel with two access regions having a low resistance—2DEG** [15, 16]. Li [15] and Ikeda [16] have reported recessed gate hybrid GaN MISHEMTs with threshold voltage 2 V, specific on-resistance of 10 mΩ cm^2 and breakdown in the kV range. With reference to Fig. 7.3, the on resistance of this device is:

$$R_{ON,\ MISHEMT} = 2R_{CONTACT} + R_{CHANNEL} + R_{SG,\ 2DRG} + R_{GD,\ 2DEG}$$

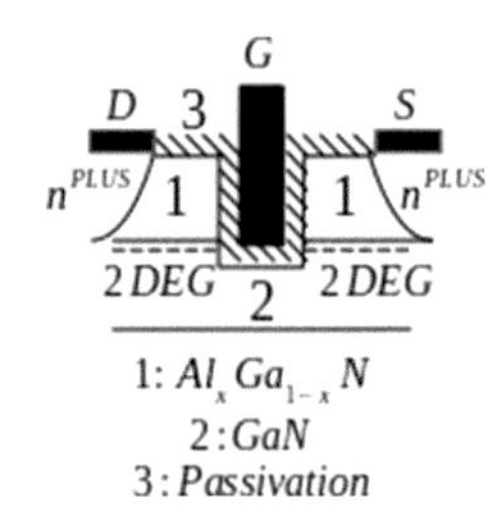

Fig. 7.3 Recessed gate hybrid MISHEMT method to convert normally on AlGaN-GaN HEMT to normally of HEMT. Dark spots are electrons

The channel resistance is directly proportional to the channel length and inversely proportional to the electron mobility.

The recessed gate region is the most important part of this normally-off transistor. The surface roughness of the recessed area, the presence of electrically active defects, and the electronic quality of the insulator/GaN interface all control the properties of this device. The channel mobility also affects the total on-resistance. Consequently, the insulator-GaN interfaces in the recessed channel of MISHEMTs ia an active research topic [13].

7.3.3 Cascode Combination of Enhancement NMOS FET and Normally-On HEMT

This is a very simple way to convert a normally on HEMT to a normally off device, by connecting an enhancement mode NMOS in the cascode configuration—Fig. 7.4.

The advantages/disadvantages of the three schemes are listed as follows:

p-GaN gate:

Advantages:

Low resistance under gate and no dielectric problems.

Disadvantage:

p-GaN etching must be optimized for low access resistance and better reliability. Gate voltage swing is limited.

Recessed gate hybrid MISHEMT

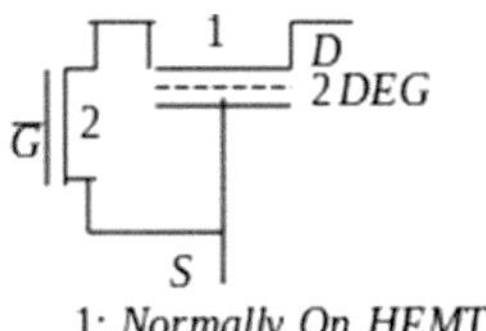

Fig. 7.4 Cascode configuration of normally on AlGaN-GaN HEMT with enhancement mode (normally off) NMOS to covert HEMT to normally off HEMT. Dashed line represents 2DEG

Advantages:

Large forward breakdown and standard device driving in applications.

Disadvantages:

Inappropriate for low voltage (<100 V) applications due to gate channel resistance.

Gate region properties (interface roughness, dielectric properties) affect device reliability and performance.

Enhancement NMOS and Normally On HEMT Cascode:

Advantages:

Stable, tried and tested silicon MOSFET technology.

Disadvantages:

Inappropriate for low voltage (<600 V) and high frequency (>1 MHz) applications. Silicon NMOS needs to be optimized for each application resulting in increased package complexity.

7.4 AlGaAs-GaAs Normally on HEMT|p (*Pseudomorphic*) HEMT and Double Heterojunction HEMT [1–19]

The structure and energy band diagram of the alumimum-gallium-arsenide—gallium arsenide (AlGaAs-GaAs) HEMT|pHEMT are shown in Fig. 7.5a–d. AlGaAs is an alloy of aluminum and gallium arsenide($Al_xGa_{1-x}As\ x \rightarrow\ mole\ fraction$) but is commonly written in its shortened form AlGaAs. The original AlGaAs GaAs HEMT exploited the ultra high electron mobility of the two dimensional electron gas (2DEG) at the heterogeneous junction interface between AlGaAs and GaAs. AlGaAs has a higher bandgap than GaAs and is n doped. So electrons are transferred from the AlGaAs layer (also called the *supply layer*) into the GaAs quantum well channel as it is energetically more stable for them. It is also known as the barrier layer due to its higher bandgap energy compared to the channel material. **But due to the small discontinuity of the conduction band energy at the AlGaAs—GaAs heterojunction interface, the immobile donor charge in the AlGaAs supply layer is also modulated by the applied gate potential**. *Consequently the channel sheet charge density is easily saturated, resulting in degraded modulation efficiency and speed of the AlGaAs/GaAs HEMT device.* **Using clever bandgap engineering tricks, if the GaAs channel is substituted with an indium gallium arsenide (InGaAs) thin film inserted between the GaAs buffer and the AlGaAs supply layer, the resulting AlGaAs/InGaAs heterostructure has a larger conduction band discontinuity**. *Now electrons are tightly confined within the InGaAs quantum well and the electron mobility of InGaAs is*

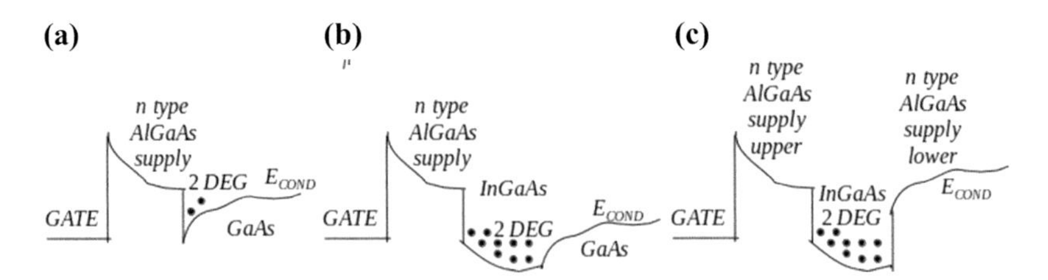

Fig. 7.5 **a–c** AlGaAs-GaAs HEMT, pseudomorhic AlGaAs HEMT and double pseudomorphic AlGaAs-GaAs HEMT

*higher compared to GaAs.*So there is a significant improvement in the modulation efficiency and performance of the HEMT device. **InGaAs has a lattice constant mismatch between AlGaAs and GaAs, but it can be grown dislocation free as long as its thickness is less than a certain critical thickness**. *Below this critical thickness value, the InGaAs layer can be compressed and distorted from its normal cubic crystalline structure to match the lattice constant of both AlGaAs and GaAs.* The strain from this lattice mismatch is contained entirely within the InGaAs layer. **As the InGaAs layer is unnaturally compressed to match the lattice constant and structure of GaAs, the device is called *pseudomorphic* HEMT**.

While the single InGaAs layer pseudomorphic HEMT is ideal for ultra high frequency RFlmicrowave applications, in order to use it for ultra high power applications, a *double pseudomorphic* HEMT structure is required. there is an n-AlGaAs supply layer not only on top but also below the channel. **The common power pHEMT structure has a InGaAs channel between the two n-AlGaAs supply layers**.

7.5 m(Me)tamorphic HEMT and MODFET

The indium gallium arsenide (InGaAs) layer that gives the pHEMT its enhanced ultra high electron mobility also introduces crystalline imperfections due to strain. ***As all the layers of a HEMT are deposited using molecular beam epitaxy, this means that the deposited layer must have the same crystalline structure as the substrate layer on which it is laid down***. So, channels with indium mole fractions >25% are not fabricated on GaAs substrates, Fig. 7.6a, b. Indium phosphide (InP) overcomes the strain induced performance limitation since it is lattice matched with 53% mole fraction InGaAs channels. But manufacturing these devices is complicated and increases manufacturing costs.

GaAs based *metamorphic* HEMT (mHEMT) technology addresses this issue by providing a properly grown buffer between the substrate and active device layers. *In the mHEMT, the device active layers are grown on a strain relaxed, compositionally graded buffer layer which transforms the lattice constant from GaAs up through InP, allowing a high In content channel layer to be grown on low cost GaAs substrates. The buffer layer*

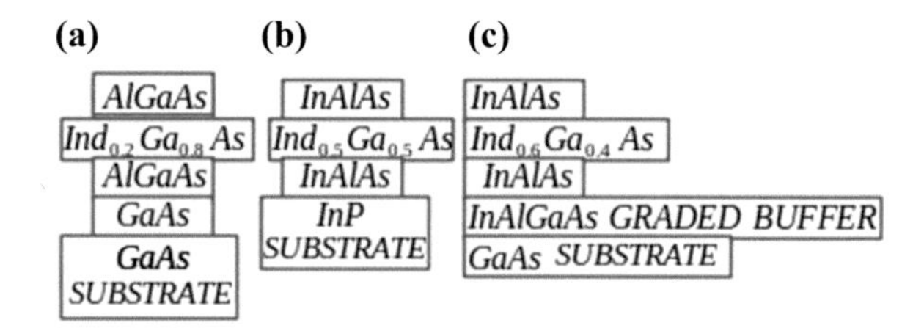

Fig. 7.6 **a–c** 30% InGaSd pHEMT (**a**), 50% InP HEMT (**b**) and 60% InP mHEMT on GaAs substrate (**c**). Compositions for the top Schottky layer InGaAs channel and substrate are shown

provides the ability to tailor the lattice constant to any indium content channel desired (20–70%), and allows the device designer an additional degree of freedom to optimize transistors for high frequency gain, power, linearity and low noise. For example, using a metamorphic buffer layer, InP-based high electron mobility transistors (53% In) can be grown on GaAs substrates for a substantial cost reduction and manufacturability improvement over InP-substrate based devices—Fig. 7.6c.

The MODFETs exceptional switching frequency arise from its structure. The wide band element is doped with donor atoms (excess electrons in its conduction band). These electrons diffuse to the adjacent narrow bandgap material's conduction band due to the available lower energy states. Immediately, a potential and attendant electric field is generated, between the materials. The electric field pushes electrons back to the wide band material's conduction band. Finally diffusion and electron drift processes balance each other, creating a junction at equilibrium. The undoped narrow band gap material now has excess majority charge carriers. The concentration of majority charge carriers (electrons) results in high switching frequencies, and the undoped low bandgap semiconductor means donor atoms do not generate unwanted scattering sites.

The secret underlying the MODFET is that the band discontinuities across the conduction and valence bands can be modified separately. Consequently, the type of carriers in and out of the device can be controlled. Therefore, a graded doping can be applied in one of the materials making the conduction band discontinuity smaller, and keeping the valence band discontinuity the same. This diffusion of carriers leads to the accumulation of electrons along the boundary of the two regions inside the narrow band gap material.

7.6 AMS-CMC GaN HEMT Model

The AMS CMC (Compact Modelling Constortium) model [19–33] for the GaN HEMT is used in all commercially available CAD (Computer Aided Design) tools as AWR Microwave Office and Agilent ADS.

The AMS CMC HEMT model has been developed based on surface potentials, and the Fermi level is:

$$E_{FERMI,\,CMC} = V_{go} - \frac{2V_{THERMAL}\ln\left(1 + e^{\frac{V_{go}}{2V_{THERMAL}}}\right)}{\frac{1}{H(V_{go,p})} + \frac{C_g e^{\frac{-V_{go}}{V_{THERMAL}}}}{qD}} \tag{7.2a}$$

where $V_{go} = V_{GS} - V_{OFF}$ V_{OFF} is the cut-off voltage and C_{GATE}, D are the gate capacitance per unit area and density of states respectively.

$V_{go,p} = V_{go} : V_{GS} \geqslant V_{THRSH}$ $V_{go,p} = V_{THERMAL} : V_{GS} < V_{THRSH}$ H contains information about the Fermi level when $V_{go} > V_{OFF}$. The gate charge is:

$$Q_{GATE} = -\int qWndx = -\int qWC_{GATE}\left(V_{GS} - V_{OFF} - E_f - V(x)\right)dx \quad 0 \le x \le L \tag{7.2b}$$

Using the surface potential concept, the drain current is:

$$I_{DS} = \frac{\mu_{effective}C_{GATE}(\psi_D - \psi_S)\left(1 + \lambda V_{DS,\,effective}\left(V_{go} + V_{THERMAL} - \psi_m\right)\right)W}{L\sqrt{1 + \theta_{SAT}^2(\psi_D - \psi_S)^2}} \tag{7.2c}$$

where λ, θ_{SAT} are channel length modulation and velocity saturation parameters respectively.

$\Psi_m = \frac{\Psi_D + \Psi_S}{2}$ and mobility degradation due to the vertical field is:

$$\mu_{effective} = \frac{U_0}{1 + U_A\left(V_{go} - \psi_m\right) + U_B\left(V_{go} - \psi_m\right)^2} \tag{7.2d}$$

where U_0, U_A, U_B are curve fitting parameters. The drain current implementation includes drain induced barrier lowering (DIBL), via the bias dependency of the cut-off voltage. The total source, drain access region resistance is implemented as a sum of the source, drain contact resistance and the bias dependent access region resistance.

Depending on the underlying mechanism, the gate current is divided into three components. The current density for the Poole–Frenkel part of the gate current is:

$$J_{Poole,\,Frenkel} = CEe^{\alpha + \beta\sqrt{E}} \quad B = \epsilon_S K \tag{7.2e}$$

where $\alpha = \frac{-\Phi_d}{V_{THERMAL}}$ Φ_d is the barrier height for electron emission from the trap state. $\beta = \frac{\sqrt{\frac{q}{\pi\epsilon_S}}}{V_{THERMAL}}$ C is a parameter dependent on the trap concentration and E is the electric field. $E = \frac{q\alpha_p - C_{GATE}(V_{go} - \Psi)}{\varepsilon_S}\alpha_P$.

The thermionic emission and trap assisted tunnelling currents are:

$$I_{THERMIONIC\ EMISSION} = W\int J_{TE0}\left(e^{\frac{V_{GATE} - \psi}{\eta V_{THERMAL}}} - 1\right)dx \quad 0 \le x \le L$$

$$J_{TE0} = A^P T_{lattice}^2 e^{\frac{-\Phi_B}{V_{THERMAL}}} \tag{7.2f}$$

$$I_{TRAPASSIST\ TUNNEL} = W \int J_{TE0} \left(e^{\frac{V_{GATE} - \psi - V_0}{\eta V_{THERMAL}}} - 1 \right) dx \quad 0 \le x \le L \tag{7.2g}$$

J_{TA0} is the reverse saturation current density, and

$$J_1 = J_{TE0} KLW \quad J_2 = J_{TA0} KLW \tag{7.2h}$$

A^P is the effective Richardsons constant, Φ_B is the Schottky barrier height, η is the ideality factor and V_0 curve is a fitting parameter.

There are two DC transfer characteristics of a field effect transistor:

- Drain source current with linearly or pulsed varied gate source voltage and constant drain source voltage.
- Drain source current with linearly or pulse varied drain source voltage and constant gate source voltage.

Both methods create a family of curves.

References

1. Angelov, I., Desmaris, V., Dynefors, K., Nilsson, P. A., Rorsman, N., & Zirath, H. (2005). *On the large signal modelling of AlGaN/GaN HEMTs and SiC MESFETs* (pp 309–311). 13th GAAS Symposium Paris.
2. Trew, R. J. (2002). SiC and GaN transistors—Is there one winner for microwave power applications? *Proceedings of the IEEE, 90*, 1032–1047.
3. Neudeck, P. G., Okojie, R. S., & Yu, C. L. (2002, June). High-temperature electronics—a role for wide bandgap semiconductors? *Proceedings of the IEEE, 90*, 1065–1076.
4. Wu, Y. F., Saxler, A., Moore, M., Smith, R. P., Sheppard, S., Chavarkar, P. M., Wisleder, T., Mishra, U. K., & Parikh, P. (2004). 30-W/mm GaN HEMTs by field plate optimization. *IEEE Electron Device Letters, 25*(3), 117–119.
5. Binari, S. C., Klein, P. B., & Kazior, T. E. (2002). Trapping effects in GaN and SiC microwave FETs. *Proceedings of the IEEE, 90*, 1048–1058.
6. Canfield, P. C., Lam, S. C., & Allstot, D. J. (1990). Modeling of frequency and temperature effects in GaAs MESFETs. *IEEE Journal of Solid State Circuits, 25*(1), 300–306.
7. Scheinberg, N., Bayruns, R., & Goyal, R. (1988). A low-frequency GaAs MESFET circuit model. *Solid State Circuit, 23*, 605–608.
8. Lee, M., et al. (1990, October). A self-backgating GaAs MESFET model for low-frequency anomalies. *IEEE Transactions on Electron Devces, 37*, 2148–2157.
9. Curtice, W. R., Bennett, J. H., Suda, D., & Syrett, B. A. (1998). Modeling of current lag in GaAs IC's. *Microwave Symposium Digest, 2*, 603–606.

10. Desmaris, V., Eriksson, J., Rorsman, N., & Zirath, H. (2004). High CW power 0.3 μm gate AlGaN/GaN HEMTs grown by MBE on sapphire. In *Material science forum,* ICSCRM 2003, (Vol. *457–460*, No. II, pp. 1629–1632).
11. Rorsman, N., Nilsson, P. A., Eriksson, J., & Zirath, H. (2004). Investigation of the scalability of 4H-SiC MESFETs for high frequency applications. In *Material science forum*, ICSCRM 2003 (Vol. 457–460, No. II, pp. 1229–1232).
12. Angelov, I., Bengtsson, L., & Garcia, M. (1996). Extensions of the Chalmers nonlinear HEMT and MESFET model. *IEEE Transactions on Microwave Theory and Techniques, 44*, 1664–1674.
13. Roccaforte F., Greco G., Fiorenza P., & Iucolano, F. (2019). An overview of normally-off high electron mobility transistors. *Materials, 12*, 1599. https://doi.org/10.3390/ma12101599
14. Uemoto, Y., Hikita, M., Ueno, H., Matsuo, H., Ishida, H., Yanagihara, M., Ueda, T., Tanaka, T., & Ueda, D. (2007). Gate injection transistor (GIT)—A normally-off AlGaN/GaN power transistor using conductivity modulation. *IEEE Transactions on Electron Devices, 54*, 3393–3399.
15. Li, Z., & Chow, T. P. (2011). Channel scaling of hybrid GaN MOS-HEMTs. *Solid-State Electronics, 56*, 111–115.
16. Ikeda, N., Tamura, R., Kokawa, T., Kambayashi, H., Sato, Y., Nomura, T., & Kato, S. (2011). Over 1.7 kV normally-off GaN hybrid MOS-HFETs with a lower on-resistance on a Si substrate. In *Proceedings of the 23rd International Symposium on Power Semiconductor Devices and IC's (ISPSD2011)*, San Diego, CA, USA, 23–26 May 2011 (pp. 284–287). https://doi.org/10.1109/ISPSD.2011.5890846
17. https://www.mpifr-bonn.mpg.de/1206300/rws_mhemt_mse.pdf.
18. https://www.microwavejournal.com/articles/print/3185-metamorphic-transistor-technology-for-rf-applications.
19. I9. Angelov, et al. (1992, December). *IEEE Transactions on Microwave Theory and Techniques.*
20. Sadi, T., et al. (2010). *IEEE Embedded Systems Devices and Computing.*
21. Long, Y., et al. (2012, October). *IEEE Transactions on Microwave Theory and Techniques.*
22. Koumdymov, A., et al. (2008, March). *IEEE Transactions on Electron Devices.*
23. Li, M., et al. (2008, January). *IEEE Transactions on Electron Devices.*
24. Cheng, X., et al. (2009, December). *IEEE Transactions on Electron Devices.*
25. Khandelwal, S., et al. (2012, October). *IEEE Transactions on Electron Devices.*
26. Khandelwal, S. (2013). Ph.D. dissertation, Department of Electronics and Telecommunication, NTNU.
27. Khandelwal, S., et al. (2013, October). *IEEE Transactions on Electron Devices.*
28. Oh, S., et al. (1980, August). *IEEE Journal of Solid-State Circuits.*
29. Ghosh, S., et al. (2015, February). *IEEE Transactions on Electron Devices.*
30. Dasgupta, A., et al. (2014). *IEEE Journal of Electron Devices Society.*
31. Dasgupta, A., et al. (2015). *IEEE Microwave and Wireless Components Letters.* Accepted March 2015
32. Thorsell, M., et al. (2009, Janury). *IEEE Transactions on Microwave Theory and Techniques.*
33. Vitanov, S., et al. (2010, October). *Solid-State Electronics.*

8 Homogeneous Bipolar|Bijunction Transistor

8.1 NPN and PNP Homogeneous Bipolar Transistors

Conceptually, two common homogeneous np junction diode can be connected in two ways—anodes (p-doped regions) joined together or the cathodes (n-doped regions joined together). The first configuration conceptually forms the NPN transistor (negative–positive–negative) while the second is the PNP (positive–negative–positive) transistor [1–15]. *For the NPN transistor, the p-doped region forms the base, with one of the other two n-doped regions the collector, and the remaining n-doped region the emitter*. The collector and emitter regions are *not interchangeable:*—the emitter region is orders of magnitude heavily doped compared to the collector region. For a PNP transistor, the common n-doped region is the base terminal. The majority charge carrier in the NPN transistor are electrons, while for the PNP it is holes. As holes are defects in the crystal lattice, rather than actual particles (electrons) the NPN transistor is much faster and easier to analyse.

8.2 Ebers Moll Model of a NPN Transistor

The Ebers Moll model [1–6] was the first attempt to analyse the bipolar|bijunction transistor. The model was improved from Eber Moll I to Ebers Moll II and Ebers Moll III, but it has been superseded by the superior Gummel Poon model that addresses the deficiencies of the Ebers Moll model. With reference to the simplest Ebers Moll model (Fig. 8.1a), considering the base-emitter junction in isolation, the current is:

A. Banerjee, *Semiconductor Devices*, Synthesis Lectures on Engineering, Science, and Technology, https://doi.org/10.1007/978-3-031-45750-0_8

$$
\begin{aligned}
I_{FWD} &= qA\left(\frac{D_E n_i^2}{L_E N_E} + \frac{D_B n_i^2}{N_B W_B}\right)\left(e^{\frac{V_{BASE-EMITTER}}{V_{THERMAL}}} - 1\right) \\
I_{ES} &= qA\left(\frac{D_E n_i^2}{L_E N_E} + \frac{D_B n_i^2}{N_B W_B}\right)
\end{aligned}
\tag{8.1a}
$$

and similarly for the base collector junction in isolation,

$$
\begin{aligned}
I_{REV} &= qA\left(\frac{D_C n_i^2}{L_C N_C} + \frac{D_B n_i^2}{N_B W_B}\right)\left(e^{\frac{V_{BASE-COLLECTOR}}{V_{THERMAL}}} - 1\right) \\
I_{CS} &= qA\left(\frac{D_C n_i^2}{L_C N_E} + \frac{D_B n_i^2}{N_B W_B}\right)
\end{aligned}
\tag{8.1b}
$$

where:

$D_B,\ D_C,\ D_E$ are the charge carrier diffusivities in the base, collector and emitter regions.

$L_C,\ L_E$ are the mean charge carrier diffusion lengths in the collector and emitter regions.

$N_B,\ N_C,\ N_E$ are the charge carrier concentrations in the base, collector and emitter regions.

$q,\ A,\ n_i,\ W_B$ are the electronic charge, junction area of cross-section, intrinsic carrier concentration and base width respectively.

From Kirchoff's current law:

$$
\begin{aligned}
&I_{EMITTER} = I_{FWD} - \alpha_{REV} I_{REV} \quad I_{COLLECTOR} = \alpha_F I_{FWD} - I_{REV} \\
&I_{BASE} = I_{EMITTER} - I_{COLLECTOR}
\end{aligned}
\tag{8.1c}
$$

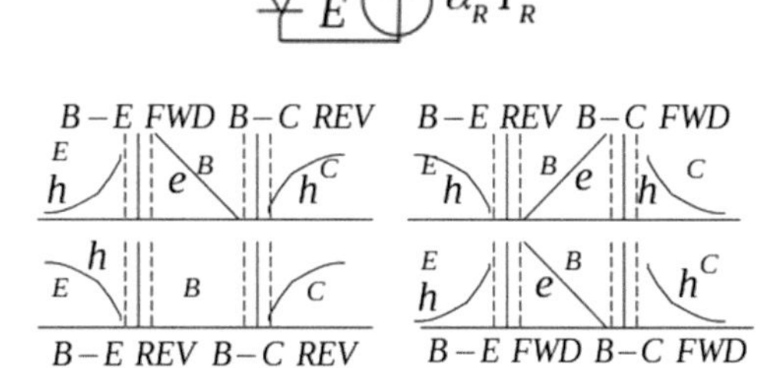

Fig. 8.1 **a** Basic Ebers Moll model of the NPN homogeneous junction transistor. **b** Charge concentrations in base (B), collector (C) and emitter (E) regions of a homogeneous junction NPN transistor under four possible bias conditions

The charge concentrations in the base, collector and emitter regions under various biasing conditions is shown in Fig. 8.1b. The analysis for a PNP transistor proceeds on similar lines, and is left for the reader.

8.3 Gummel Poon Model for a Homogeneous Junction NPN Transistors

The Gummel Poon model [1–6] includes the missing details of the simple Ebers Moll model.

- Low current drop in transistor beta (forward DC current gain) because of recombination of carriers in the base emitter junction.
- Full analysis of Early effect (base-width modulation).
- High-level injection during device saturation.
- Base emitter and base collector junction leakage currents.
- Base collector and base emitter junction capacitances.
- Base, collector and emitter terminal resistances.

The Gummel Poon model is included in SPICE [12–16] (Simulation Program with Integrated Circuit Emphasis) the gold standard electrical|electronic circuit simulation and performance evaluation tool, widely available in both open source and proprietary versions.

The equivalent circuit model for a NPN homogeneous bijunction transistor is in Fig. 8.2. Based on this equivalent circuit, the base and collector currents are listed next. Here I_{LC}, I_{LEE} are the collector and emitter leakage currents.

$$I_{CC} = \frac{I_{SS}}{q_b}\left(e^{\frac{V_{BE}}{V_{TE}}-1}\right) \quad V_{TE} = \frac{n_F k_B T_{lattice}}{q} \quad q_b = \frac{q_1}{2} + \sqrt{\frac{q_1^2}{4} + q_2}$$
$$q_1 = \frac{V_{BE}}{V_B} + \frac{V_{BC}}{V_A} + 1 \quad q_2 = \frac{I_{SS}\left(e^{\frac{V_{BE}}{V_{TE}}} - 1\right)}{I_{KF}} + \frac{I_{SS}\left(e^{\frac{V_{BC}}{V_{TC}}} - 1\right)}{I_{KR}} \tag{8.2a}$$

$$I_{EC} = \frac{I_{SS}}{q_b}\mathrm{e}^{\left(\frac{V_{BC}}{V_{TC}}-1\right)} \quad V_{TC} = \frac{n_R k_B T_{lattice}}{q} \quad I_{LEE} = \frac{I_{SE}}{q_b}\mathrm{e}^{\left(\frac{V_{BE}}{V_{TEL}}-1\right)}$$
$$V_{TEL} = \frac{n_E k_B T_{lattice}}{q} \tag{8.2b}$$

$$I_{LC} = \frac{I_{SC}}{q_b}\mathrm{e}^{\left(\frac{V_{BC}}{V_{TCL}}-1\right)} \quad V_{TCL} = \frac{n_C k_B T_{lattice}}{q} \tag{8.2c}$$

The junction capacitances in Fig. 8.2 are both nonlinear, because of depletion region and diffusion processes. Here

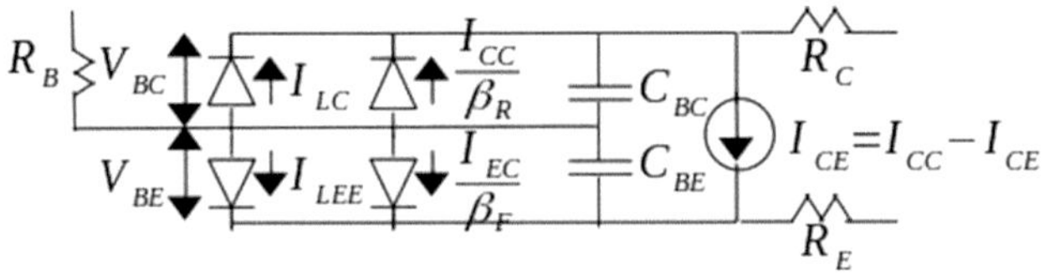

Fig. 8.2 Basic Gummel Poon model

$$C_{BE}(V_{BE}) = \frac{\tau_F \partial I_{CC}}{\partial V_{BE}} + \frac{C_{JE}}{\left(1 - \frac{V_{BE}}{V_{JE}}\right)^{m_E}} \quad V_{BE} < FCV_{JE} \tag{8.3a}$$

$$C_{BE}(V_{BE}) = \frac{\tau_F \partial I_{CC}}{\partial V_{BE}} + \frac{C_{JE}}{F_{2E}} \quad V_{BE}\left(F_{3E} + \frac{m_E V_{BE}}{V_{JE}}\right) \quad FC \geq V_{JE}$$
$$F_{2E} = (1 - FC)^{1+m_E} \quad F_{3E} = 1 - FC(1 + m_E) \tag{8.3b}$$

$$C_{BC}(V_{BC}) = \frac{\tau_R \partial I_{CE}}{\partial V_{BC}} + \frac{C_{JC}}{\left(1 - \frac{V_{BC}}{V_{JC}}\right)^{m_C}} \quad V_{BC} < FCV_{JC} \tag{8.3c}$$

$$C_{BC}(V_{BC}) = \frac{\tau_R \partial I_{CE}}{\partial V_{BC}} + \frac{C_{JC}}{F_{2C}} \quad V_{BC}\left(F_{3C} + \frac{m_C V_{BC}}{V_{JC}}\right) \quad FC \geq V_{JC}$$
$$F_{2C} = (1 - FC)^{1+m_C} \quad F_{3C} = 1 - FC(1 + m_C) \tag{8.3d}$$

The Gummel Poon model parameters, included in SPICE[12–16] are listed below:

I_{SS}	transport saturation current
β_F, β_R	ideal forward|reverse DC current gain
n_F, n_R	forward|reverse current emission coefficient
V_A, V_B	forward|reverse Early voltages
I_{KF}, I_{KR}	forward|reverse beta high current roll-off corner
I_{SC}, I_{SE}	base–collector|emitter leakage saturation current
n_C, n_E	base collector|emitter leakage current transmission coefficient
m_C, m_E	base collector|emitter np grading coefficient
V_{JC}, V_{JE}	base collector|emitter junction built-in voltage
C_{JC}, C_{JE}	base collector|emitter zero bias capacitance
τ_F, τ_R	ideal forward|reverse transit time FC-forward bias depletion capacitance coefficient
R_B, R_C, R_E	base, collector and emitter terminal resistances

There are two DC transfer characteristics of a bipolar|bijunction transistor:

- Collector emitter current with linearly or pulsed varied base emitter voltage and constant collector emitter voltage.
- Collector emitter current with linearly or pulse pulse collector emitter voltage and constant base emitter voltage.

  Both create a family of curves.

8.4 VBIC Enhancement to Gummel Poon Model

The Vertical Bipolar Inter Company (VBIC) [7–10] model was formulated entirely by the semiconductor and Computer Aided Design (CAD) industry to correct the deficiencies of the SPICE [12–16] Gummel Poon Model, while exploiting its power and flexibility. The bijunction transistor properties not addressed by the original Gummel Poon model are:

- Improved Early effect analysis.
- Inclusion of quasi-saturation.
- Parasitic substrate transistor modelling.
- Parasitic fixed (oxide) capacitance modelling.
- Avalanche multiplication analysis.
- Improved temperature dependence modelling.
- Decoupling of base and collector currents.
- Inclusion of self heating (electrothermal) effects.
- $C\infty$ continuous (smooth) modelling.
- Improved heterogeneous junction bipolar transistor (HBT) modelling.

In the VBIC formulation bipolar transistors are 4 terminal: 5 terminal transistors can be modelled by embedding the VBIC four terminal equivalent circuit model inside a subcircuit. Figure 8.3a, b show typical planar, vertical layered NPN transistors that VBIC is designed to model. The VBIC model can also be used for planar, vertical layer PNP and HBT devices, but is inappropriate for lateral BJTs. VBIC four terminal equivalent circuit embedded inside a five terminal circuit does not accurately model transistor action of the second parasitic BJT.

Compact models for circuit simulation must scale properly with device geometry. But because of the large number of available BJT layout topologies, this scaling does not work very well—this explains lack of geometry features in VBIC. Geometry scaling for VBIC is left to either pre-processing for the generation of model libraries for circuit simulation, or via scaling relations specific to a particular technology implemented either in the simulator or the CAD system.

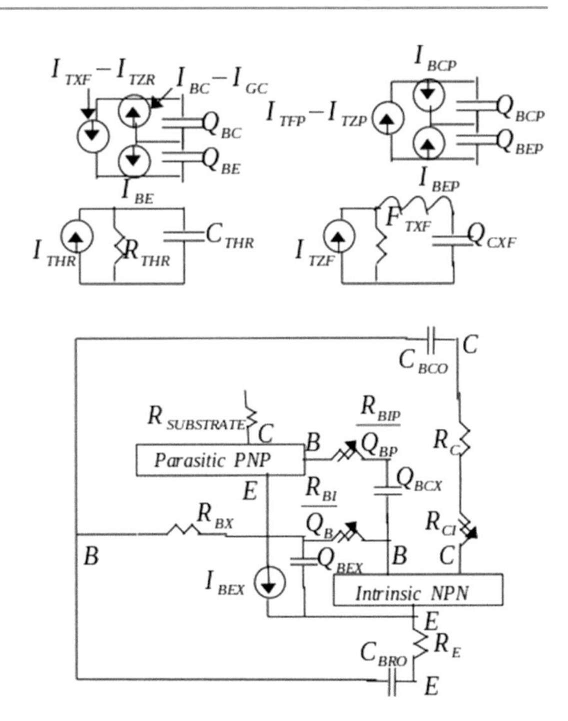

Fig. 8.3 **a–d** VBIC enhanced Gummel Poon NPN homogeneous junction transistor model parasitic PNP transistor and intrinsic NPN transistor (**a**, **b**). Thermal and phase circuits (**c**, **d**). **e** VBIC enhanced homogeneous junction NPN transistor equivalent circuit model

The equivalent circuit of a VBIC based NPN transistor consists of an intrinsic NPN transistor, a parasitic PNP transistor, parasitic resistances and capacitances, a local thermal network (used only with the electrothermal version of the model), and a delay circuit to account for excess phase for the forward transport current (Fig. 8.3a, b).

In the electrothermal version of VBIC branch currents and charges in the electrical part of the model depend on the local temperature rise, the voltage on the node dt. *The thermal-2-equivalent circuit has two nodes external to the model so that the local heating|dissipation can be connected to a thermal network. This thermal network models the thermal properties of the material in which the BJT and surrounding devices are built.*

The equivalent network (Fig. 8.3e) has fixed and bias dependent resistances, capacitances, and current sources, these circuit elements are in reality voltage controlled charge and current sources $Q(V_1, V_2, \ldots V_n)$ $I(V_1, V_2, \ldots .V_n)$. Therefore resistors and capacitors are then voltage controlled current and charge sources. Branch currents and charges that are controlled by more than one branch voltage, include transconductance and transcapacitance elements when they are linearized, to enable DC solution, AC, noise and transient simulations. The VBIC circuit parameters [6] are listed below:

- Zero|non-zero phase forward transport current.
- Non-zero phase forward charge and inductance.
- Reverse transport. forward|reverse base emitter currents.

- Intrinsic base collector, weak base collector avalanche currents.
- Forward|reverse parasitic transport currents.
- Parasitic base emitter|collector currents.
- Intrinsic|external depletion, diffusion and parasitic base emitter charge.
- Intrinsic (diffusion, depletion), external (diffusion only) and parasitic base collector charges.
- Intrinsic (modulated), external (fixed) collector resistance.
- External fixed base, fixed emitter and substrate resistances.
- Modulated intrinsic and parasitic base resistances.
- Parasitic overlap base collector|emitter capacitances.
- Thermal current, capacitance and resistances.

The details of the formulation of the VBIC model are in [6].

References

1. Sze, S. M., & Lee, M. K. (2021). *Semiconductor physics and devices*. Wiley. ISBN 9789354243226.
2. Martins, E. R. (2022). *Essentials of semiconductor device physics*. Wiley. ISBN 987-1-119-88413-2.
3. Pierret, R. *Semiconductor device fundamentals*. ISBN 10 0201543931; 13 978-0201543933.
4. Hess, K. *Advanced theory of semiconductor devices*. Wiley, IEEE Press. ISBN 100780334795; 13 978-0780334793.
5. Streetman, B. G., & Banerjee, S. K. *Solid state electronic devices*. ISBN: 9789332555082, 9789332555082.
6. Pullfrey, D. (2013). *Understanding modern transistors and diodes*. Cambridge University Press. ISBN 13 978-0521514606.
7. https://designers-guide.org/vbic/
8. https://edadocs.software.keysight.com/pages/viewpage.action?pageId=5923753
9. https://link.springer.com/chapter/10.1007/1-4020-7929-X_2
10. https://www.globalspec.com/reference/66420/203279/6-2-the-vbic-model
11. https://help.simetrix.co.uk/8.1/simetrix/simulator_reference/topics/analogdevicereference_bipolarjunctiontransistor_vbicwithoutselfheating.htm
12. https://ngspice.sourceforge.io/
13. https://www.pspice.com/
14. https://www.synopsys.com/implementation-and-signoff/ams-simulation/primesim-hspice.html
15. https://www.ti.com/tool/TINA-TI
16. https://www.analog.com/en/design-center/design-tools-and-calculators/ltspice-simulator.html

9 Metal Oxide Semiconductor Field Effect Transistor

9.1 The Long Channel MOSFET

The MOSFET [1–19] is a *minority carrier* four terminal (drain, gate, source, body|substrate) device. For a p doped substrate, electrons flow between drain and source, and this device is a NMOS. Similarly, for a n doped substrate, holes conduct current between the drain and source, and this device is a PMOS. *So, a MOSFET is either a "p-channel n-substrate" or a "n-channel p-substrate" device. Originally, MOSFETs had their gate length greater than 1 micron—"long channel" MOSFET. State-of-art MOSFETs, as used in microprocessors and computer memory integrated circuits, have gate lengths in the low nanometer range.* Short channel MOSFETs are examined in detail later in this chapter. The layered structure of a NMOS is in Fig. 9.1a. In a typical circuit, the source and substrate|body at the same potential, conventionally the ground.

There are two key DC transfer characteristics for a MOSFET.

- Drain source current with linearly or pulse varied gate source voltage and constant drain source voltage.
- Drain source current with linearly or pulse varied drain source voltage, and constant gate source voltage.

Both measurements generate a family of curves. The sequence of physical steps involved in a drain source current versus gate source voltage sweep at constant drain source voltage are:

- The applied positive gate source voltage repels holes from the substrate close to the oxide|semiconductor gate substrate interface (surface), resulting in a space charge layer, across which some of the applied gate-body voltage gets dropped.

A. Banerjee, *Semiconductor Devices*, Synthesis Lectures on Engineering, Science, and Technology, https://doi.org/10.1007/978-3-031-45750-0_9

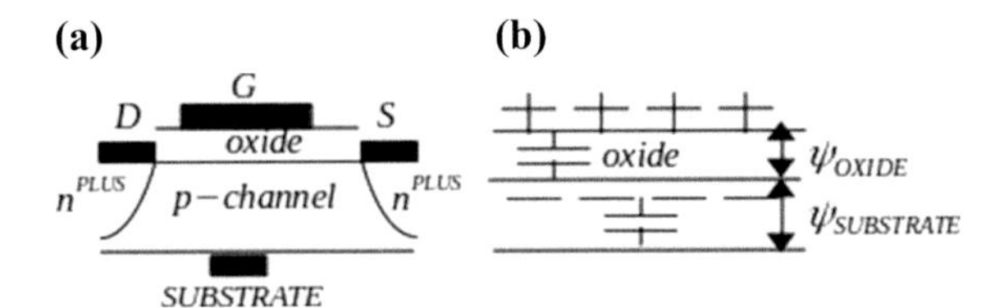

Fig. 9.1 **a**, **b**Physical structure of a silicon homogeneous junction MOSFET and charge distribution in a MOS capacitor

- The potential in the top part of the substrate also causes a voltage drop across the depleted body source pn junction. The source body diode is forward biased and electrons are injected into the body.
- The electron injection is highest at the surface corresponding to the highest substrate potential at the surface. The electrons form a thin channel at the surface, which increases exponentially with applied bias.
- This electron charge becomes dense enough to electrostatically screen the body from the gate. Gate body voltage increase are absorbed mostly in the oxide. Then the channel charge and gate bias do not satisfy any exponential relationship.

The basic, exponential current–voltage relationship of a forward-biased diode is still present at high gate bias. The relevant bias for the forward-biased source-channel diode is $(\Psi_S - V_S)$, where Ψ_S is the surface potential in the substrate|body, and V_S is the source potential. The potential-divider action of the series arrangement of the oxide capacitance and the semiconductor capacitance (Fig. 9.1b) has its effect. Then the drain current is exponentially related to the surface potential. The surface potential is approximately equal to the applied gate substrate bias.

$$\Delta\psi_{SURFACE} = \frac{\Delta V_{GATE,SUBSTRATE}}{1 + \frac{C_{SURFACE}}{C_{OXIDE}}} \quad C_{OXIDE} = \frac{\in_{OXIDE}}{t_{OXIDE}} \quad C_{SURFACE} = \frac{\Delta Q_{SURFACE} + \Delta Q_{SUBSTRATE}}{\Delta\psi} \tag{9.1a}$$

At low gate bias the charge in the semiconductor is small, so $C_{SURFACE} \ll C_{OXIDE}$, and the surface potential changes directly proportionally to the applied gate substrate bias. Then the drain current is exponentially related to the gate bias. The surface capacitance increase with increasing gate bias due to increased charge in the channel. The change in the surface potential with the applied gate substrate bias decreases. The electron charge increases exponentially with the surface potential, but no longer exponentially with the applied gate substrate bias. The boundary between the exponential and non-exponential regions is gradual, and is associated with the threshold voltage. When $V_{GATE\,SOURCE} \leq V_T$ the transistor is operating in the sub-threshold region, and the *drain current at zero gate source voltage is the OFF current*. When $V_{GATE\,SOUTCE} \geq V_T$ the drain current in this region is the ON current.

The region of the MOSFET in between the drain and source with the gate at the top and the substrate at the bottom with a metallic substrate contact is a parallel plate MOS capacitor—*gate (top plate), insulating oxide (middle), p-doped substrate (bottom plate).* The energy band diagram of the MOS capacitor is in Fig. 9.2a, b. The gate is heavily doped, n-type polysilicon, for which the Fermi level is coincident with the conduction band edge, the insulator is silicon dioxide, and the substrate is p-type silicon. T*o equilibrate the system, the Fermi level in the body must be raised, accomplished by transfer of electrons from the gate. The transferred electrons recombine with holes in the p-type body, creating a space-charge region of ionized acceptors near to the interface with the oxide. There is band-bending in this region. The gate acquires a net positive charge due to loss of electrons.* The charge difference across the oxide creates an electric field. The total potential differences across the oxide and semiconductor are Ψ_{OXIDE}, $\Psi_{SUBSTRATE}$. Their sum gives:

$$V_{BI} = \frac{\Phi_{SUBSTRATE} - \Phi_{GATE}}{q} \tag{9.1b}$$

where $\Phi_{SUBSTRATE}$, Φ_{GATE} are the respective work functions.

Work function difference between the gate and substrate surface region ensures that the substrate surface is less p-type than the rest of the bulk, i.e., it is n-type—**channel inversion.** In order to create a high electron concentration at the substrate-gate oxide interface surface, at lower positive applied gate substrate voltage than what would be required if there were no band bending at equilibrium, i.e., if the built in voltage was zero. **The no band bending condition is flat band condition** (Fig. 9.3a), achieved by applying a negative potential to the gate. The electron concentration (per unit volume) in the semiconductor is:

$$n(x) = n_i e^{\frac{\psi(x) - \Phi_{SUBSTRATE}}{V_{THERMAL}}} \tag{9.1c}$$

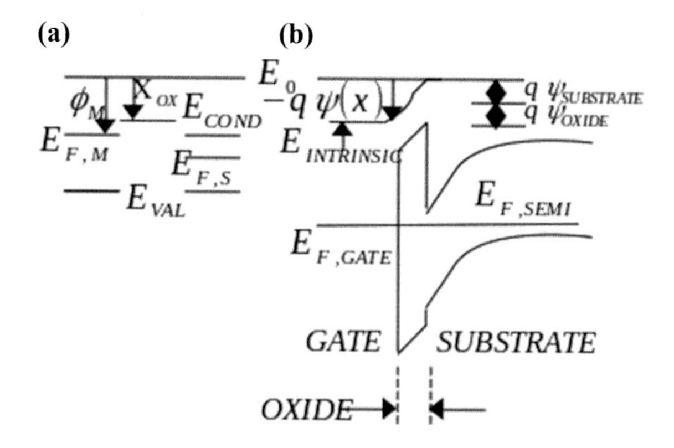

Fig. 9.2 **a**, **b** Energy band diagram for MOS capacitor with layers separated and in contact

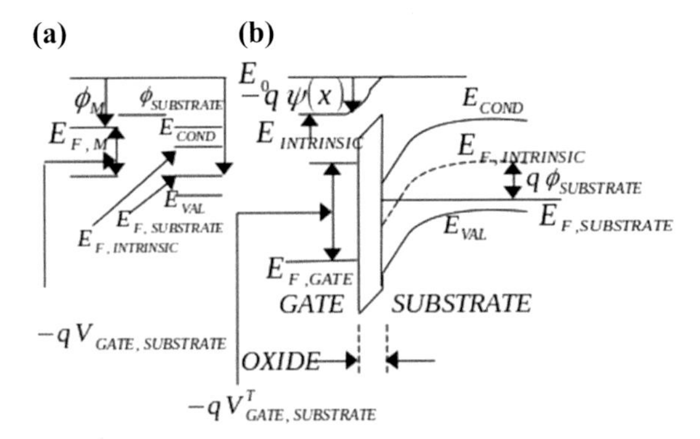

Fig. 9.3 **a**, **b** Flat band and strong inversion in MOS capacitor

Deep inside the substrate, where the acceptor doping concentration is uniform,

$$n_{SUBSTRATE} = \frac{n_i^2}{N_{ACCEPTOR}} \quad \Phi_{SUBSTRATE}$$
$$= \frac{E_{FERMI,SUBSTRATE} - E_{FERMI,INTRINSIC,SUBSTARTE}}{q} \tag{9.1d}$$

Using Eqs. 9.1c, d, e can be re-written as:

$$n(x) = N_{ACCEPTOR} e^{\frac{\psi(x) - 2\Phi_{SUBSTRATE}}{V_{THERMAL}}} \tag{9.1f}$$

With the onset of **strong inversion** (Fig. 9.3b) of the oxide substrate interface surface, the total electron charge is:

$$Q_n = \int -qn(x)dx \quad 0 \le x \le L \tag{9.1g}$$

9.2 Surface Charge Model of Long Channel MOSFET Drain Current

The MOSFET being examined is invariant in the z-direction. So Poisson's equation for the body and channel regions ignoring hole charge in the substrate's depletion|space-charge region is:

$$\frac{\partial^2 \psi}{\partial x^2} + \frac{\partial^2 \psi}{\partial x^2} = \frac{qN_{ACCEPTOR}}{\epsilon_{SUBSTRATE}} \left(1 + e^{\frac{\psi - 2\Phi_{SUBSTRATE} - V_{CHANNEL,SUBSTRATE}}{V_{THERMAL}}}\right) \tag{9.2a}$$

The electric field at the substrate surface is strongest in the y direction. A*ssuming the Gradual Channel Approximation, the variation of surface potential* Ψ *in the x direction is negligible compared to that in the y direction—the above equation now one dimensional.*

To solve (9.2a), both sides of it are multiplied by $\frac{\partial\Psi\partial y}{\partial y}$ and then integrated from deep in the substrate, where $\psi = 0$ (assuming that zero bias is applied to the substrate terminal). and $\partial\psi/\partial y = 0$, to the surface, where $\Psi = \Psi_{SURFACE}$ $\frac{\partial \Psi_{SURFACE}}{\partial y} = E_y(0)$.

The last limit is the field in the semiconductor at the oxide-substrate interface. Using this and Gauss's law the charge per unit area in the substrate is:

$$Q_{SURFACE} = \sqrt{2q \in_{SUBSTRATE} N_{ACCEPTOR}}$$
$$\sqrt{\psi_{SURFACE} + \frac{k_B T_{lattice}}{q} e^{\frac{-(2\Phi_{SUBSTRATE}+V_{CHANNEL,SUBSTRATE})}{V_{THERMAL}}}\left(e^{\frac{\psi_{SUBSTRATE}}{V_{THERMAL}}} - 1\right)} \tag{9.2b}$$

As any MOSFET operates with external voltages, for the applied gate-substrate bias,

$$V_{GATE,SUBSTRATE} - V_{FLAT\,BAND} = \psi_{OXIDE}(x) + \psi_{SURFACE}(x) \tag{9.2c}$$

Once more applying Gauss's law gives the charge per unit area of the gate electrode:

$$Q_{SURFACE}(x) = -C_{OXIDE}(V_{GATE\,SUBSTRATE} - V_{FLAT\,BAND} - \psi_{SURFACE}(x)) \tag{9.2d}$$

Careful manipulation of 9.2 a-d gives an implicit relation between surface potential and applied gate-substrate voltage:

$$\psi_{SURFACE} = V_{GATESUBSTRATE} - V_{FLATBAND} - \gamma AA$$
$$\gamma = \frac{\sqrt{2q \in_{SUBSTRATE} N_{ACCEPTOR}}}{C_{OXIDE}} \tag{9.2e}$$

where

$$AA = \sqrt{\psi_{SURFACE} + \frac{k_B T_{lattice}}{q} e^{\frac{-(2\Phi_{SUBSTRATE}+V_{CHANNEL,SUBSTRATE})}{V_{THERMAL}}}\left(e^{\frac{\psi_{SUBSTRATE}}{V_{THERMAL}}} - 1\right)}$$

Denoting inversion as the presence of electrons at the oxide-substrate interface surface, as required for a drain current, three types of inversion conditions are possible: weak, moderate and strong respectively, defined as:

$$0 < \psi_{SURFACE}(x) \leq \Phi_{SUBSTRATE} + V_{CHANNEL,SUBSTRATE}(x)$$

$$\Phi_{SUBSTRATE} + V_{CHANNELSUBSTRATE} lt; \psi_{SURFACE}(x) \leq 2\Phi_{SUBSTRATE} + V_{CHANNEL,SUBSTRATE}(x)$$

$$\psi_{SURFACE}(x) > 2\Phi_{SUBSTRATE} + V_{CHANNEL,SUBSTRATE}(x) \tag{9.2f}$$

The charge per unit area in the substrate consists of electrons near the surface and ionized acceptors in the depleted region of the substrate:

$$Q_{SURFACE}(x) = Q_{electron}(x) + Q_{SUBSTRATEIONIZED} \tag{9.2g}$$

Using the Charge Sheet Approximation that states that all electrons reside in a sheet at the substrate surface, so that there is no voltage drop through such a sheet in the y direction. **Then if any potential is applied to the substrate, voltage $\Psi_{SURFACE} - V_{SUBSTATE}$, is dropped entirely across the space-charge region of ionized acceptors.** Using the Depletion Approximation for the substrate space charge region (neglecting hole charges) gives:

$$Q_{SUBSTRATE}(x) = -\sqrt{2q\epsilon_{SUBSTRATE} N_{ACCEPTOR} \psi_{SURFACE}(x)} \tag{9.2h}$$

After applying the above expressions in 9.2h, it can be re-written as:

$$Q_{electron}(X) = -C_{OXIDE}(V_{GATE\ SUBSTRATE} - V_{FLAT\ BAND} - \Psi_{SURFACE}(X) - \gamma\sqrt{\Psi_{SURFACE}(X)}) \tag{9.2i}$$

The electron current in the two dimensional sheet is:

$$I_{eletron} = W\overrightarrow{J}_{eletron} = \mu_{eletron} W\left(\frac{Q_{eletron} d\,\psi_{SURFACE}(x)}{dx} - \frac{k_B T_{lattice} d\,Q_{eletron}}{q dx}\right) \tag{9.2j}$$

where W is the channel's width. The total drain current is:

$$\int I_{DS} dx = -\int I_{electron} dx = -W\int \mu_{electron} d\psi_{SURFACE} + \frac{k_B T_{lattice} W}{q}\int \mu_{electron} d\,Q_{electron} \tag{9.2k}$$

where $\mu_{electron}$ is the *substrate bulk electron mobility*. The electron mobility at the substrate-oxide interface is different from the substrate bulk electron mobility due to scattering. Assuming that the effective electron mobility is one dimensional without any x direction dependency, the drain current can be re-expressed as:

$$\int I_{DS}dx = \frac{\mu_{electroneffective}W}{L}\int -Q_{electron}d\psi_{SURFACE} + \frac{k_B T_{lattice}W}{q}\int dQ_{electron} \quad (9.2l)$$

$$I_{DS,DIFFUSION} = \frac{C_{OXIDE}\mu_{electroneffective}k_B T_{lattice}W}{Lq}$$
$$\left((\psi_{SURFACE}(L) - \psi_{SURFACE}(0)) + \gamma\left(\sqrt{\psi_{SURFACE}(L)} - \sqrt{\psi_{SURFACE}(0)}\right)\right) \quad (9.2m)$$

The expression for the drift current is left as an exercise for the reader.

- The electron charge $Q_{electron}(x)$ can never equal zero if current is to be maintained. The condition for drain current saturation $|(Q_{electron}(L))| \ll |Q_{SUBSTRATE}(L)|$. Thus, the channel never pinches off completely.
- As $V_{DEAIN,SUBSTRATE}$ increases beyond the value at which $|(Q_{electron}(L))| \ll |Q_{SUBSTRATE}(L)|$ then the condition $|(Q_{electron}(x))| \ll |Q_{SUBSTRATE}(x)|$ is satisfied at x values closer to the source.

This effect—decrease in effective channel length of the MOSFET, increases the drain current and is called ***channel-length modulation***.

9.3 Strong Inversion Source Reference Model Drain Current HERE

The strong inversion source reference model (simplified surface potential model) is used in SPICE [20–25] as the level 1 MOSFET model. SPICE [20–25] is the universally used electricallelectronic circuit simulation and performance evaluation software tool. It is a simplification of the surface potential model. It is based on three assumptions.

$$\psi_{SURFACE}(x) = 2\Phi_{SUBSTRATE}(x) + V_{CHANNEL,SUBSTRATE} \quad (9.3.1)$$

$$V_{CHANNEL,SOURCE}(x) = V_{CHANNEL,SUBSTRATE}(x) - V_{SOURCE,SUBSTRATE} \tag{9.3.2}$$

$$V_{CHANNEL,SOURCE}(x) << 2\Phi_{SUBSTRATE}(x) + V_{SOURCE,SUBSTRATE} \tag{9.3.3}$$

From the surface potential model the drain current is due to drift when the source is in strong inversion. Strong inversion source reference model starts with the drift portion of the drain current from the surface potential model. This expression contains the surface potentials at both the source and drain ends of the channel. From the above assumptions:

$$\begin{aligned} \psi_{SURFACE}(0) &= 2\Phi_{SUBSTRATE} + V_{SOURCE,SUBSTRATE} \quad \psi_{SURFACE}(L) \\ &= 2\Phi_{SUBSTRATE} + V_{DRAIN,SUBSTRATE} \end{aligned} \tag{9.3.4}$$

$$\begin{aligned} \psi_{SURFACE}(L) &= \psi_{SURFACE}(0) + V_{DRAIN,SUBSTRATE} \quad \psi_{SURFACE}(L) \\ &= 2\Phi_{SUBSTRATE} + V_{DRAIN,SUBSTRATE} + V_{SOURCE,SUBSTRATE} \end{aligned} \tag{9.3.5}$$

Combining 9.3.1- 9.3.5 gives:

$$V_{DRAIN,SOURCE} << 2\Phi_{SUBSTRATE} + V_{SOURCE,SUBSTRATE} \tag{9.4}$$

This inequality is used for a binomial expansion to the second order of the term $\Psi_{SURFACE}^{\frac{3}{2}}$ in the drift current expression for the drain current in the surface potential model, resulting in the expression for the drain current in the strong inversion model:

$$\begin{aligned} I_{DRAIN} = &\frac{C_{OXIDE}\mu_{electron,effective}V_{DRAIN,SOURCE}W}{L} \\ &\left(V_{GATE,SOURCE} - V_{THRSH} - \frac{mV_{DRAIN,SOURCE}}{L}\right) \end{aligned} \tag{9.4a}$$

The threshold voltage and body effect coefficient are defined as:

$$\begin{aligned} V_{THRSH} = &V_{FLATBAND} + 2\Phi_{SUBSTRATE} \\ &+ \gamma\sqrt{2\Phi_{SUBSTRATE} + V_{SOURCE,SUBSTRATE}} \\ m = &1 + \frac{\gamma}{\sqrt{2\Phi_{SUBSTRATE} + V_{SOURCE,SUBSTRATE}}} \end{aligned} \tag{9.4b}$$

This is a very simplified version of the surface potential model, based on the assumption that the drain voltage and current both reach a maximum saturation value:

$$V_{DRAIN,SATURATION} = \frac{V_{GATE,SOURCE} - V_{THRSH}}{m} I_{DRAIN,SATURATION}$$
$$= \frac{C_{OXIDE}\mu_{effective,electron}\left(V_{GATE,SOURCE} - V_{THRSH}\right)^2 W}{2Lm} \quad (9.4c)$$

9.4 Threshold Voltage and Body Effect Coefficient Relation

The threshold voltage and body effect coefficient can be related to eachother in a straightforward way. Using the same assumptions as in the SPICE [20–25] level 1 MOSFET model, the channel charge is related to the gate source, flat band, channel substrate and source substrate voltages as:

$$Q_{electron}(x) = -\left(AC_{OXIDE} + BV_{CHANNEL,SOURCE}\right)$$
$$A = V_{GATE,SOURCE}$$
$$-V_{FLAT\ BAND} - 2\Phi_{SUBSTRATE} - \gamma\sqrt{2\Phi_{SUBSTRATE} + V_{SOURCE,SUBSTRATE}}$$
$$B = 1 + \frac{\gamma}{2\sqrt{2\Phi_{SUBSTRATE} + V_{SOURCE,SUBSTRATE}}} \quad (9.5a)$$

The channel charge at the source end of the channel is:

$$Q_{electron}(0) = -C_{OXIDE}\left(V_{GATE,SOURCE} - V_{THRSH}\right) \quad (9.5b)$$

The body effect coefficient m, is introduced into the strong inversion model when terms involving $\Psi_{SURFACE}^{\frac{3}{3}}(0)$, $\Psi_{SURFACE}^{\frac{3}{2}}(L)$ are expanded in a Taylor's series, in the equation for the drift current in the surface potential model. ***The body effect coefficient denotes the effect of the body|substrate in reducing the channel charge when the drain source voltage is increased.***

In a two terminal structure, e.g., parallel-plate capacitor, there is a one to one correspondence between the charges on the two plates, represented by the surface and gate charges in the MOS capacitor. However, the presence of a third region, the body, causes a loss of channel charge in excess of the corresponding loss of gate charge. This degrades the drain current, and to compensate, m is included in the expression for the drain current in the strong-inversion model. Considering a change in the drain source voltage that changes the channel source voltage from 0 to a non-zero value at some location x in the region the changes in the respective charges are:

$$\begin{aligned} \Delta Q_{GATE} &= -C_{OXIDE} V_{GATE,SOURCE}(x) \quad \Delta Q_{SUBSTRATE} \\ &= -C_{SUBSTRATE}(x) V_{GATE,SOURCE}(x) \\ \Delta Q_{electron}(x) &= C_{OXIDE} m(x) V_{GATE,SOURCE}(x) \end{aligned} \tag{9.5c}$$

Now the body effect coefficient can be re-written as:

$$m = 1 + \frac{C_{SUBSTRATE}(x)}{C_{OXIDE}} \quad m = \frac{\Delta Q_{electron}(x)}{\Delta C_{GATE}(x)}$$

When electron mobility in the direction of the electron flow (longitudinal) direction is field dependent, the electron drift velocity is given as:

$$\frac{1}{V_{DRIFT}} = \frac{1}{\mu_{electroneffective}|E|} + \frac{1}{v_{SATURATION}} \tag{9.5d}$$

The expression for the drain current, after integrating over the entire channel, is:

$$I_{DRAIN} = \frac{C_{OXIDE} \mu_{electron,effective} V_{DRAIN,SOURCE} W}{L + \frac{\mu_{electron,effective} V_{DRAIN,SOURCE}}{v_{SATURATION}}} \left(V_{GATE,SOURCE} - V_{THRSH} - \frac{m V_{DRAIN,SOURCE}}{2} \right) \tag{9.5e}$$

In this equation, the charge appears as if it is evaluated where. $V_{CHANNEL,SOURCE} = \frac{V_{DRAIN,SOURCE}}{2}$, but the velocity is modified. If the saturation velocity was infinite, the two equations would be the same. In addition, the drain current shows a maximum value when the basic assumption of strong inversion breaks down at the drain end of the channel. The maximum values for the drain voltage and current are:

$$\begin{aligned} & I_{DRAIN,SATURATION} \\ &= C_{OXIDE} V_{SATURATION} \left(V_{GATE,SATURATION} - V_{THRSH} \right) \left(\frac{A-1}{A+1} \right) \\ & V_{DRAIN,SATURATION} = \frac{2\left(V_{GATE,SOURCE} - V_{THRSH}\right)}{m(1+A)} \\ & A = \sqrt{1 + \frac{\mu_{electroneffective}\left(V_{GATE,SOURCE} - V_{THRSH}\right)}{m L v_{SATURATION}}} \end{aligned} \tag{9.5f}$$

The sub-threshold region of operation is when the gate source voltage is less than the threshold voltage. The channel is everywhere either in weak or moderate inversion. This operating condition holds the key to two important MOSFET properties: the ratio of the onset of the ON-current to the OFF-current, and the inverse sub threshold slope. These two properties are key to operation of MOSFETs in high speed digital circuits.

$$Q_{electron} = \sqrt{2q\,\epsilon_{SUBSTRATE}\,N_{ACCEPTOR}} \left(\sqrt{\psi_{SUBSTRATE} + \frac{k_B T_{lattice} e^{\frac{\psi_{SURFACE}-2\Phi_{SUBSTRATE}-V_{CHANNEL,SUBSTRATE}}{V_{THERMAL}}}}{q}} - \sqrt{\psi_{SURFACE}} \right) \tag{9.5g}$$

In moderate inversion $\Psi_{SURFACE} < 2\Phi_{SUBSTRATE} + V_{CHANNEL,SUBSTRATE}$, the first square root term in the above equation can be expanded using the Taylor's series. Keeping terms to first order gives:

$$Q_{electron} = \sqrt{2q\epsilon_{SUBSTRATE}N_{ACCEPTOR}}k_B T_{lattice}\frac{e^{\frac{\psi_{SURFACE}-2\Phi_{SUBSTRATE}-V_{CHANNEL,SUBSTRATE}}{V_{THERMAL}}}}{2q\sqrt{\psi_{SURFACE}}} \tag{9.5h}$$

In weak inversion, the surface potential is barely influenced by channel substrate voltage. Therefore no electric field in the x direction exists $E_x \approx 0$. Then the drain current is due to diffusion. From the above equations, under the conditions $Q_{electron}(0)$ evaluated at $V_{CHANNEL,SUBSTRATE} = V_{SOURCE,SUBSTRATE}$, and $Q_{electron}(L)$ at $V_{CHANNEL,SUBSTRATE} = V_{DRAIN,SUBSTRATE}$ gives:

$$I_{DRAIN} = AA\frac{\mu_{electroneffective}Wk_B^2 T_{lattice}^2\sqrt{2q\epsilon_{SUBSTRATE}N_{ACCEPTOR}}}{2Lq\sqrt{\psi_{SURFACE}}}\left(1 - e^{\frac{-V_{DRAINSOURCE}}{V_{THERMAL}}}\right) \tag{9.5i}$$

$$AA = e^{\frac{\psi_{SURFACE}-2\Phi_{SUBSTRATE}-V_{SURFACESUBSTRATE}}{V_{THERMAL}}}$$

The dominant effect of the surface potential on the current occurs via the exponential term in the numerator, so that the accuracy is not reduced from selecting some constant value for the surface potential in the denominator. If value of the surface potential is selected to be that at the extreme end of the weak inversion $(2\Phi_{SUBSTRATE} + V_{SOURCE,SUBSTRATE})$, then the term $\frac{\sqrt{2q\epsilon_{SUBSTRATE}N_A}}{2\sqrt{\Psi_{SURFACE}}}$ can be re-written as $C_{OXIDE}(m-1)$. In sub-threshold region of operation, the electron charge is less than the substrate charge, for the body effect coefficient and the potential divider action of the oxide and body capacitances, gives:

$$\frac{d\psi_{SURFACE}}{dV_{GATE\,SUBSTRATE}} = \frac{1}{m} = \frac{2\Phi_{SUBSTRATE} + V_{SOURCE,SUBSTRATE} - \psi^{P}_{SURFACE}}{V_{THRSH} - V^{P}_{GATE\,SOURCE}} \tag{9.5j}$$

9.5 Complimentary Metal Oxide Semiconductor Field Effect Transistor (CMOS)

All modern digital integrated circuits are fabricated with this technology consisting of a n channel, p substrate enhancement mode field effect transistor (NMOS) and a p channel n substrate enhancement mode field effect transistor (PMOS). For both transistors, the charge carriers are minority carriers.

Digital logic involves charging and discharging capacitors as quickly as possible. If ΔV is the change in voltage required to denote a change in logic level (1–0 or 0–1) at a node of constant capacitance C, then the switching time is:

$$\tau = \frac{C\Delta V}{I} \tag{9.5k}$$

where I is the average current during the switching cycle. **So a successful digital logic technology needs only small voltage swings, and must be capable of supplying large currents to circuits of low capacitance.**

Fabricating a NMOS and PMOS on the same substrate is very advanced, tried, tested and universally used technology consisting of, the following core steps (Fig. 9.4a, b):

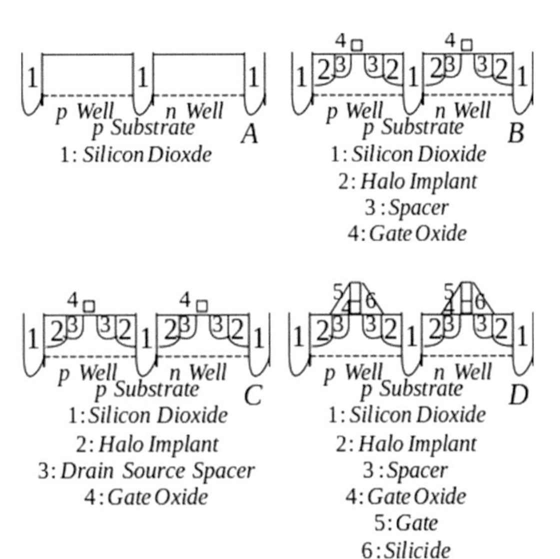

Fig. 9.4 **a** Steps A, B of CMOS fabrication process, **b** Steps C, D of CMOS fabrication process

- A: Shallow isolation trench creation, n, p well ion implantation.
- B: Gate oxidation and polysilicon layer creation, followed by source, drain extension halo layer implementation.
- C: Drain, source layers are ion implanted.
- D: Drain, gate and source passivation and silicidation.

Modern digital integrated circuit fabrication process workflows use advanced and enhanced versions of the above key steps. In CMOS technology, the threshold voltage of the PMOS is opposite, and nearly equal in magnitude, to that of the complimentary NMOS transistor.

In an CMOS inverter, the two gates of the NMOS and PMOS are are connected together, and the threshold voltages are such that in either of the logic states (HI or LO) only one of the transistors is ON, minimizing static power drain. Also, CMOS logic gates can be made with much fewer transistors than their emitter coupled logic (ECL) rival, enabling cost effective fabrication of dense, high speed logic circuits. Further, in integrated circuits, each transistor has to be connected at the top surface, and this is easy to achieve with CMOS. In CMOS, in which two of the contacts (drain, source) are closely aligned with the third contact (gate).

In CMOS logic the ON-current is the drain current when the source end of the channel is in strong inversion condition, $V_{GATE,SOURCE} > V_{THRSH}$. The drain current increases with the drain source voltage to reach a maximum value—saturation current I_{DSAT}. The supply voltage sets the upper limit on both the drain source and gate source voltages. MOSFET gate lengths have been progressively reduced to increase packing density (reduced device area for each MOSFET) reduce parasitic capacitances and improve the saturation current of the device. Shorter gate lengths L no longer has a strong effect on the saturation current. This is because *L is already sufficiently small for the lateral field E_x to attain a sufficiently high value for velocity saturation to occur over a significant part of the channel.* Consequently the inclusion of saturation gives a much smaller increase in the saturation current, as L is reduced, than is predicted by the basic SPICE Level 1 model, which does not limit the electron velocity to v_{SAT}. To attain their saturation velocity, electrons still have to be accelerated over the source side of the channel, so a high mobility is essential. T*he reduced channel length in a MOSFET increases the drain current in the linear and saturation regions of operation.*

The ON-current can be increased by both the effective electron mobility $\mu_{electron,effective}$ and by increasing the channel charge, achieved by increasing the oxide capacitance as well as the overdrive voltage ($V_{GATE,SOURCE} - V_{THRSH}$). CMOS uses unipolar power supply, increasing the gate source voltage would mean increasing the drain source voltage. But increasing the drain source voltage is limited by the necessity of keeping $E_x < E_y$. Also, a low supply voltage value is needed for CMOS enabled portable electronic devices. The alternative to increasing the overdrive voltage is to reduce the threshold voltage—with consequences for the sub threshold current. In densely packed CMOS circuits with hundreds of thousands of transistors per unit area, the current per transistor when the it is OFF ($V_{GATE,SOURCE} = 0$ for NMOS) must be very very small, else the static power drain would be very high. Unwanted power dissipation also occurs during switching.

9.5.1 Lattice Strain and Charge Mobility

The channel region of silicon based sub-micron gate length field effect transistors is mechanically strained to improve both electron and hole mobility. **The strain modifies the band structure, and calibrated strain application exploits this feature to reduce the effective mass in the desired direction of conduction.**

Hooke's Law relates the mechanical deformation of a one dimensional object to the applied force $\overrightarrow{F}_{EXTERNAL} = K$ where K is a material specific constant.

When generalized to a three dimensional object, e.g., a cube of crystalline semiconductor, normal and shear forces are introduced which deform a given side length in a given face arising from all the possible normal and shear forces. Each force per unit area is a stress and each component of deformation is a strain. The force required to produce a given strain is determined by the elastic stiffness constants of the material. In cubic crystals e.g., as Si and GaAs, there are three independent such constants. In Si, their magnitudes range from 64 to 166 Gpa (Giga Pascal), which indicates the immense pressures within a crystalline solid. The resulting strain from a given stress is determined by the elastic compliance constants of the material, in turn related to the elastic stiffness constants.

The simple one dimensional crystal model used in Chap. 1 illustrates the importance of atomic spacing on how band structure is related to crystal structure. **In short, mechanical stress|strain imposed on the crystal lattice will change atomic spacing, in turn modifying electron band structure.** For example, it is advantageous to lower the heavy hole band energy with respect to the light hole band energy. *Achieving this in real world semiconductor integrated circuits requires very careful process engineering as strain introduced in one region must be compensated else the crystal would crack and be destroyed.* Valence bands become so warped by the added strain that they cannot be classified as 'heavy' nor 'light'. Rather, as the strain splits the bands, they are referred to as 'top' and 'bottom' bands. **As holes want to occupy the higher band, it is important that the curvature of the top band be such that the effective mass in the desired direction of conduction be reduced.** One way to achieve this for a $< 110 >$ channel on a $\{001\}$ Si surface in a sub-100 nm device *is to apply a uniaxial compressive stress of about 1 GPa to the p-type channel. A common method to add such a high stress in the crystal is to etch recesses in the silicon at the source and drain regions, and then fill-in by epitaxially growing SiGe.* With a Ge mole fraction of $x \approx 0.3$, the more expansive SiGe puts the Si channel under a compressive stress of the required magnitude.

For n-channel silicon MOSFETs on $\{001\}$ substrates, a $< 110 >$ tensile stress induces a shear strain that enhances electron conduction in this direction. *Specifically, the six fold symmetry of the conduction band minima is broken, resulting in two [001] valleys moving to lower energies. The conduction band minimum moves towards the X-point as the shear strain increases. So the degeneracy of the two conduction band minima at the X-point is removed. Physically, the two transverse effective masses are no longer equal, i.e., the Brillioun zone is no longer symmetrical in directions perpendicular to the new* k_x *direction.* The beneficial effect of breaking the six fold symmetry of the conduction bands in unstrained bulk material is illustrated in Fig. 9.5a, which represents the E-k relationship for silicon near the conduction band edge E_{COND}. The valley is steeper in two of the principal directions than in the remaining orthogonal direction. The equivalent constant energy surfaces in three dimensions are shown in Fig. 9.5b. The four spheroids in the horizontal plane are called the Δ4 valleys, and the two in the perpendicular direction are called Δ2 valleys. The tensile strain breaks the six fold degeneracy of these valleys, raising the energy of the Δ4 set, and lowering the energy of the Δ2 set. For a given energy E with respect to the bottom of the conduction band in the Δ4 valleys, $E - E_{COND}$ for the Δ2 valleys is increased. The electrons seek the lower energy states, so the Δ2 valleys become heavily populated. *If the in-plane effective mass in the Δ2 valleys remains at* m_t^P. *less than the longitudinal effective mass* m_l^P, *then the electron mobility in the channel is increased.*

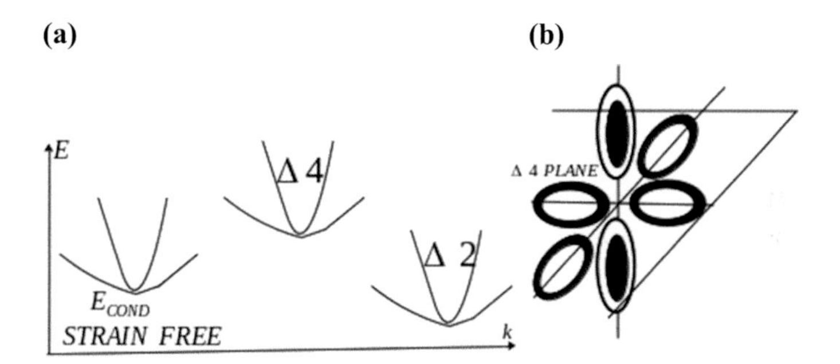

Fig. 9.5 **a** Symmetry breaking of unstrained semiconductor conduction band due to imposed lattice stress. **b** Equivalent constant energy surfaces, elliptical rings represent high energy, low occupancy Δ4 states

Fig. 9.6 Polysilicon gate energy band diagram

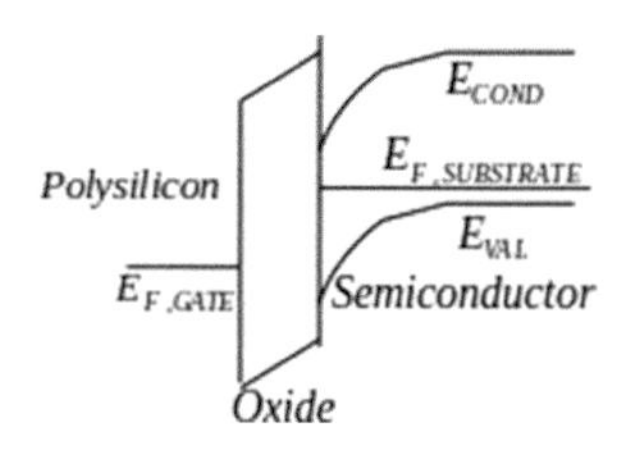

9.5.2 High k Dielectrics in Gate Capacitors

Increasing the oxide capacitance puts more charge in the channel for a given gate source voltage and increases the drain current. With gate lengths in the nanometer range, e.g., 90 nm, the oxide thickness is approximately 2 nm, leaving little room for further shrinking the oxide thickness. Thinner oxide films that satisfy predefined standards of integrity and uniformity are very difficult to fabricate. Also leakage to the gate of drain intended current increases sharply. Traditionally, the gate oxide has been silicon dioxide as the physics of the silicon–silicon dioxide interface is very well understood.

However, the relative dielectric constant of silicon dioxide is 3.9. Including nitrogen in the oxide boosts the dielectric constant to 4–5. This is an active research area, under the constraint that the properties of the silicon–silicon dioxide interface must be preserved. *For example, using a dielectric material derived from hafnium with traces of nitrogen preserves the silicon–silicon dioxide properties, but makes the dielectric later thicker, while boosting the overall dielectric constant by $\approx$15–20%, and reduces leakage current.* The ON current is increased if the dielectric layer thickness is selected as:

$$t_{high-k} < \frac{t_{silicondi\,oxide}\epsilon_{high-k}}{\epsilon_{silicondi\,oxide}}$$

9.5.3 Poly Silicon Gates and Capacitances

State-of-art MOSFETs in digital circuits have a gate stack consisting of a metal gate electrode and high-k dielectric. For long highly doped polycrystalline silicon gates have been used to counter the disadvantages of pure metal (e.g., aluminium) such as penetration into the silicon dioxide, during thermal processing prevented using metal as a mask to facilitate the self-alignment of the source and drain regions to the gate. Polycrystalline silicon does not have any such problems, but creates depletion layers at the gate oxide surface. So, an additional potential drop is introduced in the device. Therefore,

$$V_{GATE\,SUBSTRATE} - V_{FLAT\,BAND} = \psi_{POLY} + \psi_{OXIDE} + \psi_{SURFACE} \tag{9.6}$$

where Ψ_{POLY} is the potential drop inside polycrystalline silicon gate layer. Clearly, the charge in the channel, experiences less of the gate substrate voltage than it would otherwise. For state-of-art ultra high speed digital circuits, this is a problem, and so metal gates, less reactive than aluminium with high-k dielectrics need to be used. The energy band diagram of the polysilicon layer gate MOSFET is in Fig. 9.7.

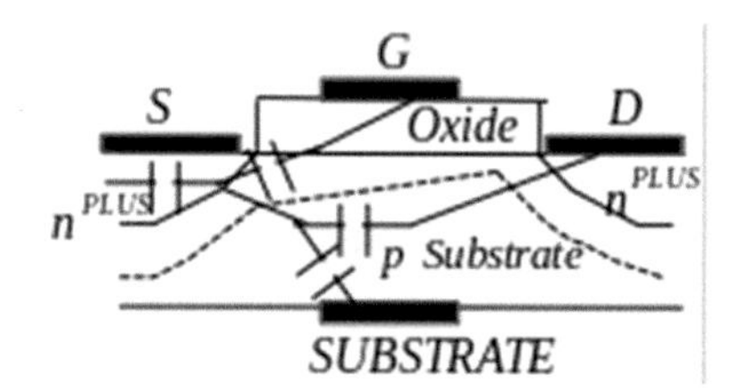

Fig. 9.7 Parasitic capacitances inside a short channel MOSFET. Dashed line shows depletion layer

9.5.4 Gate Leakage Current

As MOSFET physical dimensions are scaled down, specifically oxide thickness, electrons can flow from the channel (and overlapped part of the drain) in an NMOS to the gate. This tunnelling current (called *leakage current*) reduces the drain current. Expressions for the tunnel current are based on the assumption that the electrons in the channel form a classical two dimensional sheet where the electrons are confined to quasi-bound states. The tunnel current density is:

$$J_{TUNNEL} = \frac{q}{\hbar}\int n_{2D}(E)T(E)dE \tag{9.7a}$$

where $n_{2D}(E)$ is the electron density of the two dimensional sheet of electrons at the oxide semiconductor interface, and T(E) is the tunnelling transmission probability. If the rectangular tunnelling barrier is assumed, then an analytical expression for T(E) can be obtained in a straightforward manner. The denominator for this expression contains hyperbolic functions that involve the thickness of the tunnelling barrier, the electron effective mass in the oxide and the height of the potential barrier.

$$E_{CONDOXIDE} - E_{COND}(0) = X_{SILICON} - X_{OXIDE} \tag{9.7b}$$

where the χ's are electron affinities. Then the tunneling probability is:

$$T \approx e^{\frac{-2t_{OXIDE}\sqrt{2m^{P}_{OXIDE}(E_{CONDOXIDE}-E)}}{\hbar}} \tag{9.7c}$$

The tunnelling depends both on the thickness of the barrier and on its height. Therefore, for low tunnelling, the oxide must have a low electron affinity. For silica, $\chi = 0.9$ eV and 2.9 eV for hafnium.

9.5.5 Short Channel Effect and Threshold Voltage

The expressions presented in previous sections for the ON and sub-threshold current, indicate a lower limit for the sub-threshold current. Denoting $I_{DRAIN,ON}$ to be the current when $V_{GATE,SOURCE} = V_{THRSH}$ and $I_{DRAIN,OFF}$ when $V_{GATE,SOURCE} = 0$, then the ratio of these two currents is:

$$\frac{I_{DRAIN}(ON)}{I_{DRAIN}(OFF)} = e^{\frac{-V_{SOURCE}}{mV_{GATE\,SOURCE}}} \tag{9.7d}$$

So for a current ratio of 10^{-4} m (body effect coefficient) at its optimum value of unity indicates that the minimum threshold voltage is about 0.24 V.

Both the currents are dependent on the depth of the drain and source regions inside the substrate, and the gate length L. To include the depth into the drain current current calculations, the device equations need to be solved numerically. Analytical expressions are too complicated to extract meaningful insights from them.

The numerical calculation results show that for a given channel length the sub-threshold current increase is directly proportional to the depth of the drain, source regions. Also, for a given drain, source depth, the threshold voltage is directly proportional to the channel length.

The numerically computed drain characteristics indicate that increase in the drain, source regions enhance the drain ON current with increasing drain source voltage. This effect is not the same as channel length modulation in longer devices after the drain source voltage is increased beyond the pinch-off value. Current saturation in sub-micron channel length MOSFETs is dependent more on velocity saturation than pinch-off conditions.

These effects can be quantified by a reduction in the threshold voltage as either the junction depth is increased, or as the length of an already short channel is decreased. These MOSFET performance characteristics, applicable to short gate length devices are called ***short channel effect.***

The short channel effect is an electrostatic effect, when field lines emanate from the positively biased drain, and must terminate on negative charges: the p type substrate provides negatively charged sites, leading to a space charge region extending from the drain. This region is wider than the space charge region around the source because of the greater reverse bias at the drain substrate np junction. The drain related space charge at some location x increases the surface potential at that location. This increases both $|Q_{electron}(x)|$, $E(x)$, *resulting in drain current increase, called* ***charge sharing****. This is because the total space charge (depletion) at point x is now determined by both the gate and drain potentials. A given surface potential value at a location x can now be achieved with a lower gate source voltage than would be needed in absence of depletion due to the drain source voltage. As L decreases, the space charge region from the drain progressively encroaches on the source. Eventually, the surface potential at* $x = 0$ *is affected by the drain source voltage. The*

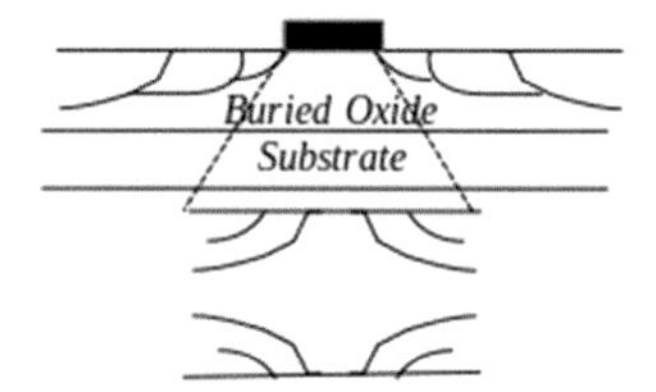

Fig. 9.8 **a**, **b** (**a**) Cross section of a thin body SOI MOSFET for SOI MOSFET dynamic logic circuits (**b**) Electric field lines just below

surface potential at x = 0 sets the potential barrier height at the source-channel junction, and thereby controls the electron flow into the junction and the drain current, called ***drain induced barrier lowering (DIBL)****.*

The effect of the short channel and drain induced barrier lowering are embodied in the effective threshold voltage given by:

$$V_{THRSH,EFFECTIVE} = V_{THRSH} + \Delta V_{THRSH}(junction_{DEPTH}, L, V_{DRAINSOURCE}) \tag{9.7e}$$

The electrostatic interaction between the drain and the channel via the depletion region in the substrate is a capacitive phenomenon, Fig. 9.8. Therefore, $|\Delta V_{THRSH}|$ is reduced by decreasing the drain source capacitor. As the sidewall of the np junction defines one of the lengths of this plate of the capacitor, the drain source depth must be be decreased. The chief drawback of this depth reduction is that the lateral access resistance to the channel from the source and drain contacts is increased. To counter this, only the part of the source and drain closest to the channel is thinned.

Alternatively, to reduce the influence of the drain on the surface charge at x = 0 is to shield the source from the field issuing from the drain. This is achieved by raising substrate acceptor concentration, but this in turn boosts the threshold voltage. This is countered by having a non-uniform substrate doping profile $N_A(x, y)$. The desired high doping is achieved deeper into the substrate using fast-diffusing dopants such as B, As and P. Doping of the sensitive surface region is achieved by lightly doping using slower diffusants such as In and Sb. The resulting doping profile is called **retrograde**. In the x direction, acceptor concentration is increased only close to the source and drain junctions. The feature is called **halo doping**. These highly doped regions are also called **pocket implants**.

9.6 Silicon on Insulator MOSFET (SOI)

Silicon MOSFETs can also be fabricated as silicon on insulator (SOI) structures (Fig. 9.9a), which have interesting properties; a layer of silicon oxide is implanted into the silicon wafer, and the CMOS FETs are then fabricated in the overlying surface layer of silicon. However, oxygen implant disturbs the crystalline structure of the silicon surface layer, and it is costly to recover the perfection of this critical region. *An easier way to construct SOI structure is to exploit Van der Waals forces to bond one silicon wafer to the oxide-coated surface of another silicon wafer. Then the thickness of the wafer with the oxide layer is carefully reduced resulting in the SPO structure. This process is known as Smart Cut, but expensive.* The source and drain regions reach through to the buried oxide, so there is no 'floor' component of parasitic capacitance at the drain|source substrate np junctions. The SOI MOSFET's speed improvement is also due to the dynamic threshold voltage effect. The dynamic threshold voltage is a result to the separation of the substrate and the other components of the MOSFET.

When the MOSFET is switched on by applying a positive gate source bias, the potential of the floating components also rise momentarily, boosting the forward bias across the source channel junction, and increasing the current. Also, when both the upper silicon layer and the buried oxide are thin, and the underlying substrate is heavily doped. The field issuing from the drain preferentially penetrates the two thin oxides and terminates on the gate and the substrate, Fig. 9.9b. The surface potential potential at the source $\Psi_{SURFACE}(0)$ is screened from the applied drain potential and the short-channel effect is reduced. The thin oxide substrate combination acts like a 'bottom' gate, making the SOI MOSFET a template for the multiple gate MOSFET, e.g.,the FinFET. The insulating layer of partially depleted SOI MOSFETs isolates devices from each other, which is vital for RF circuitry, and decouples noisy logic blocks from sensitive analogue circuitry, useful in mixed-signal applications.

There are two types of power dissipation in SOI MOSFETs—dynamic and static. The various DC leakage and sub-threshold currents in an NMOS are shown in Fig. 9.10. When the device is ON, a high drain current is required and, this current is drawn from the power supply. If the latter is more than the drain current, current leakage paths exist in the transistor.

Leakage currents are curbed by using thin oxides only where they are absolutely necessary. For example, transistors in the input/output parts of an integrated circuit do not need thick oxide layer.

Current leakage also occurs via the substrate, e.g., the usual current in a reverse biased np junction. Another current leakage mechanism is the **gate induced drain leakage** (GIDL). GIDL is similar to Zener breakdown, and arises from band-to-band tunnelling in

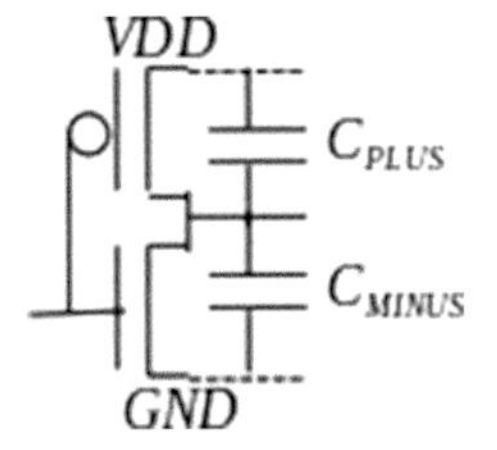

Fig. 9.9 CMOS inverter

the depletion region where the gate overlaps the drain, in np diodes with heavy doping on either side of the junction. GIDL becomes a significant current leakage mechanism because of very high fields arising from the heavy doping of the pocket implant. As the gate substrate voltage increases, the bands bend more and leakage current increases. These leakage currents exist when the transistor is ON. Current can also be drawn from the power supply when the transistor is OFF, but the drain source voltage is greater than zero. Such currents are the junction leakage current, and the sub-threshold current due to injection from the source. The latter is non-zero because the electric field from the drain affects the surface potential $\Psi_{SURFACE}(0)$.

In integrated circuits it is necessary that the OFF|ON ratio be as small as possible: i.e., switching between OFF and ON would be a vertical transition in the I_{DRAIN}, $V_{GATE,SOURCE}$ plane. In reality the transition has a finite slope due to **inverse sub-threshold slope**. *This is the change in gate source voltage needed to reduce the drain current by a factor of 10.*

$$S = \frac{\partial V_{GATE\,SOURCE}}{\partial \log(I_{DRAIN})} = 2.303 m V_{THERMAL} \tag{9.8a}$$

The common CMOS inverter (Fig. 9.10) illustrates dynamic power loss. The two capacitors C_{PLUS}, C_{MINUS} in combination represent the total capacitance C_{TOTAL} at the output node, **specifically the intrinsic and extrinsic capacitances of the transistors in both the logic gate and in the element that it drives and the interconnect capacitance**. It is charged through the PMOS, and discharged via the NMOS.. During discharge of the capacitor C_{MINUS}, v_{INPUT} is HI and the PMOS is OFF. The output goes LO and C_{PLUS} charges up. The current drawn from the power supply during this event is

$$i_{PS} = \frac{C^2(V_{DD} - V_{OUT})}{\partial t} \tag{9.8b}$$

The average power dissipated during this half-period $\frac{T}{2}$ is

$$P_{AVG} = \frac{2}{T}\int (V_{DD} - V_{OUT}) i_{PS} dt = f C^{PLUS} V_{DD}^2 \quad 0 \le t \le T \tag{9.8c}$$

where f is the clock frequency. Over one clock period, the average power dissipated per gate is:

$$P_{AVG,CYCLE} = f V_{DD}^2 (C_{MIMUS} + C_{PLUS}) = f V_{DD}^2 C_{TOTAL} \tag{9.8d}$$

The static power dissipated is:

$$P_{STATIC} = V_{DD} I_{DRAIN,SUBTHRESHOLD} \tag{9.8e}$$

9.7 Hot Electron Effect in Short Channel MOSFETs

'Hot carriers' are holes or electrons ('hot electrons') that have gained very high kinetic energy after being accelerated by a strong electric field in regions of high field intensities within a semiconductor (e.g., MOS) device. Their high kinetic energy, enables hot carriers to get injected and trapped in regions of the device where they are normally absent. The resulting space charge degrades the device's performance characteristics—'hot carrier effects'. According to the Hitachi Semiconductor Device Reliability Handbook [28], there are four commonly encountered hot carrier injection mechanisms.

- Drain avalanche hot carrier injection (DAHC)
- Channel hot electron injection (CHE)
- Substrate hot electron injection (SHE)
- Secondary generated hot electron injection (SGHE).

In the remainder of this discussion, the device being examined is the MOSFET.

Drain avalanche hot carrier (DAHC) injection causes the worst MOSFET performance and device degradation under normal operating temperature range. *DAHC occurs when a high voltage applied at the drain under non-saturated conditions* ($V_{DRAIN,SOURCE} > V_{GATE,SOURCE}$) *results in very high electric fields near the drain, which accelerate channel carriers into the drain's depletion region. Experimentally, it has been proved that the worst effects occur when the drain source voltage is twice the gate source voltage.* The accelerated channel carriers collide with silicon lattice atoms, creating dislodged electron–hole pairs in the process—impact ionization. Some of the displaced electron hole pairs also gain enough energy to overcome the electric potential barrier between the silicon substrate and the gate oxide. Under the influence of drain-gate field, hot carriers that surmount the substrate-gate oxide barrier get injected into the gate oxide layer where they might get trapped. This hot carrier injection process occurs mainly in a narrow injection zone at the drain end of the device where the lateral field is at its maximum.

Hot carriers can be trapped at the silicon–silicon dioxide interface ('interface states') or within the oxide itself, forming a space charge (volume charge) that increases with addition of more charges are trapped. These trapped charges change both the threshold voltage (Vth) and its transconductance (g_m).

Injected carriers that do not get trapped in the gate oxide become gate current. Majority of the holes from the electron hole pairs generated by impact ionization flow back to the substrate. These returning holes form a large portion of the substrate's drift current.

Excessive substrate current is an indication of hot carrier degradation. In extreme cases, abnormally high substrate current can upset the balance of carrier flow, resulting in latch-up.

Channel hot electron (CHE) injection occurs when both the gate voltage and the drain voltage are significantly higher than the source voltage, with $V_{GATE,SOURCE} > V_{DRAIN,SOURCE}$ Channel carriers that travel from the source to the drain are often driven towards the gate oxide even before they reach the drain because of the high gate voltage. CHE occurs when the substrate back bias is very positive or very negative. Carriers of one type in the substrate are driven by the substrate field toward the silicon–silicon dioxide interface. As they move toward the substrate-oxide interface, they gain additional kinetic energy from the high field in surface depletion region. They eventually overcome the surface energy barrier and get injected into the gate oxide, where some of them are trapped.

Secondary generated hot electron (SGHE) injection is the generation of hot carriers from impact ionization involving a secondary carrier that was created by an earlier incident of impact ionization. This occurs when the applied voltage at the drain is high, which is the driving condition for impact ionization. The main difference is the influence of the substrate's back bias in the hot carrier generation. This back bias results in a field that drives the hot carriers generated by the secondary carriers toward the surface region, where they further gain kinetic energy to overcome the surface energy barrier.

Hot electron effects in MOSFETs can be controlled by:

- Increasing channel length
- Drain and source should be double diffused—n^{PLUS}, n^{MINUS}
- Drain junctions must be graded
- Use self aligned n^{MINUS} regions between the channel and n^{PLUS} regions to create offset gate
- Use buried p^{PLUS} regions.

9.8 Semiconductor Industry Standard BSIM MOSFET Model

The BSIM are a family of semiconductor industry standard SPICE [22–27] MOSFET largelsmall signal device model libraries, freely available from [27]. These libraries were developed by the IGFET (Insulated Gate Field Effect Transistor) Group at the University of California Berkeley. The family of libraries consist of:

- BSIM Bulk: Bulk MOSFET library which started out as BSIM1, BSIM2,.. and after BSIM6 was renamed as BSIM Bulk.
- BSIMSOI BSIM SPICE [22–27] library for Silicon On Insulator (SOI) MOSFETs

- BSIMCMG/IMG BSIM SPICE [20–27] library for multi gate MOSFETs

The focus here is on the most commonly used library, BSIM Bulk. All BSIM libraries have been designed with the Compact Model concept, that enables easy exchange of information between the designers and foundaries. Such a model must guarantee convergence on various types of operating conditions, and the results must be accurate with minimum simulation time.

The BSIM Bulk Compact Model scheme exploits SPICE [20–27]'s built in threshold voltage model examined earlier. The threshold voltage model expresses currents in terms of voltages, with different equations for sub-threshold, linear and saturation regions of operation. Judicious use of interpolation smooths out the interface regions of the three regions of operation. The threshold voltage based model is much faster than surface potential based counterpart, avoid many iterations, and produce results as accurate as those generated with the surface potential models.

The core BSIM Bulk model has two main sub-components—the current voltage (I-V) and capacitance voltage (C-V) models. In combination these two account for all properties of sub-micron MOSFETs. The I-V component explains:

- Channel length modulation (CLM) and drain induced barrier lowering (DIBL)
- Velocity saturation and mobility degradation
- Gate induced drain leakage current (GIDL)
- Impact ionization current and direct tunnelling gate current
- Drain source and parasitic resistances

The C-V component accounts for:

- Fringe and overlap capacitances

The remainder of the BSIM Bulk model elaborates on the electronic noise, short channel, quantum and temperature related effects.

The BSIM Bulk MOSFET model has a huge set of SPICE [22–27] parameters divided into the basic model parameters, ultra high frequency (RF|microwave) parameters, temperature parameters etc., Some of the basic parameters essential for circuit design are mentioned below. The interested reader can easily download excellent free literature on this topic [27].

- Length of extracted long channel device
- Width of extracted long channel device
- Interface trap capacitance
- Sub-threshold swing factor
- Drain-bias sensitivity of sub-threshold swing

- Body-bias sensitivity of sub-threshold swing
- Coefficient of drain induced threshold voltage shift for log channel devices with pocket implant
- Vertical non-uniform doping effect on surface potential
- Threshold voltage shift due to non uniform vertical doping
- Drain induced barrier lowering (DIBL) coefficient
- DIBL exponent coefficient
- Body bias sensitivity to DIBL effect
- Low field mobility
- Effective field parameter
- Phonon|surface scattering parameter
- Coulombic scattering parameter
- Body|substrate bias sensitivity on mobility
- Saturation velocity
- Smoothing factor to drain source voltage to saturation drain source voltage transition
- Correction factor for velocity saturation
- Velocity saturation exponent for non-zero substrate source voltage
- Channel length modulation parameter
- Gate bias bias dependent modulation for channel length modulation
- Substrate current body effect coefficient
- Drain induced threshold voltage shift
- Length dependency of drain induced threshold voltage shift
- Drain source dependency of drain induced threshold voltage shift
- Source extension resistance per unit width at high gate source voltage
- Zero bias source extension resistance per unit width
- Drain extension resistance per unit width
- Zero bias drain extension resistance per unit width
- Lightly doped drain (LDD) resistance per unit width at high gate source voltage
- Zero bias LDD resistance per unit width
- Gate bias dependency of source drain extension resistance
- Body|substrate dependency of source drain extension resistance
- W dependency parameter of source drain extension resistance
- Sheet resistance
- Drain induced barrier lowering (DIBL) effect on global scaling parameters
- Body|substrate sensitivity on DIBL
- Gate source voltage dependency on early voltage
- Drain source G degradation factor due to pocket implant
- Pre-exponential coefficient for gate induced drain leakage (GIDL)
- Exponential coefficient for GIDL
- Band bending parameter for GIDL
- Pre-exponential coefficient for gate induced source leakage (GISL)

- Exponential coefficient for GISL
- Band bending parameter for GISL
- First parameter of impact ionization
- First drain source voltage dependent parameter of impact ionization current
- Parameter for Igcs and Igcd
- Parameter for gate source current (Igs)
- Drain source overlap length for Igs
- Drain source overlap length for gate drain current
- Parameter for gate drain current
- Factor for gate oxide thickness in drain source overlap regions
- Drain source dependency of Igcs and Igcd
- Exponent for gate oxide ratio
- Nominal gate oxide thickness for gate dielectric tunnelling current model only
- Flat band offset parameter
- Channel (body|substrate) doping concentration for CV
- Saturation velocity for CV
- Channel length modulation parameter for CV
- Outer fringe capacitance
- Outer fringe capacitance coefficient
- Non LDD region source gate overlap capacitance per unit channel width
- Non LDD region drain gate overlap capacitance per unit channel width
- Overlap capacitance between gate and lightly doped source region
- Coefficient of bias dependent overlap capacitance for source side
- Coefficient of bias dependent overlap capacitance for drain side
- Gate body|substrate overlap capacitance per unit channel length
- Quantum mechanical effect prefactor|switch in inversion
- Charge centroid parameter—slope of capacitance voltage curve under QME in inversion
- Charge centroid parameter—starting point for QME in inversion
- Bulk charge coefficient for for charge centroid in inversion

The BSIM Bulk model is complex. The above was just the list of basic SPICE [22–27] parameters.

9.9 Three Dimensional Transistors—State-Of-Art MOSFET Structures

As silicon based MOSFET technology has become mature, tried and tested, and most importantly synchronous logic circuit clock frequencies have increased higher and higher, the MOSFET has evolved from the initial planar layered to a three dimensional structure.

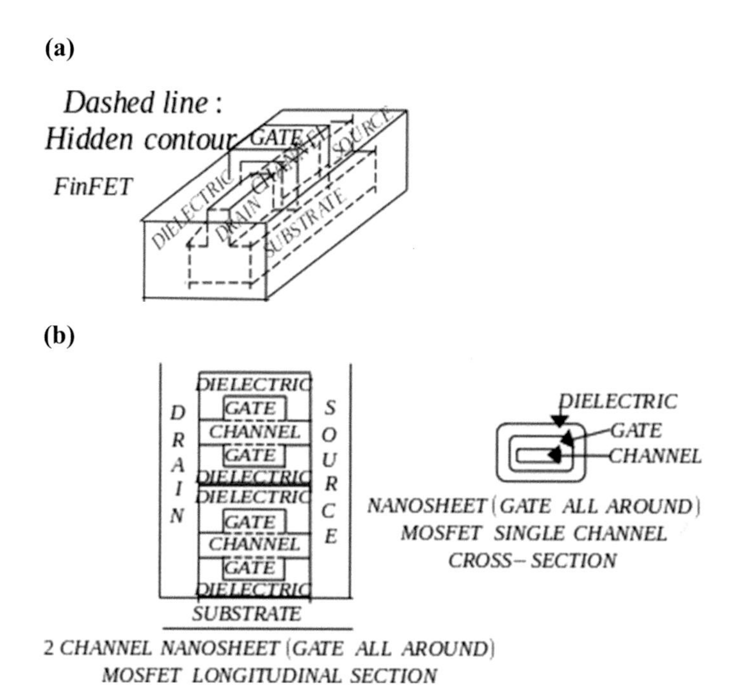

Fig. 9.10 **a** Three dimensional structure of FinFET, **b** Nanosheet (Gate All Around—GAAFET) MOSFET

This evolution is because of the necessity of curtailing unwanted effects of short channel MOSFETs, viz., leakage current. One is the *FinFET* (Fig. 9.10a) and the other is the *nanosheet MOSFET* (Fig. 9.10b). *Simultaneously, both MOSFET channel length and width have sub-micron dimensions, enabling more three dimensional MOSFETs to be packed into a single wafer die, i.e., increased packaging density.*

References

1. Sze, S. M., & Lee, M. K. (2021). Semiconductor physics and devices John Wiley and Sons. ISBN 9789354243226.
2. Martins, E. R. (2022, June). Essentials of Semiconductor Device Physics John Wiley and Sons. ISBN 987-1-119-88413-2.
3. Pierret, R. Semiconductor device fundamentals. ISBN 10 0201543931 ISBN 13 978–0201543933.
4. Hess, K. Advanced theory of semiconductor devices Wiley IEEE Press ISBN 10 0780334795 ISBN 13 978–0780334793.
5. Streetman, B. G., & Banerjee, S. K. Solid state electronic devices. ISBN: 9789332555082, 9789332555082.
6. https://eepower.com/technical-articles/an-introduction-to-cmos-technology/#.

7. Mead, C. (1972). Fundamental limitations in microelectronics—I. *MOS Technology, Solid State Electronics, 15*, 819–829.
8. Dennard, R. H., Gaensslen, F. H., Yu, H. N., Rideout, V. J., Bassous, E., & LeBlanc, A. R. (1974). Design of ion-implanted MOSFET's with very small physical dimensions. *IEEE Journal of Solid-State Circuits, SC-9,* 256–268.
9. Iwai, H. (1998). CMOS Scaling towards its Limits, IEEE, pp. 31–34.
10. 16. Pullfrey D. Understanding Modern Transistors and Diodes Cambridge University Press 2013 ISBN 13 978–0521514606.
11. https://www.ijert.org/evolution-of-cmos-technology.
12. https://ewh.ieee.org/r5/denver/sscs/Presentations/2004_12_Loke.pdf.
13. https://ieeexplore.ieee.org/document/1485588.
14. https://www.epfl.ch/labs/iclab/wp-content/uploads/2019/02/ICSES2000sl.pdf.
15. https://vlsi-backend-adventure.com/short_channel_effect.html.
16. http://www0.cs.ucl.ac.uk/staff/ucacdxq/projects/vlsi/report.pdf.
17. https://inst.eecs.berkeley.edu/~ee130/sp03/lecture/lecture27.pdf.
18. https://www.cambridge.org/core/books/abs/introduction-to-semiconductor-devices/shortchannel-effects-and-challenges-to-cmos/468FB7DF74AF708CECF03F99111D9B9C.
19. https://www.springerprofessional.de/en/short-channel-effects-in-mosfets/10709750.
20. https://ieeexplore.ieee.org/document/8705955.
21. https://siliconvlsi.com/dibldrain-induced-barrier-lowering/.
22. https://ngspice.sourceforge.io/download.html.
23. https://www.orcad.com/pspice-free-trial.
24. https://www.synopsys.com/implementation-and-signoff/ams-simulation/primesim-hspice.html.
25. https://www.ti.com/tool/TINA-TI.
26. https://www.analog.com/en/design-center/design-tools-and-ca;lculators/ltspice-simulator.html.
27. https://bsim.berkeley.edu/.
28. https://www.renesas.com/us/en/document/grl/semiconductor-reliability-handbook.

10 Noise in Semiconductor Devices

10.1 Semiconductor Noise Mechanisms [1–77]

10.1.1 Thermal Noise

"Thermal" noise is a temperature dependent common noise characteristic of any semiconductor device, caused by random variations in the position of the free electrons (Brownian motion) that induces a time varying random electromotive force at the terminals. Random thermal motion of electrons inside the semiconductor material causes temporary concentration of carriers at one end or the other. Thus one end contact will be at a more negative potential than the potential at the other end contact. A time varying polarity and magnitude voltage, called the thermal noise voltage appears. Random motion of free electrons is unavoidable at all temperatures. Thermal noise corrupts low amplitude signal output or the device is operating at RF|microwave frequencies.

The Nyquist theorem states that for a linear resistance in thermal equilibrium at temperature T, the noise current or voltage variations are independent of the conduction mechanisms, type of material and geometry|shape of the resistor. The generated noise depends only on the resistance value and its temperature T. The noise voltage spectral density (open circuit conditions) and the corresponding noise current spectral density (short circuit condition) are:

$$S(V_{NOISE}) = \frac{\overline{v}^2_{NOISE}}{\Delta f} = 4k_B T R \frac{V^2}{Hz} \quad S(I_{NOISE}) = \frac{\overline{i}^2_{NOISE}}{\Delta f} = \frac{4k_B T}{R} \frac{A^2}{Hz} \tag{10.1a}$$

The noise power spectral density (**the noise power delivered to an identical resistor R, per unit bandwidth**) and the Norton, Thevenin (*corresponding to a noiseless resistor in parallel|series with a noise source*) equivalent current and voltage are:

A. Banerjee, *Semiconductor Devices*, Synthesis Lectures on Engineering, Science, and Technology, https://doi.org/10.1007/978-3-031-45750-0_10

$$S_P = k_B T \quad \bar{i}^2_{NOISE} = \frac{4k_B T \Delta f}{R}(Norton) \quad v^2_{NOISE} = 4k_B T R \Delta f (Thevenin) \tag{10.1c}$$

- The previous four expressions are equivalent. One can be used to derive another.
- Nyquist theorem is valid for linear devices in thermal equilibrium.
- The spectral density of thermal noise is constant with respect to frequency (till frequencies at which the quantum mechanical correction is imposed)—e.g., light.
- The total available noise power of a resistor is infinite at infinite bandwidth. The quantum correction limits it to a finite value: $\frac{(\pi k_B T)^2}{6h}$

For a general impedance with non-zero complex component, the noise current spectral density is:

$$S(I_{NOISE}) = \frac{4k_B T R}{X^2 + R^2} \tag{10.1e}$$

For a linear two terminal network containing only noisy resistors, capacitors, and inductors, the noise current|voltage at its terminals can be calculated by substituting each resistor with its appropriate model. Then each individual contribution at the output is calculated and the resulting mean square values are added. An easier approach involves evaluation of the real part of the complex impedance Z seen when looking into the terminals and then application of Thevenin equivalent circuit, to get:

$$\bar{v}^2_{NOISE} = 4k_B T \int \Re Z df \quad 0 \le f \le \infty \tag{10.1f}$$

This is the Nyquist's formula. For a one-port containing only linear resistors, 10.1f reduces to:

$$\bar{v}^2_{NOISE} = 4k_B T R_{equivalent} \Delta f \quad 0 \le f \le \infty \tag{10.1g}$$

Nyquist's theorem is not valid for arbitrarily high frequency, since by integrating the expression for noise signal spectral power density over the entire frequency range results in infinite noise power. So a quantum correction is imposed. Specifically, if $\frac{hf}{k_R T} \gg 1$ the noise spectral power density is:

$$S(p) = \frac{\hbar f}{e^{\frac{\hbar f}{k_B T}} - 1} \tag{10.1h}$$

Denoting the Planck factor as:

$$H(f) = \frac{\frac{\hbar f}{k_B T}}{e^{\frac{\hbar f}{k_B T}} - 1}$$

the mean square value of the open-circuit noise voltage delivered by a resistor is re-written as

$$\overline{v}^2_{NOISE} = 4k_B T \int H(f)R(f)df \quad f_1 \le f \le f_2 \tag{10.1i}$$

For linear multiports containing some non-reciprocal devices, Thevenin's and Nyquist's theorems must be formulated differently—the previous impedance expressions must be replaced by a linear combination of some elements of the impedance matrix used to describe the non-reciprocal network. Nyquist's theorem can be applied to linear dissipative and distributed networks e.g., a transmission line or waveguide.

Nyquist's theorem cannot be directly applied to nonlinear devices|circuits. If the nonlinear system is in thermal equilibrium and the resulting variations are small, Nyquist's theorem holds in the small signal approximation of the input impedance of the circuit. Therefore, the noise model of a nonlinear resistor is in the form

$$\overline{v}^2_{NOISE} = 4k_B T \left(\frac{dV}{dI} + \frac{Id^2V}{2dI^2} \right) \Bigg|_{I=I_{CC}} \tag{10.1j}$$

- The frequency distribution of thermal noise power is uniform, at up to $f_C = 0.15 k_B T x 10^{34}\ Hz$
- The instantaneous amplitude of the thermal noise has a Gaussian distribution. The mean value of the fluctuation is zero.
- The peak factor is roughly 4 if only peaks occurring at least 0.01% of the samplingtime are considered

The Nyquist theorem based thermal noise model is valid for lightly doped semiconductors and linear resistors only.

10.1.2 Diffusion Noise

Diffusion noise arises from charge carrier velocity variations due to collision|scattering during charge carrier diffusion. Charge carrier diffusion is a nonuniform carrier distribution. If carrier density increases at one end of a semiconductor, a carrier concentration gradient is set up, causing carriers to move from high concentration region to low concentration region. The moving electrons are scattered via collisions with the lattice and|or with ionized impurity atoms. *Scattering is random, so that the instantaneous value of the diffusion current is also random, resulting in diffusion noise.*

To quantify diffusion noise, a volume of semiconductor material (e.g., n type) is partitioned into cells of dimensions Δx, Δy, Δz. *The event of any electron moving from one cell to any of the neighbouring cells is independent of any other electron moving from the same cell to any other neighbouring cell.* The diffusion noise current spectral density is:

$$S(I_{DIFFUSIONNOISE}) = \frac{4D_{electron}n(x)q^2\Delta y\Delta z}{\Delta x} \tag{10.2a}$$

Using Einstein's expression relating electron diffusivity and electron mobility, the diffusion noise spectral density can be re-written as:

$$D_{electron} = \frac{k_B\mu_{electron}T}{q} \quad S(I_{DIFFUSIONNOISE}) = \frac{4k_B\mu_{electron}n(x)\Delta y\Delta z}{\Delta x} = \frac{4k_BT}{\Delta R} \tag{10.2b}$$

where $D_{elecron}$, $\mu_{electron}$, ΔR are respectively the electron diffusivity, mobility and resistance of a cell. **Diffusion noise reduces thermal noise when Einstein's relation holds.** *While thermal noise is due to random motion of carriers, diffusion noise arises from random collisions charge carriers move through the semiconductor device.* Diffusion noise spectrum is flat.

Diffusion noise is present in all semiconductor devices, as the forward current versus forward voltage curve for each does not follow Ohm's law. A transistor operating in the linear region has weak diffusion noise current, but diffusion noise current is measurable and strong in the transistor's saturation region of operation.

10.1.3 Shot Noise

Shot noise arises as charge carriers are discrete. **All semiconductor devices contain a potential barrier. The current through the device is due to those electrons that possess sufficient energy to cross the potential barrier. The passage of charge carriers (electrons) across this barrier is a series of independent, random events.** For example, for a stream of electrons in vacuum, between two electrodes, the number of electrons crossing a reference plane will vary with each sampling time, due to both the random emission rate of electrons at the source plane, and the random distribution of individual velocities. Every time an electron crosses the reference plane, an elementary current pulse appears in the external circuit. The area of this reference plane is equal to the elementary charge q and the transit time (the average time needed to traverse the distance from the source to the reference plane). *This noise mechanism is shot noise and is due to the discrete nature of electric charge.*

Shot noise appears in all devices collecting a flow of electrical charges, under the condition that the charge carriers have ballistic trajectories with no interactions during flight. *This holds strictly when charge carrier density is low with high external electric field.* Otherwise, the randomness of their position and velocity is reduced by repulsion between identical charges or collisions with lattice atoms.

Shot noise analysis starts with the arrival of an electron at the reference plane. Ignoring transit time the instantaneous current is the sum of elementary Dirac pulses of weight q, with average value and spectral density given as:

$$I(t) = q\sum \delta(t - t_i)\overline{I}(t) = I_0 = \lambda q S(I) = 2q^2\lambda \frac{A^2}{Hz} \tag{10.3a}$$

where λ is the average number of electrons collected per second. The spectrum, computed while ignoring transit times, is white. If transit times are included a cut-off frequency $\omega_{CUTOFF} = \frac{3.5}{T_{transit}}$ is introduced. Shot noise is modelled as a noise current with mean square value:

$$\overline{i}^2_{electron} = S(I)\Delta f \tag{10.3b}$$

where Δf is the measurement system bandwidth.

The current flowing through a homogeneous np junction is due injection of minority carriers into the bulk region, followed by their diffusion and recombination. Shot noise is a combination of diffusion noise and generation-recombination noise for the minority carriers. There might also be noise contributed by emission across the junction potential barrier: the passage of each carrier is an independent random event. The junction current of a homogeneous pn junction diode is:

$$I = I_S\left(e^{\frac{qV}{k_BT}} - 1\right) \tag{10.3c}$$

where I_S V are the saturation current and applied diode voltage respectively. Assuming that the two current components generate independent random variations,

$$\overline{i}^2_{electron,total} = 2qI_S\Delta f\left(e^{\frac{qV}{k_BT}} + 1\right) = 2q\Delta f I_S(I + I_S) \tag{10.3d}$$

From the small signal low frequency conductance expression, the transconductance is:

$$g_m = \frac{dI}{dV} = \frac{qI}{k_BT} \tag{10.3e}$$

Combining the above two expressions, the total averaged shot noise current is:

$$\overline{i}^2_{electron} = 2qI\Delta f = 2k_B g_m \Delta f T \quad A^2 \tag{10.3f}$$

For a metal semiconductor junction (e.g., Schottky diode), the current flow is due to majority carriers. There are two types of carriers::

- Carriers going from metal into semiconductor, which encounter a potential barrier of height E_O, producing a current $-I_S$, which is independent of the applied voltage.
- Carriers flowing in the reverse direction, from semiconductor into metal,which encounter a barrier of height $q(\Phi_{CONTACTPOTENTIAL} - V)$, where V is the voltage applied to the metal. The overall current is the sum of these two components, the total current spectral density is

$$S(I_{total}) = 2q(I + I_S) = 2k_B g_m T\left(\frac{I + 2I_S}{I + I_S}\right) \quad (10.3g)$$

- Shot noise depends only on the DC current, not temperature. through the device, modifying the bias current represents an easy way to control the noise level. This property is useful in the design of calibrated noise sources.
- A junction diode always generates shot noise, the potential barrier is associated with its depletion layer. Since fluctuations are increasingly smoothed out as the average current increases, in forward biased junctions the noise current is much lower than predicted.
- The power distribution versus frequency is uniform—white noise up to frequencies close to the reciprocal of the transit time.
- Shot noise instantaneous amplitude distribution is Gaussian The ratio of variations to the average current can be reduced by increasing the number of electrons reaching the collector terminal.
- Shot noise could ideally be reduced if the magnitude of an individual charge could be diminished.

Shot noise is always generated in junction diodes, and its effects are observed in reverse-biased diodes or forward biased junctions operating at very low currents.

10.1.4 Generation Recombination Noise

Generation recombination noise is due to statistical variation in the population of charge carriers due to random generation|recombination, random trapping, and the release of carriers in a semiconductor.

When a covalent bond is broken, an electron–hole pair is generated, and each covalent bond breaking event requires a small amount of extra energy This energy can be supplied either thermally or by illuminating the semiconductor surface. Since the flow of energy (phonons or photons) is quantized and non-uniform, the generation of charge carriers is a random process, both in space and in time.

Opposing charge generation is simultaneous electron hole recombination. Recombination depends on the Brownian motion of carriers. Recombination is also a random process. On average, generation must balance recombination.

Traps in the bulk or on the surface of a semiconductor are very important in this noise mechanism. Electrons and holes are captured, and then released after a variable but finite time interval, additional variations in the population of charge carriers. All crystal lattice defects—impurity atoms or molecules that contaminate the surface of the semiconductor during fabrication, act as traps.

Variations in charge current density in a semiconductor material results in current fluctuations—as per Langevin's method, and is controlled by a differential equation:

$$\frac{d(\Delta N)}{dt} = H(t) - \frac{\Delta N}{\tau_0} \tag{10.4a}$$

where ΔN is the variation in the number of carriers τ_O is the average lifetime of added carriers, and H(t) is a random white noise source that controls the variation. The power spectrum of the generation recombination noise is:

$$S(p) = \frac{4\Delta N^2}{1 + \omega^2 \tau_O^2} \tag{10.4b}$$

Generation recombination noise is modelled as noise current generator with power spectral density given by (10.4b), and the mean square value:

$$\bar{i}^2_{electron} = \frac{4I^2\tau}{N\left(1 + \omega^2\tau^2\right)} \tag{10.4c}$$

where N is the average number of carriers, I is the average (DC) current through the device, and τ is the generation-recombination time constant. The level of generation-recombination noise is proportional to the square of the average (DC) current through the device. Since τ may be as brief as 1 ns, the generation recombination noise spectrum is flat up to a frequency of about 1/τ. Then it falls at about 20 dB per decade. Generation recombination noise induces variations in the conductivity of the material as the number of charge carriers is fluctuating; in contrast, for thermal noise, the number of carriers is roughly constant, but their spatial distribution fluctuates.

Generation recombination noise shows up in regions of a semiconductor material where carriers concentrations are low, e.g., in intrinsic semiconductors, lightly doped semiconductors, and in the space charge layer of every junction.

10.1.5 Flicker (1/f) Noise

The exact physical cause for flicker noise is unclear, it occurs in many non-electrical situations as well. **1/f noise is inversely frequency dependent—increasing with decreasing frequency, often extending to very low frequencies 10 – 4 Hz.**

Flicker noise is a general process in discontinuous, non-equilibrium systems e.g., DC current flow through a semiconductor device. Possible underlying causes could be either defects affecting the semiconductor lattice (e.g., unwanted impurity atoms), or to interactions between charge carriers and the surface energy states of the semiconductor, imperfect contacts (e.g., between granules of carbon in carbon resistors) etc., Flicker noise in carbon resistor is called "excess noise", as it adds to the thermal noise of the resistor—a wire wound resistor has no excess noise.

- Model A relates flicker noise to surface states, and is based on the time constants associated with the recombination process.
- Model B proposes an empirical approach relating flicker noise to bulk effects in dissipative media.
- Model C uses quantum mechanics to propose that flicker noise is conditioned photon emission each time a charge carrier collides with the lattice.
- Model D states that flicker noise is produced by the fluctuation of a parameter u, subject to the diffusion equation

$$\frac{\partial u}{\partial t} = D\nabla^2 u \tag{10.5a}$$

Here ∇ is the Laplacian operator and D is a constant. This equation is of the same form as that used to describe thermal conductivity or carrier diffusion. The solution to this equation is not an accurate representation of flicker noise, for finite dimension systems, the power spectral density has a 1/f region covering several decades. Below and

above this region the frequency dependency is of the form $\frac{1}{f^k}$, k being an integer or a real number.

Flicker noise generated in semiconductor materials e.g., GaAs is a bulk phenomenon associated with local high frequency and long range low frequency variations, the lowest frequency being limited only by the bulk volume. Perturbation theory states that any large dimension distributed system and high resistance and some capacitance will generate flicker noise when disturbed. Then the 1/f noise is modelled by a current generator with mean square value:

$$\bar{i}^{\,2}_{electron} = \frac{K I^{\alpha} \Delta f}{f^n} \tag{10.5b}$$

where K is a device dependent constant; α is a constant ($0.5 \leqslant \alpha \leqslant 2$), n = 1 for pink noise characterized by constant power per octave (n could be 2 due to variations in the Earth's rotational speed, and 2.7 for galactic radiation). Δf is the bandwidth of the measurement system and I is the DC current through the device.

Flicker noise in homogeneous semiconductors can be represented using a parameter α such that the power spectral density of a resistance R is:

$$S_R = \frac{\alpha R^2}{f N} \tag{10.5c}$$

where f the measurement frequency, and N the total number of free charge carriers. α corresponds to the normalized contribution to the relative noise of a single electron, per unit bandwidth (independent of contributions of other electrons). An initial trial average value of 0.002 was proposed for α [1–77]. α also depends on the quality of the crystal—in a impurity free|perfect material its value could be 2 or 3 orders of magnitude lower.

- The probability distribution function of flicker noise amplitudes is not Gaussian.
- The spectrum increases with decreasing frequency, never reaching frequency zero.
- Flicker noise is prominent whenever electric currents are due to a very small number of charge carriers.

In bipolar transistors flicker noise is generated due to lattice defects or unwanted impurities present in the emitter base region. Carriers are trapped, then released randomly. The associated time constants are long and this explains why the noise spectral density is more important at low frequency. It has been determined that flicker noise current in a bipolar transistor is:

$$\bar{i}^2_{electron} = \frac{K_1 I^\alpha \delta f}{f^\beta} \tag{10.5d}$$

K_1 is a constant weakly dependent on temperature; I is the forward current through the device; δf is the bandwidth centred on f: ($0.5 \leqslant \alpha \leqslant 2$) and $\beta \approx 1$.

Flicker noise is absent in carbon resistor, while for an integrated resistor the mean flicker noise voltage is:

$$\bar{v}^2 = \frac{K_R R_{SHEET} \delta f V_{DC}^2}{A_R f} \tag{10.5e}$$

R_{SHEET} is the sheet resistance, A_R is the area of the resistor, and K_R is a technological constant. For a diffused or ion-implanted resistor, its value is approximately

$5x10^{-24}S^2cm^2$, while for thick-film resistors, it is roughly 10 times greater. For integrated gallium arsenide resistors, flicker noise is proportional to the square of the applied voltage.

In a field effect transistor, the drain current shows variations with a 1/f spectrum, due to fluctuations in the population of charge carriers in the channel—a result of carrier trapping in the surface states situated at the silicon–silicon dioxide interface. For a MOSFET the surface state density at the Fermi level is the only parameter that influences flicker noise. So to reduce MOSFET flicker noise, the surface state density in the vicinity of the Fermi level must be reduced. Flicker noise increases with decreasing temperature, as the density of surface states increases toward the conduction band. For MOS transistors operating in strong inversion, flicker noise does not depend on the gate bias, because the surface potential varies very slowly with gate charge. *Thus the only way to lower the noise level is to modify the device geometry.*

Heterogeneous junction devices have higher levels of flicker noise. Any inhomogeneity in the lattice structure or materials as well as contamination during processing, poor ohmic contacts, etc. can increase the flicker noise. Flicker noise is low in devices with homogeneous structure and a significant volume of material for instance a 1-W resistor is less noisy than a 0.25-W resistor of the same value and type.

In modern integrated circuits, reduction of physical dimensions of transistors increases flicker noise. The effects of flicker noise are not limited to low frequency, since it can be up-converted by an existing nonlinearity, as in active mixers, frequency dividers, and voltage-controlled oscillators.

10.1.6 Burst (Popcorn) Noise

The underlying physical reason behind burst (popcorn) noise is not clear. Burst noise occurs in planar diffused devices, tunnel diodes, bipolar transistors, integrated circuits and film resistors. Material contamination by heavy metal atoms during processing, or crystallographic damage of regions close to junctions may cause burst noise. Bursts appear commonly in small dimension devices operating in high currentlvoltage density conditions, in single trap activity in a region with few free carriers e.g., the depletion region of the emitter junction.

For a bipolar transistor, burst noise is modelled as a current generator with mean square value:

$$\bar{i}^2_{electron} = \frac{K_1 I_B \Delta f}{1 + \frac{\pi^2 f^2}{4a^2}} \tag{10.6a}$$

where a is the burst rate, K_1 is a constant, and I_B is the transistor base current.

Burst noise corrupts bipolar transistor collector current at a random rate, between several hundred per second to one every few minutes, with variable burst widths: between several microseconds and a few minutes. For a particular device under test, the amplitude remains the same. Bursts can overlap. The burst noise spectrum is proportional to $\frac{1}{f^2}$.

- Burst noise is observed at low frequencies.
- The instantaneous burst noise amplitude is 2 to 100 times that of thermal noise. For a device under test, this amplitude is constant, as it is determined only by the specific defect of the junction.
- Burst noise can appearldisappear at random.
- Among all devices fabricated in a single run, this noise might affect only a few, indicating poor quality.

State-of-art semiconductor device manufacturing technologies, in ultra clean fabrication facilities have eliminated this problem.

10.1.7 Avalanche Noise

Avalanche noise is the result of carrier multiplication due to impact ionization in a reverse-biased np junction. In a reverse biased np junction with a large applied bias, the electric field in the depletion region is enhanced, such that minority carriers (holes in the n region and electrons in the p region) are accelerated hard and their energy is increased. A chance collision with a neutral lattice atom generates one or more electron–hole pairs by impact—*impact ionization.*

Due to positive feedback, these pairs of impact generated carriers are accelerated and undergo collisions themselves producing additional pairs of charge carriers. Crystal structure imperfections result in microplasma generation in low-volume regions and the current through the junction is increased. Uncontrolled avalanche process leads to breakdown.

Every carrier crossing the junction induces an elementary current pulse. Total current is a superposition of all elementary current pulses, which varies with the number of carriers traversing the junction potential barrier per second. The movement of avalanche generated carriers across the junction is more complicated, since the carriers are multiplied by a factor M—random variable of both space and time. The probability of generating new pairs by collision is different for electrons and holes and breakdown does not occur simultaneously. Applying some simplifying assumptions, (e.g., M is independent of position x) the mean square avalanche noise current and its corresponding spectral density are:

$$\bar{i}^2 = 2q\overline{M}^2\Delta f\left(I_{hole}(0) + I_{electron}(w) + 2qA\int g(x)dx\right) \quad S(I)$$
$$= 2\overline{M}^2\left(I_{hole}(0) + I_{electron}(w) + 2qA\int g(x)dx\right) \quad 0 \le x \le \infty \qquad (10.7a)$$

w is the width of the depletion region and g(x) is the number of charge pairs generated at coordinate x per unit volume per unit time, and A is the junction area. With the adopted simplifications, the avalanche noise spectrum is white.

10.7a applies to reverse voltages higher than 8 V, where the breakdown mechanism comes into play. For reverse voltages less than 5 V, breakdown is the result of the Zener effect. If the reverse voltage lies between 5 and 8 V, the power spectral density is:

$$S(I) = 2q\left(I_{hole}(0) + I_{electron}(w) + 2qA\int g(X)dX\right) \quad X = \left(1 + f\left(M^{P}\right)\right) \tag{10.7b}$$

$f\left(M^{P}\right)$ is a function of the averaged ionized rates. In this case, primary carriers are generated by the tunnel effect, and the secondary carriers by collision.

Avalanche noise has a typical waveform with several levels separated by a few milliVolts. At start there is fast random switching among all levels. As the reverse current increases, the highest level becomes dominant. The avalanche noise spectrum is white. Avalanche noise mainly occurs in reverse-biased diodes operating at more than 8 V.

10.2 Microwave Noise

As transistors are nowadays increasingly operating at microwave frequencies (100 s of MHz—10 s of GHz) the types of noise specific to this frequency range is analysed in detail, using accurate small signal scattering parameters (S parameters) analysis.

10.2.1 Hybrid π Bipolar Transistor Microwave Noise Model

The noise free and noisy hybrid π common emitter bipolar transistor microwave noise equivalent circuits [24] are in Figs. 10.1a, b respectively. The emitter junction is conductive, and generates *shot noise* on the emitter. The emitter current is a combination of the base and collector currents (I_B, I_C),

Collector reverse current also generates *shot noise*. The emitter, base, and collector are made of semiconductor material and each have a finite resistance resulting in *thermal noise*. The base resistance is value higher than both the collector and emitter resistances, So the thermal noise contribution of the collector, emitter resistances can be neglected in the initial analysis. Three sources are introduced in a noiseless transistor: noise due to variation in the DC bias current ($I_{B,N}$), the DC collector current ($I_{C,N}$), and the thermal noise of the base resistance. The signal source would also have internal conductance and thereby generate noise and its susceptance affects the noise level through noise tuning. In silicon transistors, the collector reverse current ($I_{CO,B}$) is very small, so the noise ($I_{CO,N}$) generated by this current can be neglected. The mean square value of the above noise generator in a narrow frequency interval Δf is given by:

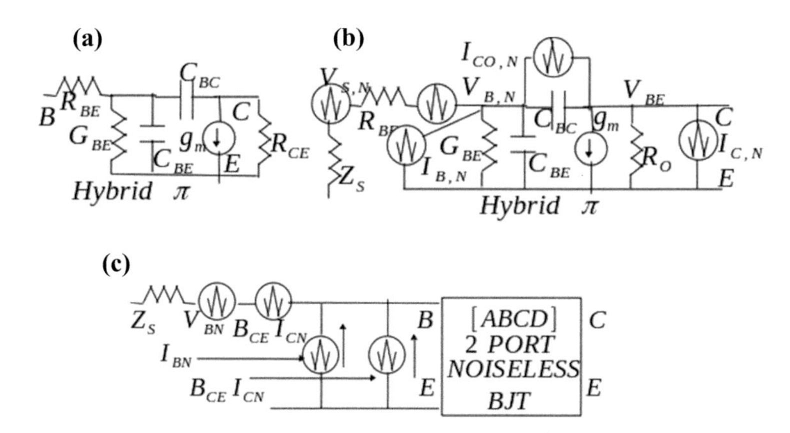

Fig. 10.1 **a**, **b** Hybrid π common|grounded emitter noiseless and noisy bipolar transistor microwave noise equivalent circuit. **c** BJT noise sources transformed to input, with noiseless intrinsic transistor

$$\bar{i}^2_{B,N} = 2qI_B\Delta f \quad \bar{i}^2_{C,N} = 2qI_C\Delta f \quad \bar{i}^2_{CO,N} = 2qI_{CO,B}\Delta f \quad \bar{v}^2_{B,N} = \bar{v}^2_{S,N} = 4k_BTR_B\Delta f \tag{10.8abcd}$$

The corresponding current|voltage noise power spectral densities are:

$$S(i_{C,N}) = \frac{\bar{i}^2_{C,N}}{\Delta f} = 2qI_C = 2k_Bg_mT \quad S(i_{B,N}) = \frac{\bar{i}^2_{B,N}}{\Delta f} = 2qI_B = \frac{2k_Bg_mT}{\beta} \quad S(v_{B,N}) = \frac{\bar{v}^2_{B,N}}{\Delta f} = 2k_BTR_B \quad S(v_{S,N}) = \frac{\bar{v}^2_{S,N}}{\Delta f} = 2k_BTR_S \tag{10.8efgh}$$

where $R_B,\ R_S$ are the base and source resistances.

10.2.2 Generalization of the Noisy Bipolar Transistor Equivalent Circuit

The equivalent circuit model for the noisy silicon bipolar transistor [24] (Fig. 10.1b) is generalized to a two port network with all the noise current|voltage sources transformed to the input (Fig. 10.1c) and a noiseless two port network for the transistor. The noise free two port transistor is represented in the [ABCD] matrix form, whose elements are:

$$A_{CR} = \frac{1}{1 - \frac{g_m}{j\omega C_{BC}}} \quad B_{CE} = \frac{-1}{g_m} \quad C_{CE} = A_{CE}(g_m + g_{BC} + j\omega C_{BE}) \quad D_{CE} = \frac{\frac{1}{R_{BE}} + j\omega(C_{BE} + C_{BC})}{g_m - j\omega C_{BC}} \tag{10.8ijkl}$$

The DC forward current gain and the transition frequencies are:

$$\beta = g_e(f)R_{BE} \quad f_T = \frac{g_e}{2\pi(C_{BC} + C_{BE})}$$

In silicon transistors the collector reverse current is very small, and so the noise generated by this current can be neglected. **The *noise factor F* is the ratio of the total mean square noise current and the thermal noise generated from the source resistance.** The total noise is that obtained from the entire network, defined as (with reference to Fig. 10.1c):

$$v_{N,NETWORK} = V_{BN} + I_{BN}\left(R_S + R_B^P\right) + AA\left(R_S + R_B^P\right) + j(I_{BN}X_S + AAI_{CN}) - I_{CM}R_E \quad AA = \frac{1}{\beta} + \frac{jf}{f_T} \tag{10.8m}$$

The total nose and noise factor are:

$$V_{N,TOTAL} = V_{N,SOURCE} + V_{N,NETWORK} \quad F = \frac{\overline{V}^2_{N,TOTAL}}{V^2_{N,SOURCE}} = \frac{V^2_{N,NETWORK} + V^2_{N,NETWORK}}{V^2_{N,SOURCE}} \tag{10.8no}$$

The expression for the noise factor F can be re-written in terms of the A, B, C, D parameters examined previously. This is left as an exercise for the reader. The noise factor expression, in the special case when the source reactance is zero, is:

$$F = 1 + \frac{1}{R_S}\left(R_B^P + \frac{R_E}{2}\right) + \frac{1}{2\beta R_E R_S}\left(\left(R_B^P + R_S\right)\left(\left(2R_E + R_S + R_B^P\right) + \left(R_B^P + R_S\right)\left(\frac{1}{\beta} + \frac{\beta f}{f_T}\right)\right)\right) \tag{10.8p}$$

It can be simplified further for the following conditions:

$$F = \frac{1}{R_S}(\left\langle R_B^P\right\rangle + < \frac{\left(R_B^P + R_S\right)^2}{2\beta R_E} > + < \frac{R_E}{2} + \frac{\left(f\left(R_B^P + R_S\right)\right)^2}{2\beta f_T^2} > \quad \beta \gg 1 \quad \omega C_{BC}R_E \ll 1 \tag{10.8q}$$

where the contribution of the first, second and last terms are due to the base resistance, base current, and collector current respectively.

10.3 T Equivalent Circuit for Noisy Bipolar Transistor

The T equivalent circuit model for the bipolar transistor at microwave frequencies [24] is in Fig. 10.2. C_{TE}, Z_S are respectively the emitter junction capacitance and the complex source impedance.

The T-configuration is simpler than the hybrid π model for minimum noise current calculation. Formulating the noise correlation matrix with the base collector capacitance C_{BC}, is easy in the hybrid π topology. Silicon bipolar transistor noise can be modelled with three noise sources—base resistor thermal noise, forward biased base emitter junction resistance shot noise and and collector junction noise. The mean square values of noise sources in a narrow frequency range Δf are:

$$\overline{e}_E^2 = 4k_B R_E T \Delta f \quad \overline{e}_B^2 = 4k_B R_B T \Delta f \quad \overline{e}_S^2 = 4k_B R_S T \Delta f \tag{10.9abc}$$

$$\left(\overline{I}_{CP} e_E^P\right) = 0 \quad \alpha = \frac{\alpha_0}{1 + \frac{jf}{f_n}} \quad \beta = \frac{\alpha}{1-\alpha} \quad g_E = \frac{1}{R_E} \quad R_E = \frac{k_B T}{q I_E} \tag{10.9d}$$

where the base resistance thermal noise voltage source is e_B. The shot noise voltage source e_E is generated by the forward-biased emitter base junction R_E. The collector noise current source I_{CP} comes from the collector partition, which is strongly correlated to the emitter–base shot noise.

The noise factor by definition is the ratio of the input signal to noise ratio to the output signal to noise ratio. Alternatively, the noise factor can be re-written as:

$$F = \frac{\frac{S_{INP}}{N_{INP}}}{\frac{S_{OUT}}{N_{OUT}}} = \frac{N_{OUT}}{GN_{INP}} = \frac{N_{OUT}}{BGk_B T} \quad F = \frac{\overline{I}_L^2}{\overline{I}_{LO}^2} \tag{10.9ef}$$

where i_{LO} is the value of i_L due to the source generator noise e_S only, and i_L is the total load current or the collector current (AC short-circuited current) due to all the noise sources. These are expressed as:

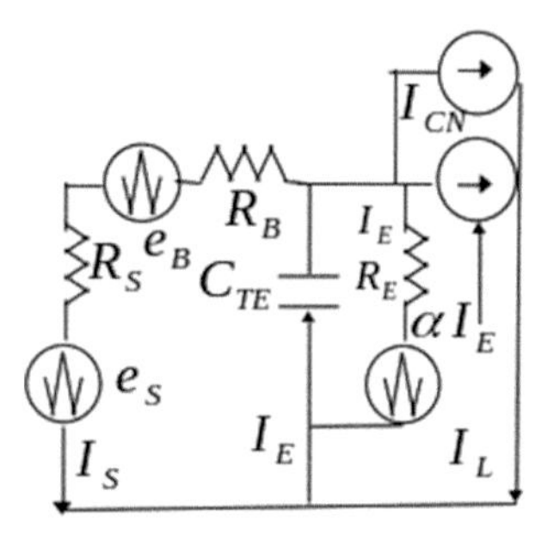

Fig. 10.2 T equivalent circuit model for noisy silicon bipolar transistor

$$I_L = \frac{\alpha\Big(e_B + e_S + e_E(1 + j\omega C_{TE}(R_B + Z_S)) + \frac{I_{CP}}{\alpha}((1 + j\omega C_{TE}R_E)(R_S + Z_S) + R_E)\Big)}{(1 + j\omega C_{TE}R_E - \alpha)(R_B + Z_S) + R_E} \tag{10.9g}$$

$$I_{LO} = \frac{\alpha e_S}{(1 - \alpha + j\omega C_{TE}R_E)(R_B + Z_S) + R_E} \tag{10.9h}$$

The expression for the noise factor is evaluated easily, and simplified in the source impedance zero case. It is then used for determining the minimum noise factor and the generator thermal noise are expressed as ($\Delta f = 1$ Hz):

$$\begin{aligned} &\overline{e}_E^2 = 4k_B R_E T \quad \overline{e}_S^2 = 4k_B R_S T \quad \overline{e}_B^2 = 4k_B R_B T \quad R_E = \frac{k_B T}{qI_E} \quad I_{CP} \\ &= 2qI_E\big(\alpha_0 - |\alpha|^2\big) \quad \alpha = \frac{\alpha_0}{1 + \frac{jf}{f_B}} \end{aligned} \tag{10.9i}$$

10.4 Noisy Gallium Arsenide (GaAs) Field Effect Transistors

A noise model of a grounded source GaAs field effect transistor (FET) [24] with the noise sources at the input and output is in Fig. 10.3a, b. This configuration is the field effect transistor equivalent of the common emitter bipolar transistor, examined earlier. The mean square values of the noise sources in the narrow frequency range Δf are:

$$\begin{aligned} &\overline{I}_D^2 = 4k_B g_m PT\Delta f \quad \overline{I}_G^2 = \frac{4k_B C_{GS}^2 \omega^2 RT\Delta f}{g_m} \quad \left(\overline{I}_G I_D^P\right) \\ &= 4j\omega C_{GS}k_B T\left(C\sqrt{PR}\right)\Delta f \quad S(I_D) = 4k_B g_m PT \end{aligned} \tag{10.10abcd}$$

$$\begin{aligned} &S(I_G) = \frac{4k_B C_{GS}^2 \omega^{21} RT}{g - m} \quad S\left(I_G I_D^P\right) = -4jC_{GS}\omega k_B T\left(C\sqrt{PR}\right) \quad C = \frac{-j\left(\overline{I}_G I_D^P\right)}{\sqrt{I_D^2 I_G^2}} \\ &P = \frac{\overline{I}_D^2}{4g_m k_B T}/hz \quad R = \frac{g_m \overline{I}_G^2}{4C_{GS}^2 k_B \omega^2 T}/Hz \end{aligned} \tag{10.10defg}$$

where C, P, R are curve fitting parameters. The values of these three parameters for MESFET are: C: 0.6–0.9 P: 1.2 R: 0.4

Y parameters are used to determine the noise parameters of the GaAs field effect transistor. First the transistor's equivalent circuit is transformed to an equivalent circuit in which all the output noise sources are transformed to the input, with a noiseless transistor attached to the output of the noise source circuit (Fig. 10.4a, b). With respect to Fig. 10.4a,

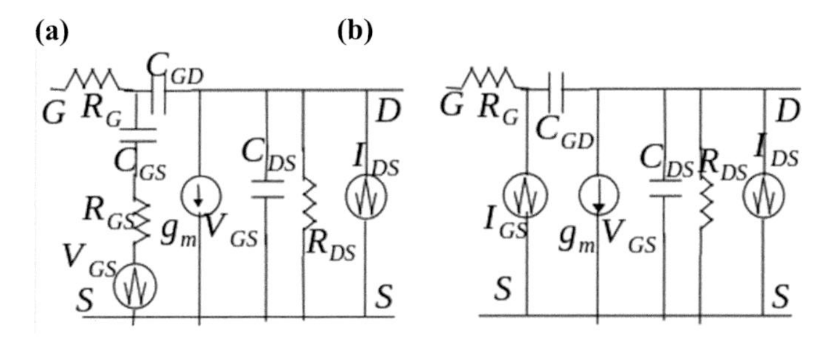

Fig. 10.3 **a**, **b** GaAs field effect transistor with noise sources and noise sources transformed to input and output noise sources only

the noise matrix is:

$$
\begin{aligned}
&[C_Y]_{FET} = [C_{Y11} C_{Y12} C_{Y21} C_{Y22}] \quad C_{Y11} = \frac{4k_B C_{GS}^2 \omega^2 T}{g_m} \\
&C_{Y12} = -4jk_B \omega C_{GS} C\sqrt{PR}T \\
&C_{Y21} = -4jAC_{GS} k_B C\sqrt{PR}T \quad C_{Y22} = 4g_m k_B PT
\end{aligned} \tag{10.10h}
$$

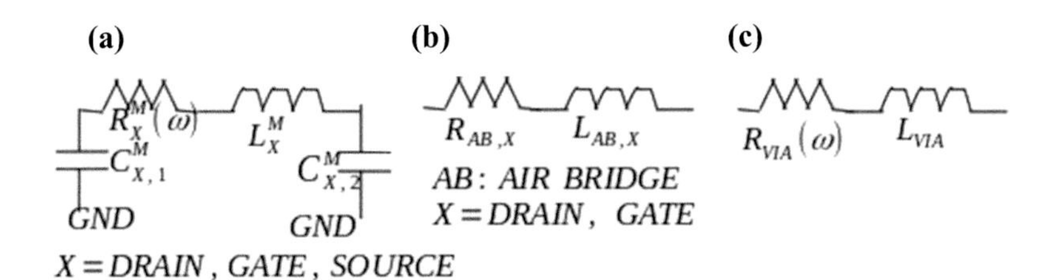

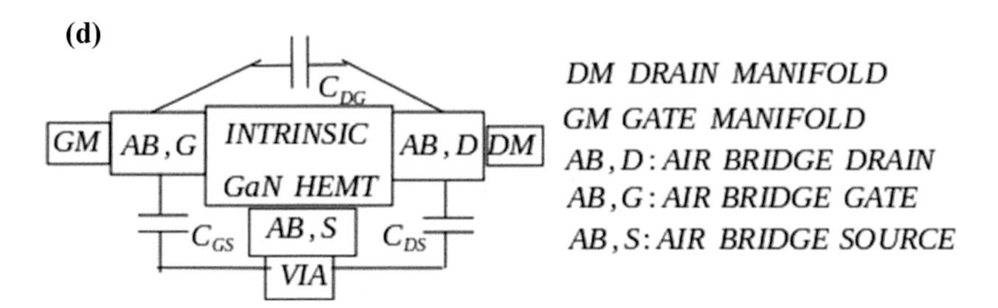

Fig. 10.4 **a** Lumped element model for drain, gate connection manifold **b** drain, gate air bridge lumped parasitic element noise circuit **c** source connection via lumped parasitic element noise circuit **d** GaN HEMT connected to printed circuit board—intrinsic noiseless HEMT and peripheral noise generating sub-circuits

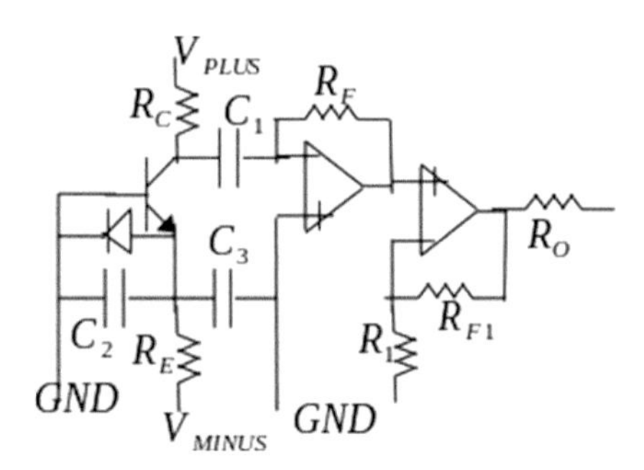

Fig. 10.5 Test bench for measuring bijunction transistor base spreading resistance

After the noise source transformation, the noiseless GaAs field effect transistor can be expressed in terms of its ABCD matrix, and this in turn can be related to the noise matrix before the transformation.

The transformed transistor noise matrix, is expressed in terms of the noise matrix of the nontransformed noise matrix is:

$$\begin{aligned}[C_a]_{TR} = [T][C_Y]_{TR}[T]^P_{TR} &= \left[\overline{e}_N\overline{e}_N i^P_N \overline{e}^P_N \overline{i}_N i^P_N\right] \ [T] \\ &= [0 B_{CS} 1 D_{CS}] \ \ [T]^P_{TR} = \left[0 1 B^P_{CS} D^P_{CS}\right]\end{aligned} \tag{10.10i}$$

where B_{CS}, D_{CS} are the elements of the noiseless transistor's ABCD matrix, and [T] is the transformation matrix.

After substituting all the values for the elements of the respective matrices, followed by considerable manipulation, a number of intermediate parameters are generated:

$$\begin{aligned}CONST &= \frac{g_{DS} + j\omega(C_{GS} + C_{GD} + C_{DS})}{j\omega C_{GD} - g_m} \quad C_{uu^P} = 4k_B T(AABB) \\ AA &= \frac{g_m P(1 + CONST R_S(j\omega C_{GD} - g_m))}{j\omega C_{GD} - g_m} \\ BB &= \frac{1 + CONST(j\omega C_{GD} - g_m)}{(j\omega C_{GD} - g_m)}\end{aligned} \tag{10.10jkl}$$

$$\begin{aligned}C_{ui^P} = \frac{4j\omega k_B C_{GS} C\sqrt{PR}T(1 + CONST(j\omega C_{GD} - g_m))}{j\omega C_{GD} - g_m} + A1 \quad A1 = PBBCONST^P \\ B1 = CONST^P\left(g_m PCONST - j\omega C_{GS} C\sqrt{PR}\right)\end{aligned} \tag{10.10mn}$$

$$C_{u^P} = 4k_B T\left(C_{GS}\left(\frac{C_{GS}\omega^2 R}{g_m} + j\omega(j\omega C_{GD} - g_m)CONSTP\right)\right) + B1 \tag{10.10o}$$

Now using the intermediate parameters, the noise parameters for the GaAs field effect transistor, e.g., noise resistance, optimized Y parameter matrix, minimum noise factor etc., are as below (10.10 p, q, r, s)

$$R_N = \frac{C_{uu^P}}{2k_BT} \quad Y_{OPT} = \sqrt{\frac{C_{u^P}}{C_{uu^P}} + j\left(\Im\left(\frac{C_{u^P}}{C_{uu^P}}\right)\right)^2} + j\Im\left(\frac{C_{ui^P}}{C_{uu^P}}\right)$$
$$= G_{OPT} + jB_{OPT} \quad \Gamma_{OPT} = \frac{Y_{OPT} - Y_0}{Y_{OPT} + Y_0} \tag{10.10pqrs}$$
$$F_{NOISE,MIN} = 1 + \frac{C_{ui^P} + C_{uu^P}Y_{OPT}}{k_BT}$$

These expressions can be simplified assuming that the drain gate capacitance is neglected. This is left as an exercise for the reader.

10.5 Noisy Heterogeneous Junction Bipolar Transistor (HBT)

Using identical arguments as for the noisy homogeneous junction bipolar transistor, the primary RF noise sources in a SiGe heterogeneous bipolar transistor HBT [25–37] are the noises associated with the DC base and collector currents and the thermal noise of the base resistance.

The power spectral densities (PSD) of the base and collector current noises have the characteristics inherent to shot noise. The shot noise current is assumed to be a Poisson distributed stream of an elementary charge q. These charges need to overcome a potential barrier, and thus flow in a completely uncorrelated manner. In a bipolar transistor, the base current shot noise is $2qI_B$ arising from the flow of base majority holes across the emitter–base junction potential barrier. *As the hole current overcoming the emitter base barrier is determined by the minority hole current in the emitter* I_B*, it reappears again in the base shot noise current expression.* Similarly, the collector current shot noise results from the flow of emitter majority electrons over the emitter base junction potential barrier, and has a spectral density $2qI_C$.

Any DC current through any np junction has shot noise, and the collector current passing through the collector–base junction has the value $2qI_C$ for the shot noise. T*he transition of carriers across a reverse biased collector–base junction (for low noise amplification), is a drift process. A DC current passing through such a junction alone does not have intrinsic shot noise. The collector current shows shot noise as the electron current injected into the collector–base junction from the emitter already has shot noise.*

The emitter current shot noise has two independent components, one due to electron injection into the base, and the other due to hole current injection into the emitter, both with "shot" like characteristics.

$$S_{INJ,E,E} = \frac{i^2_{elecron,emitter}}{\Delta f} = 2qI_C \quad S_{INJ,H,E} = \frac{i^2_{hole,emitter}}{\Delta f}$$
$$= 2qI_B < i_{electron,emitter} i^2_{hole,emitter} >= 0 \tag{10.11abc}$$

The collector current shot noise is a delayed version of the emitter electron injection induced shot noise:

$$i_{C,SHOT} = i_{electron,emitter} = i_{injected,electron,emitter} e^{-j\omega\tau_{electron}} \tag{10.11d}$$

where $\tau_{electron}$ is the transit time of the electrons of the emitter injected shot noise current.

The averaged values of the shot noise currents are:

$$\begin{aligned} &< i^2_{NOISE,C,SHOT} > = 2qI_C\Delta f < i^2_{NOISE,E,SHOT} >= 2qI_E\Delta f < i_{NOISE,C,SHOT} i^P_{NOISE,E,SHOT} > \\ &= 2qI_C\Delta f e^{j\omega\tau_{electron}} \end{aligned} \tag{10.11efg}$$

For the widely used common emitter configuration of bipolar transistors, the above expressions for the averaged currents are re-written as: HERE HERE

$$\begin{aligned} &< i^2_{NOISE,BASE} > =< i^2_{NOISE,E,SHOT} > + < i^2_{NOISE,C,SHOT} > -2\Re(< i_{NOISE,E,SHOT} i^P_{NOISE,C,SHOT} >) \\ &= 2q(I_B + 2I_C\Delta f(1 - \Re e^{j\omega\tau_{electron}})) \end{aligned} \tag{10.11h}$$

$$\begin{aligned} &< i^2_{NOISE,COLLECTOR} >= 2qI_C\Delta f \\ &< i^P_{NOISE,B} i_{NOISE,C} >= 2qI_C\Delta f(e^{-j\omega,\tau_{electron}} - 1) \end{aligned} \tag{10.11i}$$

This transport shot noise model is also called the unified model, because it can be reduced to the conventional SPICE [73–77] noise model by setting electron noise current transit time to zero, or when $\omega << \frac{1}{\tau_{electron}}$. This model enables accurate modelling of both experimental noise data and hydrodynamic noise simulation data in SiGe HBTs.

Minority carrier velocity fluctuations also contribute to shot noise. Velocity variations lead to current density fluctuations, which propagate towards the terminals. The collector current shot noise originates from the neutral base, which is close to the physical explanation underlying the transport noise model.

Noise factor is a noise performance characteristic. It is defined as **the ratio of the input signal-to-noise (SNR) to the output signal-to-noise (SNR)**. *Often, the noise figure, defined as the logarithm of the noise factor to base 10, is used*. For a source termination admittance

$$Y_S = G_S + jB_S \quad F = F_{MIN} + \frac{R_B + \frac{1}{g_m}}{G_S}\left(Y_S - Y_{S,OPT}\right) \quad R_N = R_B + \frac{1}{g_m} \tag{10.11k}$$

where $Y_{S,OPT}$ is the optimum source admittance, and R_N is the noise resistance—noise parameters. Noise figure reaches its minimum when the source is noise matched. The source resistance determines the sensitivity of noise figure to deviations from the optimum source admittance. The smallest possible noise figure and smaller source resistance are obviously desired. The noise parameters are intrinsic properties of the linear two port network, as the S|Y|Z parameters. They are functions of the equivalent input noise voltage, current and their correlation, denoted as $S_{i,n}$, $S_{v,n}$, $S_{i_n v_n^P}$. Under certain simplifying assumptions, the equivalent input noise current|voltage can be approximately obtained as:

$$S_{I,NOISE} = 2q\left(I_B + \frac{I_C}{|H_{21}|^2}\right) \quad S_{VOLT,NOISE} = 4k_B R_B T + 2q\left(I_B R_B^2 + \frac{I_C}{|Y_{21}|^2}\right) \quad S_{I,V,NOISE} = 2q\left(I_B R_B + \frac{I_C Y_{11}}{|Y_{21}|^2}\right) \tag{10.11l}$$

Y, H refer to the conventional small signal parameters. They can be expressed using equivalent circuit parameters β, collector current, base emitter, base collector capacitances and transconductance. For a given collector current, a higher β reduces the base current and $S_{i,n}$, which is also inversely proportional to H_{21}. A high transition frequency also in increases H_{21} and reduces $S_{i,n}$. The inherent high β and high transition frequency of SiGe HBTs enables these devices to have low input noise current. For the same β, the additional bandgap engineering leverage in SiGe HBTs allows higher base doping than in homogeneous junction implanted-base Si BJTs, which reduces the input noise voltage. The real and imaginary parts of $Y_{S,OPT}$ are:

$$G_{S,OPT} = \sqrt{\frac{g_m}{2\beta R_N} + \frac{\omega^2 C_I^2}{2g_m R_N}\left(1 - \frac{1}{g_m R_N}\right)} \quad B_{S,OPT} = \frac{-\omega C_I}{2g_m R_N} \tag{10.11o}$$

$$G_{S,OPT} = \sqrt{\frac{g_m}{2\beta R_N} + \frac{\omega^2 C_I^2}{2g_m R_N}\left(1 - \frac{1}{g_m R_N}\right)} \quad B_{S,OPT} = \frac{-\omega C_I}{2g_m R_N} \tag{10.11p}$$

where $C_I = C_{BC} + C_{BE}$. The imaginary part of the source admittance is negative. Therefore series inductor at the base is thus needed for noise matching of the imaginary part. As the collector current dependence of C_I through the transconductance, it is observed that:

- The real part of the optimum source admittance increases with the collector current and frequency.
- When the diffusion capacitance dominates the combined capacitance, the imaginary part of the source admittance becomes independent of the collector current.
- The absolute value of imaginary part of the source admittance increases with frequency. The minimum noise figure is:

$$F_{MIN} = 1 + \frac{1}{\beta} + \sqrt{g_m R_B}\sqrt{\frac{1}{\beta} + \frac{f^2}{f_T^2}} \tag{10.11q}$$

when $f = \frac{f_T}{\sqrt{\beta}}$ the minimum noise figure changes from white noise (no frequency dependency) to $\frac{10dB}{decade}$ frequency dependency. Also, a low base resistance is important to reduce the minimum noise figure when $f > \frac{f_T}{\sqrt{\beta}}$.

RF semiconductor devices are also affected by low frequency 1/f noise, and is influenced by the SiGe HBT's size and biasing current. The value of such noise can be high near DC. 1/f noise is a problem for low noise, low frequency analog circuits, e.g., amplifiers used in zero intermediate frequency direct conversion receivers. This noise can also be upconverted to phase noise that degrades signal integrity of frequency translations. Although base and collector currents have flicker noise, the part due to the base current dominates. The low frequency base noise current spectrum for a SiGe HBT shows both the contribution of the 1/f noise and the shot noise component $2qI_B$ component. The performance characteristic for flicker noise is the corner frequency, is defined by the intercept of the 1/f component and the shot noise level. At higher base current values, the shot noise contribution cannot be observed easily, but inferred from the corner frequency value.

The flicker noise spectrum is:

$$S_{Iu} = \frac{K_F I_B^{\alpha}}{f} \tag{10.11r}$$

where K_F and α correspond to the KF and AF SPICE [73–77] model parameters. $\alpha = 1$ indicates carrier mobility fluctuations, and $\alpha = 2$ for carrier number variations for typical SiGe HBTs its value is approximately 2, and varies only slightly with SiGe profile and collector doping profile (2 ± 0.2).

If 1/f noise is assumed only to be a function of the number of minority carriers injected into the emitter, means the same base current for a SiGe HBT and its homogeneous junction Si counterpart.

1/f noise performance is characterized by the corner frequency, defined as the frequency at which the 1/f noise equals the shot noise level. Assuming $\alpha = 2$,

$$f_{CORNER,\frac{1}{f}} = \frac{KI_B}{2qA_E} = \frac{KJ_{CORNER}}{2q\beta} \quad J_{CORNER} = \frac{I_C}{A_E} \quad \beta = \frac{I_C}{I_B} \quad S_{\Phi} \propto \frac{S_u}{4\pi^2 f^2} \tag{10.11r}$$

Minimizing voltage controlled oscillator (VCO) and frequency synthesizer phase noise, is a key concern for designers of these two circuits. **Phase noise is a result of variations in time period of a periodic time varying current, voltage waveform. For a target frequency is f, if perturbations in the time period is less than the time period corresponding to the specified|target frequency f, the actual frequency is higher than f, and vice-versa. In the spectrum, the higher frequency appears to the right of f, and the lower frequency to the left of f.** *Consequently the spectrum has decaying tails on both sides of the center frequency f.* For an arbitrary signal u, the corresponding phase noise Φ and the PSD of φ

is related to the PSD of the physical noise u by

$$\Phi(t) \propto \int u(\tau)s(t-\tau)d\tau \quad -\infty \leq \tau \leq t \quad S_{\Phi} \propto \frac{S_u}{4\pi^2 f^2} \tag{10.11s}$$

10.6 Noisy Gallium Nitride (GaN) High Electron Mobility Transistor (HEMT) Model

A widely used noise model [40–54] for the common low noise GaN HEMT consists of a noiseless semiconductor device (intrinsic GaN HEMT) embedded inside a network of capacitors, inductors and resistors. These passive circuit components are the parasitic capacitors, inductors and resistors resulting from the way a surface mount HEMT is soldered to its printed circuit board. The surface mount HEMT soldered to a printed circuit board is ideal for very accurate electromagnetic solver analysis.

Common commercially available GaN HEMT has metal "fingers" drain, gate connections, and the source is a metal pad underneath the device, The corresponding drain, gate connectors on the printed circuit board are called *drain-gate manifold*. The HEMT source connection is a metal pad underneath the device that is soldered to a printed circuit board via. Also, there is a very small (fraction of a millimeter gap between the periphery of the HEMT and the locations of the drain-gate connections with the printed circuit board—*air bridge*. **At RF|microwave frequencies, each of these external (to the intrinsic semiconductor device) physical structures act as a network of parasitic capacitors, inductors and resistors (Figs. 10.4a, b, c and 10.5).**

With reference to Figs. 10.4a, b, c and 10.5 the admittance(Y) matrix of the drain, gate manifold network is:

$$\begin{gathered}
[Y]_{X,M} = [Y11 Y_{12} Y_{21} Y_{22}] \quad Y_{11} = j\omega C_{1,X,M} + \frac{1}{R_{X,M} + j\omega L_{X,M}} \\
Y_{12} = \frac{-1}{R_{X,M} + j\omega L_{X,M}} \quad Y_{22} = j\omega C_{2,X,M} + \frac{1}{R_{X,M} + j\omega L_{X,M}} \\
Y_{21} = \frac{-1}{R_{X,M} + j\omega L_{X,M}} \quad X{:}DRAIN, GATE
\end{gathered} \tag{10.12a}$$

The series resistance of the drain-gate manifold (Fig. 10.4a) is frequency dependent (skin effect)

$$R_{X,M}(\omega) = R_{DC,X,M} + \sqrt{\omega}(1+i)R_{RF,X,M} \quad X{:}DRAIN, GATE \tag{10.12b}$$

The drain-gate manifold model parameters are extracted from electromagnetic solver simulations—$R_{DC,X,M}$, $R_{RF,X,M}$ are determined by linear regression:

$$-\Re\left(\frac{1}{Y_{12}, X, M}\right) = R_{DC,X,M} + \sqrt{\omega} R_{RF,X,M} \quad X{:}DRAIN, GATE \tag{10.12c}$$

The drain, gate manifold parasitic capacitance, inductance are extracted using identical linear regression techniques:

$$C_{1,X,M} = < \frac{\Im\left(Y_{11,X,M} - \frac{1}{R_{X,M}(\omega)+j\omega L_{X,M}}\right)}{\omega} > C_{2,X,M}$$
$$= < \frac{\Im\left(Y_{22,X,M} - \frac{1}{R_{X,M}(\omega)+j\omega L_{X,M}}\right)}{\omega} > \tag{10.12d}$$

$$L_{X,M}(\omega) = < \Im(\frac{-1}{Y_{21,X,M}}) - \frac{R_{RF,X,M}}{\sqrt{\omega}} > \quad X{:}DRAIN, GATE \tag{10.12e}$$

The impedance parameters(Z) related to the source printed board connector via, are:

$$Z_{VIA} = R_{VIA,DC} + \sqrt{\omega} R_{VIA,RF} + j\left(\omega L_{VIA}(\omega) + \sqrt{\omega} R_{VIA,RF}\right) \tag{10.12f}$$

The via inductance obtained using linear regression is:

$$L_{VIA}(\omega) = \frac{\Im Z_{VIA} - R_{VIA,RF}}{\omega} \tag{10.12g}$$

The impedance matrix(Z) corresponding to the air barrier between the drain, gate manifolds and the physical boundary of the HEMT is:

$$Z = [Z_G + Z_S \; Z_S \; Z_S Z_D + Z_S] \tag{10.12h}$$

where each impedance has a real and imaginary component as:

$$Z_X = R_X + j\Omega L_X + \frac{1}{j\omega C_X} \quad X = DRAIN, GATE, SOURCE \tag{10.12i}$$

Linear regression, applied to the source impedance Z_S gives:

$$R_S = < \Re Z_S > \quad \omega \Im Z_S = \omega^2 L_S - \frac{1}{C_S} \tag{10.12j}$$

Substituting for the real and imaginary parts of the source impedance in the the other impedance terms of 10.12 l, followed by careful linear regression on experimental impedance data allows the extraction of the real and imaginary parts of the other two impedances.

Defining the intermediate parameter C_{COMP} and a subsequent $\Delta - Y$ transformation enables the drain, gate and source capacitances for the air barrier network to be denoted as:

$$C_{COMP} = C_D C_S + C_G C_S + C_D C_G \quad C_{DS} = \frac{C_{COMP}}{C_G} \quad C_{DG} = \frac{C_{COMP}}{C_S} \quad C_{GS} = \frac{C_{COMP}}{C_D} \tag{10.12k}$$

The AMS-CMC GaN HEMT model examined at the end of Chap. 7 includes both high and low frequency noise models. The low frequency (flicker) noise model probability spectral density (PSD) function is (10.12 l):

$$S_{i,FLICKER} = \frac{I_{DS}^2 KLP_1(f)}{C_{GATE}^2} \left(C_{GATE}\Gamma_1 V_{THERMAL}\left(\frac{1}{n_D} - \frac{1}{n_A}\right) + AA + BB + 0.5\Gamma_3\left(N_D^2 - n_S^2\right)\right) \tag{10.12l}$$

$$AA = (\Gamma_1 + C_{GATE}\Gamma_2 V_{THERMAL})\ln\left(\frac{n_D}{n_S}\right)$$

$$BB = (\Gamma_2 + C_{GATE} V_{THERMAL})(n_S - n_D)$$

where $P_1(f) = \frac{k_B T}{f^{EF} L^2 W}$ EF, Γ_1, Γ_2, Γ_3, n_D, n_S are curve fitting parameters and drain, source charge carrier densities respectively.

Similarly, the high frequency PSD is:

$$S_{i,THERMAL} = \frac{4k_B T}{I_{DS} L_{EFF}^2} \left(\frac{qC_{GATE}\mu_{effective} W}{\sqrt{1 + \theta_{SAT}^2 (\psi_D - \psi_S)^2}} \right)^2 \left(V_{go}\left(V_{go}(\psi_D - \psi_S) + \left(\psi_D^3 - \psi_S^3\right)\left(\frac{1}{V_{go}} - 1\right)\right)\right) \tag{10.12m}$$

10.7 Transistor Noise Factor|Figure Measurement with Laboratory Equipment

Transistor noise factor, or its more commonly used variation noise figure [55–72] is the key noise performance metric—the noise factor|figure concept can be applied to a circuit, sub-circuit or a transistor. *While the discussions in the previous sections elaborate on the analytical models and expressions for estimating the noise probability spectral densities of various transistor noise mechanisms, they cannot be used directly.* Therefore the noise factor|figure, which can be measured accurately with sensitive test equipment is so important.

The basic definition of noise factor is the ratio of signal to noise power at the input divided by the signal to noise power at the output. **So the noise factor of a two port network is the decrease|degradation in the signal-to-noise ratio as the signal traverses**

the network from the input to the output port—the concept is inapplicable for one port networks (e.g., oscillator). *A real world transistor, unlike an ideal device, adds some extra internal noise and degrades the signal-to-noise ratio. A low noise factor means that very little noise is added by the two port network.* Noise factor is independent of the modulation format and of the fidelity of modulators and demodulators.

- Noise factor is different from gain. Once noise is added to the signal, subsequent gain amplifies signal and noise together and does not change the signal-to-noise ratio.
- The degradation in a two port network's signal-to-noise ratio is dependent on the temperature of the input signal source that excites the network—the source thermal noise:

$$F = \frac{\frac{S_I}{N_I}}{\frac{S_O}{N_O}} = \frac{\frac{S_I}{N_I}}{\frac{GS_I}{N_{ADD}+GS_I}} = \frac{GN_I + N_{ADD}}{GN_I} = \frac{N_{ADD} + Gk_BT_O\Delta f}{Gk_BT_O\Delta f} \quad (10.13a)$$

where $G,\ S_I,\ S_O, N_I,\ N_O$ are respectively the two port network gain, input|output signal and noise, $G,\ N_{ADD}, T_O = 290K$ are the gain, added noise and reference temperature respectively.

- The thermal noise power spectral probability distribution is $k_BT_O\Delta f$ where T_O is a reference temperature (290 K) and Δf is the bandwidth.
- Noise factor is a function of frequency, but independent of bandwidth (so long as the measurement bandwidth is sufficiently narrow to resolve variations with frequency).
- *The linearized form of the noise factor is the noise figure*
- For a cascaded, interconnected n stage network, for which each stage is a two port network,the total noise factor is:

$$F = F_1 + \frac{F_2 - 1}{G_1} + \frac{F_3 - 1}{G_1G_2} + \ldots + \frac{F_n - 1}{G_1G_2\ldots G_{n-1}} \quad (10.13b)$$

where F_i G_i are the noise factor and gain of the ith stage.

Device gain is a key parameter in noise calculations. When an input power of $k_BT_O\Delta f$ is used in these calculations, it is the maximum that can be delivered to a matched load. A large input impedance mismatch decreases the actual power delivered to the device. If the gain of the device is defined as the ratio of the actual power delivered to the load to the maximum power available from the source, the mismatch loss present at the input can be neglected, as it is included in the gain definition. This definition of gain is transducer gain, G_T When cascading devices, mismatch errors arise if the input impedance of the device differs from the load impedance. *So the total gain of a cascaded series of devices* ***does not*** *equal the product of the gains.*

Available gain G_A is routinely quoted as a transistor parameter, it is the gain that results when a given source admittance Y_S drives the device and the output is matched to the load. It is very common to use insertion gain,G_I or the forward transmission coefficient,S_{21}^2, is the quantity specified or measured for gain in a 50 Ω system. If the measurement system has low reflection coefficients and the device has a good output match the cascade noise figure equation given earlier gives accurate results. With poor output match or measurement system with significant mismatch errors results in an error between the actual system and calculated performance characteristics. e.g.,, the output impedance of the first stage is different from the 50 Ω source impedance used with a second stage characterized for noise figure, the noise generated in the second stage could be altered. Fortunately, the second stage noise contribution is reduced by the first stage gain so that errors involving the second stage can be ignored. The complete analysis of mismatch effects in noise calculations is very complicated requiring understanding the dependence of noise figure on source impedance.

Noise figure is a simplified model of the actual noise generated in a transistor, which can have multiple noise contributors: thermal noise, shot noise, generation-recombination noise and partition noise. The effect of source impedance on these noise generation processes is very complex. **The noise figure that results from a measurement is influenced by the match of the noise source with the match of the measuring instrument; the noise source is the source impedance for the transistor, which itself is the source impedance for the measuring instrument.** *The actual noise figure performance of the transistor will be determined by the match of other system components in the circuit in which it is a part of.*

Designing low noise RF|microwave amplifiers requires tradeoffs between the gain of a stage and its corresponding noise figure—requiring knowledge of how the RF transistor's gain and noise figure varies as a function of the source impedance|admittance. *The minimum noise figure does not have to occur at either the system impedance, or at the conjugate match impedance that maximizes gain.*

10.7.1 Transistor Noise Parameters [55–72]

The effect of impedance mismatch is fully understood only with two characterizations of the device under test- *noise figure measurement* and *gain measurement. S-parameter measurement* ***only*** *enables calculation of the available gain in a perfectly matched device.* Optimum noise figure estimation requires a special tuner—for which noise factor—source impedance dependency is:

$$F = F_{MIN} + \frac{4R_{NOISE}}{Z_0}\left(\frac{|\Gamma_{OPT} - \Gamma_{SRC}|^2}{|1 + \Gamma_{OPT}|^2(1 - |\Gamma_{SRC}|^2)}\right) \tag{10.13c}$$

where Γ_{SRC} is the source reflection coefficient.

The three transistor noise parameters are F_{MIN}, R_{NOISE}, Γ_{OP}.

The available gain of the transistor when driven by a specified source impedance, can be calculated from its S-parameters measured with a network analyzer.

$$G_A = \frac{|S_{11}|^2(1 - |\Gamma_{SRC}|^2)}{|1 - S_{11}\Gamma_{SRC}|^2\left(1 - \left|S_{22} + \frac{S_{12}S_{21}\Gamma_{SRC}}{1-S_{11}\Gamma_{SRC}}\right|^2\right)} \tag{10.13d}$$

When the source reflection coefficient Γ_{SRC} is plotted on a Smith chart corresponding to a set of fixed gains, *gain circles* are created—a convenient format to display the relation between source impedance and gain.

10.7.2 Effect of Bandwidth

Noise factor|figure is independent of bandwidth. *The key assumption behind noise measurements is that the device under test has an amplitude-versus-frequency characteristic that is constant over the measurement bandwidth.* **So noise measurement bandwidth should be less than the device bandwidth.** State-of-art measurement|test equipment can tackle this issue.

The bandwidth defining system sub-circuits e.g., a receiver, will be the intermediate frequency or the detector will have a bandwidth much narrower than the RF circuits. Only then noise factor|figure is a valid parameter to characterize the noise performance of the RF circuitry. Otherwise, noise figure may still be used as a figure of merit for comparisons. Complete system signal-to-noise (SNR) ratio will require the input bandwidth as a parameter.

10.7.3 Noise Figure Measurement—Linearity, Noise Equivalent Temperature

Noise factor|figure measurements depend on a key property of linear two-port networks, noise linearity. **The noise power out of a device is linearly dependent on the input noise power or temperature.** Using the slope (of the output power versus source temperature curve) and a reference point, the output power corresponding to a noiseless input power (intercept of curve on the power axis), can be estimated. Then the noise figure or effective input noise temperature can be calculated. To ensure noise measurements linearity, all automatic gain control (AGC) circuitry (working in a feedback loop) must be deactivated.

Often *effective input noise temperature,* T_{EQ}, is used to quantify the noise performance of a device instead of the noise factor|figure, (NF). T_{EQ} is the equivalent temperature of a source impedance into a perfect (noise-free) device that would produce the same added noise N_{ADD}—defined as:

$$T_{EQ} = \frac{N_{ADD}}{k_B G \Delta f} = T_O(F - 1) \tag{10.13f}$$

Noise factorlfigure measurement requires a noise source with a calibrated output noise level, denoted as *excess noise ratio* (ENR units dB). A noise source is a circuit that provides two known levels of noise, e.g., widely used reverse biased low-capacitance diode driven into avalanche breakdown with a constant current. A 0 dB ENR noise source produces a 290 K temperature change between its on and off states. ENR is not the "on" noise relative to $k_B T \Delta f k_B T \Delta f$

$$ENR(dB) = 10\log\left(\frac{T_{HOT} - T_{COLD}}{T_O}\right) \tag{10.13g}$$

10.7.4 Y Factor Method for Noise Factor|Figure Measurement

The Y-Factor method is the basic noise factorlfigure measurement scheme (automatedlmanual) performed internally in a noise figure analyzer. Using a noise source, this method allows the determination of the internal noise in the device under test (e.g., transistor) and the noise factorlfigure or effective input noise temperature.

With a noise source connected to the device under test, the output power is measured with the noise source on and the noise source off (N_{ON} N_{OFF}). The ratio of these two powers is the Y-factor. Output power measurement device could be a power meter, spectrum analyzer, or special internal power detector inside noise figure meterslanalyzers. The relative level accuracy is important. State-of-art noise figure analyzers have very linear internal power detectors, that can measure input power level changes very accurately. The absolute power level accuracy of the measuring device is not important since a ratio is measured ($Y = \frac{N_{ON}}{N_{OFF}}$).

The calibrated ENR of the noise source is a reference level for input noise. So an equation for the noise added by the device under test itself (internal noise), N_{ADD} can be derived. State-of-art noise figure analyzers determine this internal noise by modulating the noise source between the on and off states and applying internal calculations.

$$N_{ADD} = G_1 k_B T_O \Delta f \left(\frac{ENR}{Y - 1} - 1\right) \tag{10.13h}$$

An expression for the total "system noise" factor F_{SYS} can be estimated, based on these estimated noise parameters. *System noise factor includes the noise contribution of each of the individual sub-circuits of the system. The noise generated in the measuring instrument is included as a second stage contribution.* If the device under test gain is large

($G_1 \gg G_2$), the noise contribution from this second stage will be small. The second stage contribution can be removed from the calculation of noise figure if the noise figure of the second stage and the gain of the device under test is known. The device gain is not needed to find the system noise factor.

$$F_{SYS} = \frac{ENR}{Y-1} \quad F_{SYS} = \frac{ENR - Y\left(\frac{T_{COLD}}{T_O - 1}\right)}{Y-1} \quad F_{SYS} = \frac{ENR\left(\frac{T_{COLD}}{T_O}\right)}{Y-1} \tag{10.13ijk}$$

When the noise figure is much larger than the ENR, the device noise masks the noise source output. So the Y-factor is approximately 1. Accurate measurement of small ratios is difficult. The Y-factor method is not used when the noise figure is more than 10 dB above the ENR of the noise source, depending on the measurement. When the noise source cold temperature is not 290 K, the system noise factor expression is modified to 10.13 j.

This expression is inapplicable for semiconductor noise sources, as the hot and and cold temperatures are interlinked. Since the physical noise source is at the cold temperature T_{COLD} the internal attenuator noise due to T_{COLD} is added both when the noise source is on and off—effectively the noise change between the on and off state remains constant ($T_{HOT} - T_{COLD}$). This is most important for low ENR noise sources when $T_{HOT} < 10T_{COLD}$. An alternate equation is used to correct for this case 10.13 k.

10.7.5 Signal Generator Twice Power Method

In absence of accurate noise sources and devices with very high noise figure such that the Y factors can be very small and difficult to accurately measure, the signal generator twice power method is used. First, the output power is measured with the device input terminated with a load at a temperature of approximately 290 K. Then a signal generator is connected, providing a signal within the measurement bandwidth. The generator output power is adjusted to produce a 3 dB increase in the output power. If the generator power level and measurement bandwidth are known the noise factor is calculated easily. The gain of the device under test gain is not required.

$$F_{SYS} = \frac{P_{GENERATOR}}{k_B T_O \Delta f} \tag{10.13l}$$

The results of this measurement are not very accurate. The noise bandwidth of the power-measuring device must be known. Noise bandwidth, Δf, is a calculated equivalent bandwidth, with a rectangular spectral shape with the same gain bandwidth product as the actual filter shape. The output power must be measured on a device that measures true power. This is essential, as a combination of noise and a CW signal is present. Thermal based power meters measure true power very accurately but often require amplification to read a low noise level and will require a bandwidth-defining filter. Spectrum analyzers have good sensitivity and a well-defined bandwidth but the detector may respond differently to CW signals and noise. Absolute level accuracy is not needed in the power detector since a ratio is being measured.

10.7.6 The Direct Noise Measurement Method

This scheme for noise factor measurement of high noise figure devices, involves device output power measurement with an input termination at a temperature of approximately 290 K. If the gain of the device and noise bandwidth of the measurement system is known, the noise factor can be estimated with a very sensitive power meter. In addition, the gain of the device under test must be known and the power detector must have absolute level accuracy.

$$F_{SYS} = \frac{N_0}{G k_B T_O \Delta f} \tag{10.13m}$$

Each of the noise factor estimation schemes examined so far measure the total system noise factor, starting with the noise source and ending with the measurement system. However the goal is to measure the noise factor of the device under test only. *From the cascade noise factor equation it is clear that if the device under test gain is large, the contribution of the measurement system will be small, and can be ignored for preliminary estimates.* The noise figure of a high gain device under test can be directly measured with the previously discussed methods. When a low gain device is tested, or the highest possible accuracy is needed, a correction is required, provided the gain of the device under test and the noise figure of the system are known.

$$F_{SYS} = F_1 + \frac{F_2 - 1}{G_1} \quad F_1 = F_{SYS} - \frac{F_2 - 1}{G_1} \tag{10.13n}$$

A simple sequence of measurements are used to achieve this goal.

- The noise source is connected to the measurement instrument, The noise power levels corresponding to the noise source switched on|off are measured. These measured values N_{ON}, N_{OFF} are used to calculate the system noise factor F_{SYS} with the Y factor method.
- The device to be tested is now added to the measurement test bench along with the noise source. The noise power levels with the noise source switched on and off ($N_{ON,DUT}$, $N_{OFF,DUT}$) are measured. The gain of the device under test is:

$$G_{DUT} = \frac{N_{ON,DUT} - N_{OFF,DUT}}{N_{ON} - N_{OFF}}$$

- The system noise factor is then computed from these measured noise power values, and then the device noise factor is then easily computed.

Noise is a series of random electrical impulses. Any noise measurement device|system estimates the mean noise level at the output of the device. The actual noise figure of the device under test is calculated using these measured noise power levels, with some correction terms. *As the time required to calculate the true mean noise level is infinite, averaging is performed over some finite time period. The difference between the measured average and the true mean will fluctuate and give rise to a repeatability error.*

For small variations, the deviation is proportional to $\frac{1}{\sqrt{<t>}}$ so that longer averaging times produce better averages. The average includes more events and so is closer to the true mean. The variation is also proportional to $\frac{1}{\sqrt{<\Delta f>}}$. Larger measurement bandwidths produce a better average because there are more noise events per unit of time in a large bandwidth, thereby improving the average value. **Noise figure must be measured with the widest possible bandwidth narrower than that of the device under test.**

10.7.7 Noise Figure Measurement Equipment

The noise figure analyzer is designed to measure noise figure using the Y factor method. It consists of a receiver with an accurate power detector and a circuit to power the noise source. It allows ENR entry and displays the calculated noise figure value corresponding to the frequency it is tuned to, along with gain.

The versatile signal|spectrum analyzer can be used to measure noise figure, using any of the methods examined previously. They are ideal for measuring high noise figure devices using the signal generator or direct power measurement method. The variable resolution bandwidths allow measurement of narrow band devices.

Like a spectrum analyzer, a network analyzer is also very flexible test equipment that can perform a variety of tasks, e.g., noise figure measurement, gain analysis, in addition to their primary task of network analysis.

As a network analyzer uses the same internal signal receiver for network analysis as well as for noise figure measurement, there are minor performance limits. Often the receiver is a double side band (DSB) type, where noise figure is measured at two frequencies and an internal correction is applied. When a wide bandwidth measurement is made, an error may be introduced if the analyzer noise figure or gain is not constant over this frequency range. When narrow bandwidth measurement is used to measure narrow-band devices, the unused frequency spectrum between the upper and lower side-band does not contribute to the measurement and a longer measurement time is needed to reduce jitter.

Network analyzers are designed to primarily measure the S-parameters of the device under test. S-parameter data can reduce noise figure measurement uncertainty by offering mismatch correction, providing a more accurate gain measurement of the device so that the second stage noise contribution can be subtracted with more precision. But mismatch also affects the noise generation in the second stage which cannot be corrected for without knowing the noise parameters of the device. Likewise, if there is a impedance mismatch between the noise source and device under test input, appropriate corrections are essential. Noise parameter measurements require a tuner and additional firmware. The resulting measurement system can be complex and expensive. Error correction in a network analyzer is for gain measurements and calculation of available gain.

A complete noise parameter measurement requires a test set along with software, a vector network analyzer and a noise analyzer to make a series of measurements, to measure the noise parameters of a semiconductor device. The estimated noise parameters can be used to calculate the minimum device noise figure, the optimum source impedance, and the effect of source impedance on noise figure. The test set has an adjustable tuner to present various source impedances to the device under test. Bias is provided to semiconductor devices that may be tested. A noise source is included for noise figure measurement,. at different source impedances. The corresponding source impedances are measured with the network analyzer. From this data, the complete noise parameters of the device can be calculated.

The complete device S-parameters are also measured so that gain parameters can also be determined. Complete noise figure measurement is complicated and time consuming process, since a number of accurate measurements have to be made.

Power meters and true RMS voltmeters can be used to measure noise figure with any of the methods described earlier. Some computerlmanual calculations are needed. Being broadband devices, they need a filter to limit their bandwidth to be narrower than that of the device under test. Such a filter will be fixed in frequency and allow measurements only at this frequency. Power meters are often used to measure receiver noise figures where the receiver has a fixed IF frequency and much gain. The sensitivity of power meters and voltmeters is usually poor but the receiver may provide enough gain to make measurements. If additional gain is added ahead of a power meter to increase sensitivity, temperature drift and oscillations must be avoided.

10.7.8 Laboratory Test Bench for Bijunction Transistor Input Noise Measurement

The mean-square equivalent noise equivalent input noise of a resistively loaded BJT amplifier (with zero small-signal impedance from both base—ground and emitter—ground) is measured over a narrow frequency band Δf centered at frequency f is expressed as:

$$\overline{v}^2_{inputnoise} = \left(4k_BTr_x + \frac{I_C r_x^2}{\beta}\left(2q + \frac{K_f}{f}\right) + 2qI_C\left(\frac{r_x}{\beta} + \frac{V_{THERMAL}}{I_C}\right)^2\right)\Delta f \tag{10.14a}$$

where, r_x is the base spreading resistance (Ohms), β is the small-signal current gain (dimensionless), I_C is the DC collector current, I_B is the DC base current, K_f is the flicker noise-coefficient, and f is the frequency at which the mean-square noise voltage is measured. If the noise measurement is made at a frequency f where the flicker noise can be ignored, the expression for the mean-square equivalent input noise becomes:

$$\overline{v}^2_{inputnoise} = \left(4k_BTr_x + 2qI_C\left(\frac{r_x^2}{\beta} + \left(\frac{r_x}{\beta} + \frac{V_{THERMAL}}{I_C}\right)\right)\right)\Delta f \tag{10.14b}$$

Accurate measurement of the base spreading resistance requires the circuit as shown in Fig. 10.5. Assuming that the operational amplifier is ideal and that the thermal noise in the feedback resistor, R_F can be ignored, the output noise is:

$$\overline{v}^2_{inputnoise} = \left(1 + \frac{R_{F1}}{R}\right)^2 \left(\frac{1}{\frac{r_x}{\beta} + \frac{V_{THERMAL}}{I_C}}\right)^2$$
$$R_F^2\Delta f\left(4k_BTr_x + r_x^2\left(2qI_b + \frac{K_f I_b}{f}\right) + \frac{2qI_C}{\left(\frac{r_x}{\beta} + \frac{V_{THERMAL}}{I_C}\right)^2}\right) \tag{10.14c}$$

If the measurement is made at a large frequency, the flicker noise component can be neglected. The base spreading resistance satisfies:

$$\left(\frac{1}{\beta}\left(\frac{A}{\beta} - k_BI_C\Delta f\right)\right)r_x^2 + \left(\frac{2AVTHERMAL}{\beta I_C} - 4k_BT\Delta f\right)r_x^2$$
$$+ A\left(\frac{V_{THERMAL}}{I_C}\right)^2 = 0 \quad A = \frac{\overline{v}^2_{noiseoutpur}}{A_1^2R_F^2} - 2qI_C\Delta f \tag{10.14d}$$

Thus 10.14d is solved to determine the base spreading resistance using the measured value of the output noise. Only the positive solution for the base spreading resistance is used since the negative value has no physical meaning. A more exact solution may be obtained by directly solving 10.14c. The coupling capacitor C_1 prevents DC current from the transistor from flowing into the feedback resistor while forcing the entire signal component of the collector current to flow though this feedback resistor. The op amp inverting terminal is at a virtual ground so that the signal component of the collector voltage is zero which eliminates the Early effect. The capacitor C_2 is a bypass capacitor which places the emitter at signal ground. Both these capacitors are electrolytic, chosen to be large so that the low frequency noise spectra is not altered.

10.7.9 MOSFET Noise Sources—Thermal, Flicker

The common noise sources in a MOSFET are:

- channel thermal noise
- 1/f noise
- resistive polysilicon gate material noise
- distributed substrate resistance noise
- shot noise from the leakage current of the drain source reverse diodes

Under normal operating conditions, only the flicker and thermal noises are relevant. Other noise sources are relevant in low temperature operating conditions.

A MOSFET in normal working conditions has an inverse resistive channel between the drain and the source. *Applying a gate voltage induces minority carriers to collect and form the conducting channel between the drain and the source. This conducting layer is at the interface of the gate oxide and the substrate.* With zero drain source voltage, the channel is a homogeneous resistor. The noise in the channel is:

$$i_{D,THN}^2 = 4k_BT_O \quad V_{DS} = 0 \quad i_{D,THN}^2 = \frac{4k_BT\mu^2W^2}{I_{DS}L^2}\int Q_n^2(V(x))dV \quad 0 \leqslant V \leqslant V_{DS} \quad V_{DS} \neq 0$$
$$Q_n(x) = C_{OXIDE}\left(V_{GS} - V_{THRSH}(x) - V(x)\right) \tag{10.15a}$$

Assuming that the effect of position dependence of both the channel potential V(x) and the threshold voltage is small, the above equation can be integrated to give:

$$i_{D,THN}^2 = \frac{8C_{OXIDE}k_BTW}{3L}\left(\frac{3V_{DS}(V_{GS} - V_{THRSH}) - 3(V_{GS} - V_{THRSH})^2 - V_{DS}^2}{2(V_{GS} - V_{THRSH}) - V_{DS}}\right) \tag{10.15b}$$

A MOSFET has three regions of operation linear, saturation point and saturation region, defined as: $V_{DS} < V_{GS} - V_{THRSH}$ $V_{DS} = V_{GS} - V_{THRSH}$ $V_{DS} > V_{GS} - V_{THRSH}$. At the **saturation point**, 10.15b is simplified to:

$$i_{D,THN}^2 = \frac{8C_{OXIDE}k_BTW(V_{GS} - V_{THRSH})}{3L} = \frac{8g_mk_BT}{3} \tag{10.15c}$$

In the **saturation region** 10.15c cannot be used. However, experimentally it has been found that 10.15c is a good approximation, as long as the device shows a good saturation. This is because the cut-off region near the drain is much smaller then the resistive reverse channel which is responsible for the noise. This expression predicts the thermal noise in the channel without the substrate effect. In practice the thermal noise is higher. This is a result of thermal noise depending on the channel potential V(x). The integral in Eq. 10.15a is now hard to calculate, so that:

$$i_{D,THN}^2 = 4k_B \gamma g_m T \tag{10.15d}$$

The factor γ is a complex function of the basic transistor parameters and bias conditions, evaluated using numerical analysis of experimental data. For modern CMOS processes with oxide thickness in the order of 50 nm or less, and with a lower substrate doping of about $10^{13} - 10^{16}\ cm^{-3}$ the factor γ is between 0.67 and 1.

The current noise in the channel also generates noise in the gate through the gate-channel capacitance. The gate noise is due to the capacitative coupling frequency depending. The gate noise is approximately:

$$i_{G,THN}^2 \approx \frac{16\pi^2 f^2 k_B T}{5 g_m} \approx \frac{4 f^2 k_B g_m T}{f_T^2} \quad f_T \approx \frac{g_m}{2\pi C_{GS}} \tag{10.15e}$$

Flicker or 1/f noise is observed in all kinds of devices, from homogeneous metal film resistors to semiconductor devices and even in chemical concentration cells. Because 1/ f noise is spread over the components, there is a misconception that there is fundamental physical mechanism is behind it. Experimental evidence suggests that several internal physical mechanisms combine to generate 1/f noise, with the MOS transistor having the highest 1/f noise of all active semiconductors, due to its surface conduction mechanism. Although there are several physical models competing to explain the 1/f noise in a MOS-FET, they are all based on Hooge's empirical mobility fluctuation model and the carrier density|number fluctuation model first introduced by McWhorter.

In the Mobility Fluctuation model the 1/f noise is referred to as $\Delta\mu$-1/f noise. The model is described by the Hooge empirical equation (only for homogeneous junction devices):

$$\frac{i_{FLICKER}^2}{I_{SC}^2} = \frac{\alpha_I}{f N} \tag{10.15f}$$

where α_I is Hooge's 1/f noise parameter, N is the number of free carriers and I_{SC} is the short circuit current. Experimental data shows that this expression is valid for a number of common semiconductor devices, with $\alpha_I \approx 2x10^{-3}$. *Deriving this expression from fundamental principles results in expressions that do not predict values for the flicker noise current accurately.* The discrepancies arise as the electron mobility in a particular region of MOSFET operation is an effective value (Mathieson's rule) controlled by the various scattering processes occurring in the device at the time of measurement. The Mobility Fluctuation model 1/f noise voltage spectrum is:

$$v_{FLICKER,MOBILITYFLUCTUATION}^2 = \frac{\alpha_I \mu_f q (V_{GS} - V_T)}{2 C_{OXIDE} f \mu_{effective} L W} \tag{10.15g}$$

The Number Fluctuation model denotes the 1/f flicker noise as Δn-1/f noise. *The fluctuation in the number of mobile carriers is caused by the random trapping and de-trapping of the mobile carriers. The traps are located at the silicon–silicon dioxide interface within the gate oxide. This causes a signal with a Lorentzian generation-recombination spectrum. Superposition or a large number of these signals with the proper time constant result in a 1/f-noise spectrum.*

According to this model, the Δn-1/f noise is proportional to the effective trap density near the quasi-Fermi level of the inverse carriers, as verified by a large number of experiments. This model explains the 1/f noise in the MOSFET's weak inversion region, where the relative 1/f noise current has a plateau. At the silicon–silicon dioxide interface in the gate oxide (oxide traps) additional energy states exist. These states and traps communicate randomly with the free charges in the channel. The flicker noise voltage spectrum, as per the Number Fluctuation model is:

$$v^2_{FLICKER,NUMBERFLUCTAUTION} = \frac{K_F}{2C^2_{OXIDE} f \mu LW} = \frac{K_f}{2C^2_{OXIDE} f LW} \tag{10.15h}$$

where K_f is an experimentally determined parameter. The discrepancies between the mobility fluctuation and number fluctuation models are:

- The Mobility Fluctuation model predicts that the flicker noise spectrum is directly proportional to DC voltages ($V_{GS} - V_{THRSH}$) but the Number Fluctuation flicker noise spectrum is independent of DC voltages.
- The Mobility Fluctuation model predicts that the flicker noise spectrum is inversely proportional to oxide capacitance, while the Number Fluctuation model states that the flicker noise spectrum is inversely proportional to the square of the oxide capacitance.

This does not mean that both or one of the model(s) are incorrect—actual flicker noise is a result of a combination of these two mechanisms.

References

1. Mohammad, S. N., Salvador, A. A., & Morkoc, H. (1995). Emerging gallium nitride based devices. *Proceedings of the IEEE, 83,* 1306-1355.
2. Yoder, M. N. (1997). Gallium nitride: Past, present and future, in International. *Electron Devices Meeting Technical Digest. 3–12.*
3. Shur, M. S. (1998). GaN based transistors for high power applications. *Solid State Electronics, 42*(12), 2131–2138.
4. Mishra, U. K., Wu, Y. F., Keller, B. P., Keller, S., & Denbaars, S. P. (1998). GaN based microwave power HEMT. In Proceedings. of International Physics of Semiconductor Devices Workshop, pp. 878–883.

5. Bernd-Ulrich, H. (1995). Klepser, Crispino Bergamaschi and Mathias Schefer, Analytical Bias Dependent Noise Model for InP HEMT's. *IEEE Transactions On Electron Devices, 42*(11), 1882–1889.
6. Felgentreff, T., Olbrich, G., & Russer, P. (1994). Noise parameter modeling of HEMTs with resistor temperature noise sources. *IEEE MTT-S Digest,WE3C-3,* pp. 853–856
7. Hikaru Hida, Keichi Ohata, & Yasuyuki Suzuki (1986, May) A new Low-Noise AlGaAs/GaAs 2 DEG FET with a surface undoped layer. *IEEE Transactions On Electron Devices,* 33(5), 601–607.
8. Hsu, S. S. H., & Pavidis, D. (2001). Low noise AlGaN/GaN MODFETs with high breakdown and power characteristics. *GaAsIC Symposium*, pp. 229–232.
9. Lee Sunglae, & Webb Kevin, J. (2004, June 6-11). Numerical noise model for the AlGaN/GaN HEMT, IEEE MTT-S International Microwave System Digest.
10. De Jaeger, J. C., Delage, S. L., & Cordier, Y. (2005). Noise Assessment of AlGaN/GaN HEMTs on Si or SiC Substrates: Application to X-band low Noise Amplifiers, 13th GAAS Symposium-Paris, p. 229.
11. Deng, J., Werner, T., & Shur, M. S. (2001). Low Frequency and Microwave Noise Characteristics of GaN and GaAs-based HFETs GaAs Mantech.
12. Pucel, P. A., Haus H. A., & Statz, H. (1975). Advances in electronics and electron physics. New York: Academic, pp. 195–265.
13. Nuttinck, S., Gebara, & Harris, M. (2003). High-frequency noise in AlGaN/GaN HFETs. *IEEE Microwave and Wireless Components Letters, 13,* 149–151.
14. Brookes, T. M. (1986, January). The noise properties of high electron mobility transistor. *IEEE Transactions on Electron Devices, 33*, 52–57.
15. Anwar, A. F. M., & Kuo-Wei Liu. (1994, November). A noise model for high electron mobility transistors. *IEEE Transactions on Electron Devices, 41*(11), 2087–2092.
16. Kamei, K. et al. (1985). Extremely low—noise 0.25 μm gate HEMT. In Proceedings of 12th GaAs Related compounds conference, pp. 541–546.
17. Sullivan, Y. T., Asbeck, S. G. J., Waung, P. M., Qiao, C. D., & Lau, S. S. (1997). Measurement of piezoelectrically induced charge in GaN/AlGaN heterostructure field effect transistor. *Applied Physics Letters, 71*(9), 2794–2796.
18. Tyagi, R. K., Ahlawat, A., Pandey, M., & Pandey, S. (2007). An analytical two dimensional model for AlGaN/GaN HEMT with polarization effects for high power applications. *Microelectronics Journal, 38*, 877–883.
19. Chang, C. S., & Fetterman, H. R. (1987). An analytical model for HEMT using new velocity field dependence. *IEEE Transactions on Electronic Devices, 34*, 1456–1462.
20. Tyagi, R. K., Ahlawat, A., Pandey, M., & Pandey, S. (2008, December). A new two-dimensional C–V model for prediction of maximum frequency of oscillation (fmax) of deep submicron AlGaN/GaN HEMT for microwave and millimeter wave applications. *Microelectronics Journal, 39*(12), 1634–1641.
21. Statz, H., Haus, H. A., & Pucel, R. A. (1974). Noise characteristics of gallium arsenide field effect transistor. *IEEE Transactions on Electron Devices, 21,* 549–562.
22. Wu Lu, Jinwei Yang, M.Asif Khan, & Ilesanmi Adesida. (2002). AlGaN/GaN HEMT on SiC with over 100 GHz ft and Low Microwave Noise. *IEEE Transactions on Electron Devices, 48*(3), 581–585.
23. Kumar, V., Lu, W., Schwindt, R., Kuliev, A., Simin, G., Yang, J., Khan, M. A., & Adesida, A. (2002). AlGaN/GaN HEMT in SiC with fT of over 120GHz. *IEEE Electron Device Letters, 23,* 455–457.
24. Rohde, U. L., Poddar, A. K., Böck, G. (2005). The design of modern microwave oscillators for wireless applications. John Wiley & Sons, Inc., ISBN 0-471-72342-8.

25. Cressler, J., & Niu, G. (2003). Silicon-Germanium Heterojunction Junction Bipolar Transistors. Artech House.
26. Van der Ziel, A. (1955, November). Theory of shot noise in junction diodes and junction transistors. *Proceedings of the IRE, 43,* 1639–1646.
27. Niu, G., Cressler, J., Ansley, W., Webster, C., & Harame, D. (2001, November). A unified approach to RF and microwave noise parameter modeling in bipolar transistors. *IEEE Transactions on Electron Devices, 48,* 2568–2574.
28. Moller, J., Heinernann, B., & Herzel, F. (2002). An improved model for high-frequency noise in BJTs and HBTs interpolating between the quasi-thermal approach and the correlated-shot-noise model. In Proceedings of the IEEE BCTM, pp. 228–231.
29. Sakalas, P., Schroter, M., Zampardi, P., Zirath, H., & Welse, R. (2002). An improved model for high-frequency noise in BJTs and HBTs interpolating between the quasi-thermal approach and the correlated shot noise model. In IEEE MTT-S International Microwave Symposium Digest, pp. 2117–2120.
30. Sakalas, P., Schroter, M., Scholz, R., Jiang, H., Racanelli, M. (2004). Analysis of microwave noise sources in 140 GHz SiGe HBTs, in IEEE RFIC Digest, pp. 291–294.
31. Niu, G., Xia, K., Sheridan, D., & Harame, D. (2004). Experimental extraction and model evaluation of base and collector current RF noise in SiGe HBTs. In IEEE RFIC Digest, pp. 615 – 618.
32. Jungemann, C., Neinhus, B., Meinerzhagen, B., & Dutton, R. (2004, June). Investigation of compact models for RF noise in SiGe HBTs by hydrodynamic device simulation. *IEEE Transactions on Electron Devices, 51,* 956–961.
33. van Vliet, K. M. (1972, October). General transistor theory of noise in PN junction like devices—i. three-dimensional green's function formulation. *Solid State Electronics, 15,* 1033–1053.
34. Cui, Y., Niu, G., & Harame, D. (2003). An examination of bipolar transistor noise modeling and noise physics using microscopic noise simulation. In Proceedings of the IEEE BCTM, pp. 183–186.
35. Niu, G. (2004). Bridging the gap between microscopic and macroscopic theories of noise in bipolar junction transistors. In Technical Digest of IEEE Topical Meeting on Si Monolithic Integrated Circuits in RF Systems.
36. Jungemann, C., Neinhus, B., & Meinerzhagen, B. (2002, July). Hierarchical 2-d DD and HD noise simulations of Si and SiGe Devices – part i: Theory. *IEEE Transactions on Electron Devices, 49,* 1250-1257
37. Jungemann, C., Neinhus, B., & Meinerzhagen, B. (2002, July). Hierarchical 2-d DD and HD noise simulations of Si and SiGe devices – part ii: Results. *IEEE Transactions on Electron Devices, 49,* 1258–1264.
38. Haus, et al. (1960). Representation of noise in linear two ports. *Proceedings of the IRE, 48,* 69–74.
39. Niu, G., & Cressler, J. (2002). Noise-gain tradeoff in RF SiGe HBTs. *Solid State Electronics, 46,* 1445–1451.
40. Bary, L., Cibiel, G., Ibarra, J., et al. (2001, May). Low frequency noise and phase noise behavior of advanced SiGe HBTs, in Technical. Digest of IEEE RFIC Symposium.
41. Rudolph, M., Behtash, R., Doerner, R., Hirche, K., Wurfl, J., Heinrich, W., & Trankle, G. (2007, January). Analysis of the survivability of GaN low-noise amplifiers. *IEEE Transactions on Microwave Theory Techniques, 55*(1), 37-43.
42. Colangeli, S., Bentini, A., Ciccognani, A, Limiti, A., Nanni, A. (2013, October). GaN-based robust low-noise amplifiers. *IEEE Transactions on Electron Devices 60*(10), 3238–3248.
43. Nuttinck, S., Gebara, E., Laskar, J., Harris, M. (2003, April). High-frequency noise in AlGaN/GaN HFETs. *IEEE Microwave Wireless Component Letters 13*(4), 149-151.

44. Lee, S., Webb, K. J., Tialk, V., & Eastman, L. F. (2003, May). Intrinsic noise equivalent-circuit parameters for AlGaN/GaN HEMTs. *IEEE Transactions Microwave Theory Techniques, 51*(5), 1567-1577.
45. Sanabria, C., Xu, H., Palacios, T., Chakraborty, A., Heikman, S., Mishra, U., & York, R. (2005, Febuary). Influence of epitaxial structure in the noise figure of AlGaN/GaN HEMTs. *IEEE Transactions. Microwave Theory Techniques, 53*(2), 762-769.
46. Liu, Z. H., Arulkumaran, S., Ng, G. I., Xu, T. (2009, June). Improved microwave noise performance by SiN passivation in AlGaN/GaN HEMTs on Si. *IEEE Microwave Wireless Component Letters 19*(6), 383-385.
47. Boglione, L. (2013, December). Considerations on the 4NT 0 /T m ratio and the noise correlation matrix of active and passive two-port networks. *IEEE Transactions on Microwave Theory Techiques, 61*(12), 4145-4153.
48. Rudolph, M., Doerner, R. (2014, March). Bias-Dependent Pospieszalski Noise Model for GaN HEMT Devices. In Proceedings. 2014 German Microwave Conference (GeMIC), Aachen, Germany, pp. 1–4.
49. Colangeli, S., Bentini, A., Ciccognani, W., Limiti, E. (2014, September-December). Polynomial noise modeling of silicon-based GaN HEMTs. *International Journal Numerical Model, 27*(5–6), 812–821.
50. Nalli, A., Raffo, A., Vannini, G., D'Angelo, S., Resca, D., Scappaviva, F., Crupi, G., Salvo, G., & Caddemi, A. (2014, October). A scalable HEMT noise model based on FW-EM analyses, in Proceedings. 9th European Microwave Integrated Circuits Conference. Rome, Italy, pp. 1420–1423.
51. GH25–10 User Guide for the DK 2.3, United Monolithic Semiconductor, France (2014)
52. Resca, D., Raffo, A., Santarelli, A., Vannini, G., & Filicori, F. (2009, Febuary). Scalable equivalent circuit FET model for MMIC design identified through FW-EM analyses. *IEEE Transactions Microwave Theory Techniques, 57*(2), 245-253.
53. Resca, D., Santarelli, A., Raffo, A., Cignani, R., Vannini, G., Filicori, F., Schreurs, D. D. M. P. (2008, April). Scalable nonlinear FET model based on a distributed parasitic network description, IEEE Transactions. *Microwave Theory Techniques, 56*(4), 755-766.
54. Resca, D., Santarelli, A., Raffo, A., Cignani, R., Vannini, G., Filicori, F., & Cidronali, A. (2006, September). A distributed approach for millimetre-wave electron device modelling. In Proceeding 1st European. Microwave Integrated Circuits Conference, Manchester, UK., pp. 257–260.
55. Dambrine, G., Cappy, A., Heliodore, F., & Playez, E. (1988, July). A new method for determining the FET small-signal equivalent circuit. *IEEE Transactions Microwave Theory Techniques, 7*(36), 1151-1159.
56. Anderson, R. W. S-Parameter Techniques for Faster, More Accurate Network Design, Hewlett-Packard Application Note 95–1.
57. Beatty, Robert W. (1964, June). Insertion Loss Concepts, Proceedings of the IEEE, pp. 663–671.
58. Boyd, D. (1999, October). Calculate the Uncertainty of NF Measurements. Microwaves and RF, p. 93.
59. Chambers, D. R. (1983, April). A Noise Source for Noise Figure Measurements, Hewlett-Packard Journal, pp. 26–27.
60. Davenport, W. B. Jr., & Root, W. L. (1958). An Introduction to the Theory of Random Signals and Noise, McGraw-Hill Book Co., Inc, New York, Toronto, London.
61. Description of the Noise Performance of Amplifiers and Receiving Systems. (1963, March). Sponsored by IRE subcommittee 7.9 on Noise. Proceedings of the IEEE, pp. 436–442.
62. Friis, H. T. (1944, July). Noise Figures of Radio Receivers, Proceedings of the IRE, pp. 419–422.

63. Fukui, H. (1966, June). Available Power Gain, Noise Figure and Noise Measure of Two-Ports and Their Graphical Representations. IEEE Transactions on Circuit Theory, pp. 137–143.
64. Fukui, H. (Ed.). (1981). Low Noise Microwave Transistors and Amplifiers. IEEE Press and John Wiley & Sons, New York.
65. Gupta, M. S. (1971, December). Noise in Avalanche Transit-Time Devices, Proceedings of the IEEE, pp. 1674–1687.
66. Haitz, R. H., & Voltmer, F. W. (1966, November 15). Noise Studies in Uniform Avalanche Diodes. Applied. Physicd. Letters, pp. 381–383.
67. Haitz, R. H., & Voltmer, F. W. (1968, June). Noise of a self sustaining avalanche discharge in silicon: Studies at microwave frequencies. *Journal of Applied Physics,* 3379–3384.
68. Haus, H. A., & Adler, R. B. (1958, August). Optimum Noise Performance of Linear Amplifiers, Proceedings of the IRE, pp. 1517–1533.
69. Hines, M. E. (1966, January). Noise Theory for the Read Type Avalanche Diode, IEEE Transactions on Electron Devices, pp. 158–163.
70. Kanda, M. A. (1977, August). Statistical Measure for the stability of solid state noise sources. *IEEE Transactions on Microwave Throry and Techniquws*, pp. 676–682.
71. Kanda, M. (1976, December). An improved solid-state noise source. *IEEE Transactions on Microwave Theory and Techniques*, pp. 990–995.
72. Kuhn, N. J. (1963, November). Simplified Signal Flow Graph Analysis, Microwave Journal, pp. 59–66.
73. Kuhn, N. J. (1984, June). Curing a Subtle but Significant Cause of Noise Figure Error, Microwave Journal, p. 85.
74. https://ngspice.sourceforge.io/.
75. https://www.orcad.com/orcad-free-trial.
76. https://www.synopsys.com/implementation-and-signoff/ams-simulation/primesim-hspice.html.
77. https://www.ti.com/tool/TINA-TI.
78. https://www.analog.com/en/design-center/design-tools-and-calculators/ltspice-simulator.html.

11 Semiconductor Device Manufacturing Technologies

11.1 Creating High Vacuum

This discussion starts with the standardized vacuum units, followed description of how to create ultra high vacuum conditions [1–5].

11.1.1 Vacuum Units

The Pascal (named after the French mathematician|physicist Blaise Pascal) is the official SI unit for vacuum pressure—widely used in physical sciences. One Pascal is the force of one Newton per square meter acting perpendicular to a surface. It is easily converted to other common vacuum units.

1 Pa = 0.01 mbar = 0.0075 Torr = 7.5 micron(mTorr) = 0.0075 mm Hg = 0.000145 PSI

Another common metric vacuum unit is millibar (mbar). Millibar, related to the bar pressure unit, originated as a pressure measurement unit in meteorology. One mbar is equal to 100 Pa or one hPa(Hectopascal). It is related to the common pressure units as:

1 mbar = 100 Pascal = 1 hPa = 0.75 Torr = 750.0 mTorr = 0.75 mm Hg = 0.145 PSI *One Torr is the pressure equivalent of 1 mm of mercury at 0 Celsius.*

11.1.2 How to Create Vacuum

Fore vacuum pumps extract air from the chamber they are connected, to the atmospheric pressure. Also, a fore vacuum pump must be able to support secondary pumps. There are two types of fore vacuum pumps:

A. Banerjee, *Semiconductor Devices*, Synthesis Lectures on Engineering, Science, and Technology, https://doi.org/10.1007/978-3-031-45750-0_11

1. Dry running diaphragm, scroll and screw pumps.
2. Oil sealed rotary vane pumps.

The multistage Roots vacuum pumps are used widely to create ultra low vacuum in semiconductor processing equipment. The single stage Roots vacuum pump is in Fig. 11.1a.

Versatile multistage Roots pumps are dry vacuum pumps used in low, medium, high and ultra-high vacuum systems to produce "dry" conditions. The single-stage Roots pump is used as a booster pump in combination with several types of fore pumps (e.g., screw, liquid ring pumps and rotary vane pumps) to boost pumping speeds. Multistage Roots pumps do not need any fore pump and can operate from atmospheric pressure. A multistage Roots pump can consist of up to eight stages with sets of rotors on a shared shaft. Roots pumps are suitable where a dry and clean atmosphere is essential—ideal for interiors of ultra low vacuum semiconductor processing equipment chambers.

Each single stage Roots pump has two counter-rotating interconnected lobed rotors within a stator housing or casing. Gas enters through an inlet flange located perpendicular to the rotors and is then "isolated" between the rapidly counterrotating rotors and the stator. The compressed gas is then expelled via the exhaust port.

11.1.3 How to Measure Vacuum

Vacuum levels referenced in semiconductor industry processes are classified as below:

$$Atmospheric\ 760\ Torr\ Low\ Vacuum\ 760 - 25\ Torr$$
$$Medium\ Vacuum\ 25 - 10^{-3}\ Torr\ High\ Vacuum\ 10^{-3} - 10^{-9}\ Torr$$
$$Ultra\ \ High\ Vacuum\ 10^{-9} - 10^{-12}\ Torr\ Extreme$$
$$High\ Vacuum\ < 10^{-12}\ Torr$$

So special vacuum gauges are required to measure vacuum levels, each optimized to a specific vacuum range. With reference to the ultra clean environment used in semiconductor device manufacture, only those vacuum level measurement devices that operate in dry conditions are allowed. The first is the capacitance manometer gauge (Fig. 11.1b).

The capacitance manometer gauge measures vacuum levels from atmospheric pressure to 10^{-5} Torr depending on the given sensor applied. As seen from Fig. 11.1b the key component of this gauge is the diaphragm. *A thin diaphragm will distort easily at a low input pressure, while a thicker one will not respond to that low pressure, since the capacitance value of a parallel plate capacitor is inversely proportional to the plate separation.* A capacitance sensor operates by measuring the change in electrical capacitance from the distortion of a sensing diaphragm compared to some fixed capacitance electrodes. In some designs

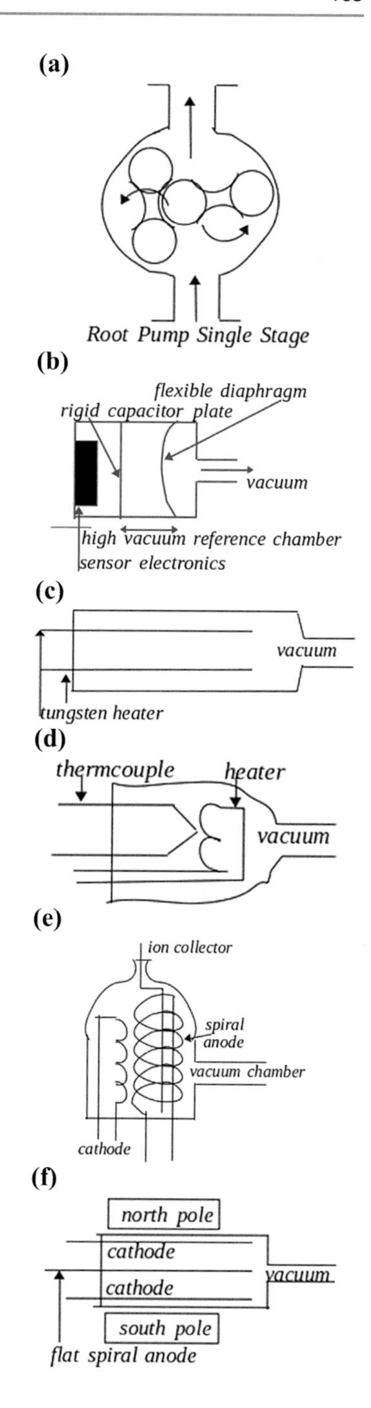

Fig. 11.1 **a** Single stage Roots pump. **b** Capacitive manometer. **c** Pirani gauge. **d** Thermocouple vacuum gauge. **e** Hot cathode vacuum gauge. **f** Inverted magnetron cold cathode ionization gauge

a variable DC voltage is applied to keep the sensor's Wheatstone bridge in a balanced condition. The amount of voltage required is directly related to the pressure.

The Pirani gauge (Fig. 11.1c) is a thermal conductivity gauge used to measure pressure in a vacuum chamber. The gauge is able to give a pressure reading due to a heated metal filament suspended in the vacuum chamber. Gas molecules in the system collide with the wire allowing it to emit heat and cool. As the vacuum is pumped down, there are fewer gas molecules to collide with the wire and the wire heats up. The wire's internal electrical resistance increases and a circuit attached to the wire detects the change in resistance. With appropriate calibration, the circuit can directly correlate the amount of resistance to the pressure in the vacuum chamber. The Pirani gauge is used to measure pressures between 0.5 to 10^{-4} Torr. Each device needs calibration to obtain accurate readings depending on the thermal conductivity and the heat capacity of the gas.

Pirani gauges can be constant current|resistance. The constant current gauge has a power supply to provide a fixed amount of energy to the metal filament. A current change implies a corresponding resistance change. The varied resistance is proportional to the pressure in the vacuum. The constant resistance gauge has a power supply which varies the current based on the constant resistance. The variation in the current is proportional to the pressure in the vacuum.

The thermocouple gauge (Fig. 11.1d) is very similar to a Pirani gauge, which has an internal heater adjacent to the thermocouple. The voltage across the thermocouple is measured, and calibrated to correspond to the chamber pressure.

Ultra high vacuum levels can only be measured with some form of ionization gauge. These come in two varieties, cold|hot cathode and measure pressures in the range 10^{-2} 10^{-10} Torr. They sense pressure indirectly by measuring the electrical ionic current produced when gas is bombarded with electrons. *Decreasing pressure decreases the available number of ions for the electrical ionic current—i.e., decreasing ionic current means higher vacuum.*

The hot cathode ionization gauge (Fig. 11.1e) has three electrodes acting as a triode, where the cathode is the filament. The three electrodes are a collector, plate, a filament and a grid. Electrons emitted from the filament move to and fro around the grid before finally entering the grid. During these movements, some electrons collide with a gas molecule, ionizing it, to form a ion|electron pair. **The number of these ions is proportional to the gas molecule density multiplied by the electron current emitted from the filament, and these ions enter into the collector to form the ion current**. *Since the gas molecule density is proportional to the pressure, the pressure is estimated by measuring the ion current.*

There are two types of cold cathode ionization gauges—the Penning gauge and the inverted magnetron, also known as the redhead gauge. The inverted magnetron (Fig. 11.1f) exploits the physical fact that ion production rate by a stream of electrons in a vacuum system is dependent on pressure and the ionization probability of the residual gas.

There are two parallel connected cathodes. An anode is placed midway between them. The cathodes are metal plates or shaped metal bosses. The anode is a loop of flattened metal wire, the plane of which is parallel to that of the cathode. A high voltage is applied between the anode and cathode halves. A external permanent magnet applies a magnetic field between cathode halves and the anode. Electrons emitted from the cathode travel in helical paths, eventually reaching the anode, thus increasing the amount of ionization occurring within the gauge. Normally the anode is operated at about 2kV, giving rise to a direct current caused by the positive ions arriving at the cathode. The pressure is proportional to the magnitude of the direct current produced. The pressure range covered by this gauge is from as low as 10^{-7} Torr. The simple rugged design makes this vacuum gauge ideal for a number of common industrial systems.

11.2 Photolithography

Photography, or "writing with light" is one of the key processing steps in semiconductor device fabrication. *Unlike all other semiconductor device manufacturing technologies to be examined in the next several sections, photolithograpy and wet etching does not need any vacuum conditions.* For a long time, state-of-art semiconductor devices were planar. For example, the drain and source regions of a field effect transistor must be at the right locations, with respect to the gate region. Photolithography enables accurate alignment of these regions, using appropriately designed masks.

Photolithography is the process of transferring geometric shapes and patterns from a mask, to a smooth, clean surface. It uses a ultraviolet light to image the mask on to a surface coated with chemicals whose molecular structure changes (de-polymerize|polymerize) when irradiated with ultraviolet light. Figure 11.2a shows a general photolithography system.

Photolothography consists of the following steps.

- Surface cleaning
- Spin coating with photoresist
- Baking at low temperature to dry|harden the deposited phororesist
- Mask alignment, exposure and development
- Hard baking and post process exposure.

Successfully imprinting a pattern on the wafer with photolithography involves tightly controlling the resolution tolerances|limits for the final imprinted pattern. With reference to Fig. 11.2a, *smaller distances on the illuminated mask translate to large distances on the projection plane (projection pupil and lens).* In standard optical microscopes, the detector sees the light in the far field region. So, connecting the reciprocal wavevector space (k-space) with the real space:

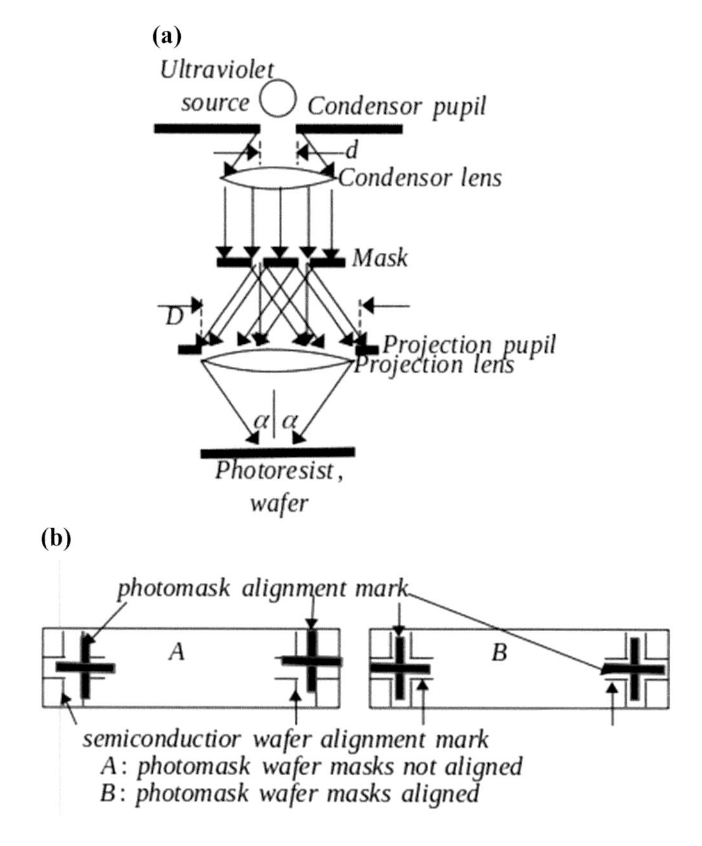

Fig. 11.2 **a** General ultraviolet light based photolithography system. **a** Photolithography mask, wafer alignment

$$\begin{gathered} K^2 = \omega^2 \mu_0 \epsilon = K_x^2 + K_y^2 + K_z^2 \sqrt{K_x^2 + K_y^2} < \tfrac{n\omega}{C} \Rightarrow \left|K_{par,\max}\right| = \tfrac{2\pi n}{\lambda} \\ Resolving\ power = \tfrac{\lambda}{2n} = \tfrac{\lambda_{EFFECTIVE}}{2} = Diffration\ Limit \end{gathered} \tag{11.1a}$$

where n is the lens material refractive index. In a real optical system, the spatial cutoff frequency is controlled by the size of the lens, quantitatively described by its numerical aperture NA. With reference to Fig. 11.2a,

$$\begin{aligned} Numerical\ Aperture\ (NA) = n\sin(\alpha)\frac{k_{par,\max}}{k} = \sin(\alpha) \Rightarrow k_{par,\max} \\ = \frac{2\pi NA}{\lambda}\ Resolving\ power = \frac{\lambda}{2\ NA} \end{aligned} \tag{11.1b}$$

The photoresist is applied using a spinning chuck (3000–6000 rpm) with the wafer held in position by applying a vacuum from underneath. The final photoresist thickness (1–2 μm) is controlled by its viscosity and the applied rotational speed. The spinning process introduces unwanted nanometer scale striations, edge beads etc., The photoresist coated wafer is then baked to evaporate the photoresist solvent, harden the phororesist and minimize the nanometer scale photoresist surface imperfections. Subsequent baking using hot plate ensures fast, uniform photoresist solvent evaporation. Two types of photoresist are used:

Positive photoresist:
Exposure to ultraviolet light de-polymerizes (*breaks up long chain molecules to individual ones*) the exposed photoresist regions. The de-polymerized photoresiat is soluble in the photoresist developer. These exposed areas are subsequently washed away, exposing the underlying material. Unexposed areas of the photoresist are insoluble to the photoresist developer. Positive photoresist creates an identical copy of the pattern, which is exposed as a mask on the wafer.

Negative photoresist:
The photoresist in areas exposed to the ultraviolet light polymerize (*individual phororesist molecules join to form long chain molecules*). The polymerized photoresist molecules are insoluble in phororesist developer solution. The photoresist in regions unexposed to ultraviolet light are removed with photoresist developer solution.

The key transfer of the mask pattern to the photoresist treated wafer surface, requires accurate alignment of the mask and the wafer. *A photolithography mask, also called a photomask, is a plate of opaque material, with transparent geometrical patterns—only the transparent areas allow the incident light to pass through, transferring the mask pattern to the photoresist coated semiconductor wafer surface.*

Sophisticated computer programs generate the mask pattern. The computer generated data is transferred to a chrome coated square quartz plate with electron beam lithography, with the quartz plate coated with appropriate photoresist. A beam of electrons moves over the mask in a vector|raster pattern. Only over those spots where the chrome needs to be removed, the electron beam is switched on. When the photoresist on the mask is exposed, the chrome can be etched away, leaving a clear path for the light in the stepper/scanner systems to travel through.

Photomask alignment [8] is another very difficult task, involving sub-micron tolerances. Commercial semiconductor device fabrication facilities use high end computer vision enabled equipment to eliminate all errors. Figure 11.2b shows how the process works in a very simple case.

After the wafer and mask have been exposed to the ultraviolet light, the mask is removed and the wafer is baked for a predefined duration. Then the wafer is treated with the appropriate photoresist developer (depending on whether the photoresist used was positive or negative). Deionized water wash removes traces of photoresist and developer solution. The wafer is now ready for the next processing stage.

Although photolithography is meant to transfer two dimensional images from the mask to the wafer, it is a three dimensional image, because of the intermediate air. *So the expected sharp contrast between light intensity in the bright and dark areas of an image is reduced because of light intensity gradient.* The Normalized Image Log-Slope (NILS) method is used to quantify the aerial image quality, and minimum acceptable NILS values are calculated using empirically determined constants. The *depth of focus* (DOF) is the vertical distance over which the image remains in focus. It is dependent on the wavelength of light used, and the numerical aperture of the imaging system. A high DOF is essential for the entire resist layer to be properly exposed during photolithography.

These key parameters need to be adjusted to transfer patterns with smaller and smaller feature sizes, to the wafer. **To reduce the feature size either the wavelength of the exposing light must be decreased or numerical aperture of the projection optics must be increased. The DOF must be sufficient to ensure accuracy and precision in the feature size through the entire thickness of a resist**.

The *phase shift* technique is used to enhance edge contrast in the image being patterned, eliminating defects from diffraction limitations at sub-wavelength patterning. Phase shift masks with different thicknesses at different sections of the pattern on the mask, which change the phase of the transmitted light, are used to achieve this goal.

Initially, photolithography was a "dry" process. *Immersion lithography bypasses the feature size limitations of dry lithography by replacing air with water as the medium between the optical system and the substrate.* Water with a refractive index of 1.44 increases the value of NA beyond 1.0, leading to a reduction in minimum single-exposure feature size to about 40 nm when using 193 nm light. Immersion lithography techniques increase the amount of light that can reach the resist (increasing the resolution) and change the phase of the light so that it improves DOF. So single exposure immersion lithography is the only patterning technique at design nodes down to 45 nm.

11.3 Deep and Extreme Ultraviolet Photolithography

Deep ultraviolet(DUV) technology for photolithography is based on projection optics since the pattern on the photomask is much larger than the final pattern developed on the photoresist. The optical system in a 193 nm photolithography machine is a *catadioptric system,* using both refractive(lens) and reflective(mirrors) for directing and controlling the light beam from the laser. *A catadioptric optical system can handle a broad bandwidth of the source laser light while limiting chromatic aberration.* Refractive elements in the optical system are fabricated from either synthetic fused silicalcalcium fluoride, both with low absorption coefficient of 193 nm light. Photomasks (or reticles) in these systems are typically made from fused silica with chrome patterns. In a step-and-scan photolithography process, a slit of light is scanned across one or more dies patterned on the reticle. The light reproduces the part of the pattern on the reticle that is illuminated on the wafer, at much reduced feature size because of passage through the reduction lens. Simultaneous (and highly precise, accurate, and repeatable) movement of both the reticle and the wafer is used to produce the full image of the die on the wafer. After patterning a die, the next die area is positioned for patterning. The alignment and positioning of masks is fully automated using computer vision techniques.

With relentless trend in gate length reduction for microprocessor MOSFETs, the Rayleigh and diffraction limit conditions dictate that light of smaller wavelength be used. Extreme ultraviolet (wavelength 13.5 nm) lithography, although far more complicated than deep ultraviolet lithography, addresses this issue.

- There is no material that allows radiation at 13.5 nm to pass through it, so lenses are ruled out. The ultraviolet beam is controlled with mirrors.
- Ordinary mirrors do not work. Special mirrors constructed with alternating layers of materials with different dielectric constant(commonly molybdenum and silicon|beryllium) are used.
- These special mirrors reflect light via constructive interference, whose wavelength exactly matches the spacing between the layers. So, constructing these mirrors is very difficult and expensive.
- A 4X reduction system consisting entirely of these special mirrors, as used in standard optical lithography systems is easily constructed, so that mask tolerances need not be much better than in current systems.

An EUV machine uses an infrared laser to excite xenon atoms to plasma state, which upon relaxation emits ultraviolet light at 13.5 nm. The rest of the machine consists of condenser and beam size reduction mirrors to pattern the photoresist coated wafer (Fig. 11.3). Some EUV machines use tin plasma to generate the EUV radiation. Such a machine is more complicated, as the used tin vapor has to be extracted constantly.

11.4 Ion Implantation

A semiconductor material either has excess electrons(n type) or holes(p type). However most base materials for semiconductor devices (e.g., silicon) are electrically neutral|intrinsic semiconductor, so that appropriate impurity atoms must be inserted into the crystal lattice to make, e.g., silicon n type. The common impurity atoms are boron, phosphorus (n type) or arsenic (p type). The semiconductor industry technique for introducing these impurities is ion implantation [6, 7]. The advantages of ion implantation over the now obsolete thermal diffusion process are:

- It produces isotropic doped regions with very tight impurity dosage tolerances.
- The physical dimensions of the doped regions can be tightly controlled.

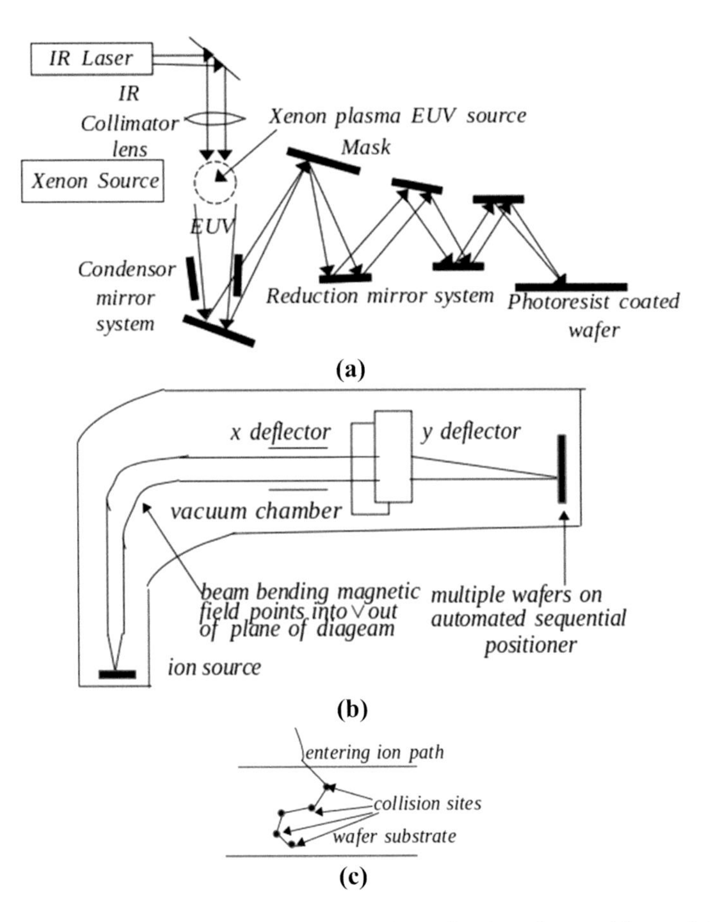

Fig. 11.3 **a** Extreme ultraviolet lithography system. **b** Ion implantation machine. **c** Collision sites for ion implanted into wafer

- The ion beam impinging the semiconductor wafer can be moved very precisely over the wafer surface. Ion implantation is a low temperature process.

The ion implanter machine is a particle accelerator (Fig. 11.4a), and the interior is maintained at very high vacuum to prevent unwanted impurities from contaminating the wafer.

Ions are extracted from the ion source using electromagnetic fields. The ion beam from the ion source is directed into a mass analyzer where the beam is focused and bent through a right angle. The radius of the bend is determined by a combination of electromagnetic field characteristics and the mass to charge ratio of the ions. Consequently, only ions of a particular mass to charge ratio is selected to exit the mass analyzer using a movable aperture (or an electromagnetic lens), other different ions that may originate from the

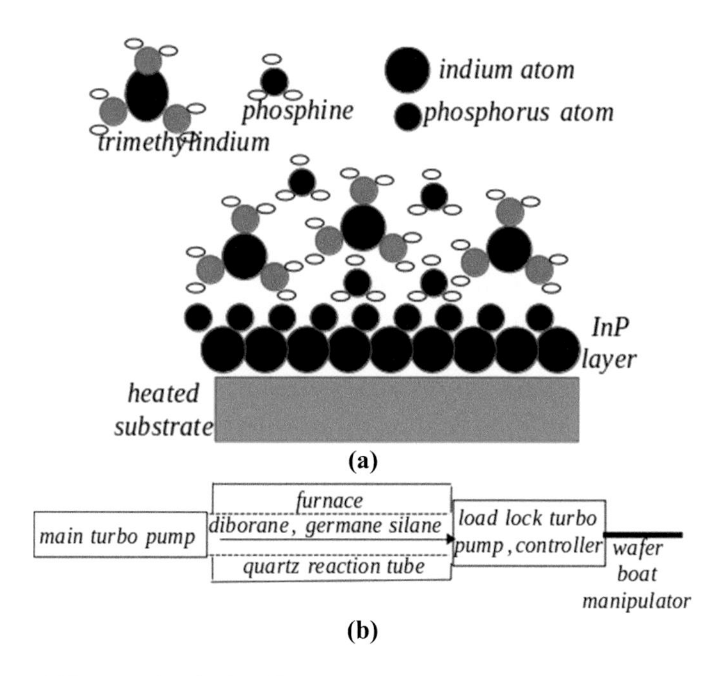

Fig 11.4 **a** MOCVD reaction. **b** UHVCVD apparatus

ion source are blocked. The beam of selected ions is then accelerated to high energies, ranging from sub-keV to MeV values (eV = electron Volt) and the high energy ion beam is steered using electromagnetic fields to impinge the semiconductor surface.

When a dopant ion strikes the wafer surface, it penetrates into the substrate crystal matrix to a depth proportional to its energy and angle of incidence. *But the newly entered ion does not immediately displace a crystal lattice atom and occupy the displaced atom's lattice site.* The newly entered ion collides with and is scattered by lattice atoms and a number of these collisions comes to a halt in an interstitial site. (Fig. 11.4b). So the concentration profile for dopant atoms versus penetration into the substrate is Gaussian. So, the ion implanted wafers need to be "activated".

Implanted ion activation is achieved by annealing the ion implanted wafer. Annealing also repairs any damage done to the wafer crystal matrix by collisions with the high-energy dopant ions and flatten the dopant distribution profile—uniform doping. Once substituted into the lattice, the dopant will act as either a donor or acceptor depending on its electronic structure.

11.4.1 Ion Implantation Masks and Photoresists

Ions are implanted only into ***specific regions of the wafer.*** **To prevent the incident ions from penetrating unwanted regions, those regions have to be covered with an appropriate resistive material, that can be removed after ion implantation is completed. In addition, to demarcate the regions that need to be covered with the ion implant resist material, a mask is necessary.**

A n ionized atom of mass M, after being accelerated over a potential U has velocity v (using purely classical physics) $V = \sqrt{\frac{2nqU}{M}}$, This expression implies that for a single ionised heavy atom e.g., arsenic after an acceleration voltage of 10 keV which is comparatively low for ion implantation, has a speed of about 160 km/s, and for a boron ion accelerated with 500 keV, the velocity is around 3000 km/s (approximately 1% of the light speed), for which the above classical view is still a valid approximation.

The entering ions suffer collisions. Inelastic collisions of electrons result in excitation|ionization of the wafer substrate atoms. In the case of ion energies of several 10 eV, which are too low for these processes, i.e., about 1 per thousand of the original energy, a charge exchange between the ion and the solid can take place during the short term formation of "quasi-molecules" of the entering ions substrate atoms.

The collisions with the atomic nuclei of the solid are elastic with a scattering cross-section increasing with decreasing ion velocity. Inelastic collisions can also occur in which the atoms of the host matrix are displaced in the microstructure of the solid, producing point defects in crystalline structure.

The material used as ion implantation mask must be sufficiently thick to block ions from impinging the wafer surface areas covered by it. The lateral resolution and sidewall steepness of this material must satisfy tight tolerances. As the substrate might heat up during ion implantation, the mask material must not soften, i.e. have a high melting temperature. The necessary ion implantation mask material film thickness increases with the ionic energy. The higher the density ρ and average atomic number Z of the ion implantation mask material used, the lower its necessary film thickness to absorb the incident ions. The ρ and Z parameters of common ion implantation mask materials as phenol, epoxy or acrylic-resins differ only to a small extent.

11.5 Dry and Wet Etching—Anisotropic and Isotropic

The goal behind photolithography is to demarcate regions of the semiconductor wafer surface for the next processing step—e.g., etching. Semiconductor wafer surfaces are etched using either wet or dry etching [9]. *As wet etching uses very corrosive chemicals, the masking material, e,g., for silicon wafers is silicon dioxide or silicon nitride.*

When a semiconductor wafer is treated with a corrosive liquid or vapor, the resulting wafer material removal may be anisotropic (uniform in one direction) or isotropic (uniform in all three directions. The material removal occurs *only* from those regions not protected by the ion implantation mask material. Material removal rate for wet etching is faster than corresponding rates for many dry etching processes. Wet etching rates can be changed by varying temperature or the concentration of the active etching species.

Liquid etching chemicals (etchants) etch away|remove crystalline materials at different rates depending upon which crystal face is exposed to the etchant. In silicon, this results in very high anisotropy. Some of the common wet etching agents for silicon are potassium hydroxide(KOH), ethylene diamine pyrocatechol(EDP), or tetramethylammonium hydroxide(TMAH). Etching a (100) silicon wafer results in a pyramid shaped etch pit, with flat and angled (54.7°) etched walls.

To achieve isotropic wet etching, a mixture of hydrofluoric acid, nitric acid, and acetic acid(HNA) is used for silicon. The concentrations of each etchant determines the etch rate. Silicon dioxide|nitride is used as a masking material against HNA. **During the reaction, the material is removed laterally at the same rate as that for material being removed in the downward direction—isotropic etching.** Wet chemical etching is generally isotropic even though a mask is present since the liquid etchant can penetrate underneath the mask. Wet chemical etching is not used if directionality is very important for high-resolution pattern transfer.

Dry etching can be of two types dry gas reaction or plasma based. In dry etching, plasmas or etchant gasses remove the substrate material, using using high kinetic energy of particle beams, chemical reaction or a combination.

Physical dry etching requires high kinetic energy(ion, electron, or photon) beams to etch off the substrate atoms. When the high energy particles knock out the atoms from the substrate surface, the material evaporates after leaving the substrate. There is no chemical reaction and only unmasked material is removed.

Chemical dry etching(vapor phase etching) **does not** use liquid reactants or etchants. Etching is achieved via a chemical reaction between etchant gases to corrode the silicon surface. The chemical dry etching is isotropic and is highly selective. Anisotropic dry etching has the ability to etch with finer resolution and higher aspect ratio than isotropic etching. Directional nature of dry etching blocks undercutting. The ions that are used in chemical dry etching are tetrafluoromethane, sulphur hexafluoride, nitrogen trifluoride, chlorine gas, or fluorine.

Reactive ion etching(RIE) the semiconductor industry's choice for fast, controlled high resolution etching. It uses both physical and chemical mechanisms to achieve high levels of resolution. The high energy collision between the high energy ions and the wafer atoms helps to dissociate the etchant molecules into more reactive species. Cations are produced from reactive gases which are accelerated with high energy to the substrate and chemically react with the silicon. The typical RIE gasses for Si are carbon hexafluoride, sulphur hexafluoride and boron chloride.

11.6 Chemical Vapor Deposition—CVD, APCVD, LPCVD, PECVD, MOCVD

Chemical vapor deposition [10] and its enhanced variants as Atmospheric Pressure CVD(APCVD), Low Pressure CVD(LPCVD), Phase Enhanced CVD(PECVD), metal oxide chemical vapor deposition(MOCVD) and ultra high vacuum chemical vapor deposition(UHVCVD) are very widely used in semiconductor fabrication *to deposit layers of material with extremely tight tolerances on thickness*. Before CVD can be used to deposit a layer of some material on a wafer, photolithography is used to demarcate those areas that need to get coated. Appropriate masking materials cover those wafer regions that do not need to be coated.

CVD is a fundamental semiconductor device manufacturing technology, each form customized for a different processing step. The different methods for CVD rely on differing process parameters e.g., different chemistries, substrate materials, temperatures, pressures, and deposition durations. However, all CVD processes utilize two main procedures:

- Decomposition reaction of a gaseous compound
- Combination reaction of one or more of those elemental parts on a substrate material.

For the decomposition and combination reactions to proceed (to generate the desired final result), the parameters that control the chemical reactions must be adjusted as required for the bonds to break and reform, similar to how water evaporates in hot or low-pressure air.

The three common chemical vapor deposition techniques used in semiconductor processing are: Atmospheric Pressure CVD(APCVD). Low Pressure CVD(LPCVD) and Phase Enhanced CVD(PECVD). While APCVD requires high operating temperatures (1000 °C). LPCVD requires a high vacuum chamber (10–1000 Pa) and moderately high temperature (lower than the ~ 1000 °C of APCVD). The required chemical reactions occur faster, and the vacuum chamber can be used to process a number of wafers at the same time, reducing processing costs.

PECVD is used for low temperature (100–400 °C) low volume chemical vapor deposition. During this process, cold plasmas are injected into the reaction chamber to boost the electron temperature of the chemical that is being deposited on the wafer. By changing the pressure, the cold plasma uses the energy of these electrons to quickly dissociate the molecules of the reactive gases. A layer of uniform thickness (of the material to be deposited), is laid down on the exposed (not covered by mask) wafer surface. For example, deposition of 3C and 6H silicon carbide(SiC) layers on silicon. A number of separate but related processes exist to achieve this task, and some allow n, p type dopant atoms to be included in the new layer to be deposited.

11.6.1 Metal Organic Chemical Vapor Deposition (MOCVD|OMVPE|MOVPE)

Metal organic vapor phase epitaxy (MOVPE), or organometallic vapor phase epitaxy (OMVPE) or metalorganic chemical vapor deposition (MOCVD) [11], is a chemical vapor deposition technique to deposit single or polycrystalline thin films on appropriate substrates. **Unlike MBE (Molecular Beam Epitaxy) which physically deposits layers on a substrate, MOCVD involves chemical reactions between the reactant gases and the substrate material to create the new layers**. MOCVD does not require high vacuum conditions, and the reaction chamber is first purged to high vacuum and then the reactant gases are pumped in to eliminate contaminants from polluting the layers to be deposited. The working pressure inside the chamber is 10–760 Torr. MOCVD is a preferred technique to fabricate devices which include thermodynamically metastable alloys—optoelectronic devices.

MOVPE|MOCVD|OMVPE uses reactants in their gaseous form. Ultra pure reactant gases in predetermined proportions are injected into the reaction chamber. The substrate is heated to a precalculated high temperature. The gas molecules coming in contact with the heated substrate react, depositing layer(s) of the reaction products on the substrate. *Crystal growth is favored by the surface reactions taking place on the substrate surface.* These reactions involve organic|metalorganics and hydrides which contain the required chemical elements for producing the desired layer on the substrate surface. This technique is used to deposit layers of compound semiconductor containing groups of the periodic table II, III, V and VI. Figure 11.5 illustrates how indium phosphide(InP) is deposited on a substrate, in three steps.

- Trimethyl indium and phosphine gases are injected into the purged reaction chamber in predetermined proportions.
- As the chamber contains only the reactants and the substrate, heating the surface triggers pyrolysis amongst the reactant gases, in contact with the wafer.
- The reaction product is deposited on the wafer.

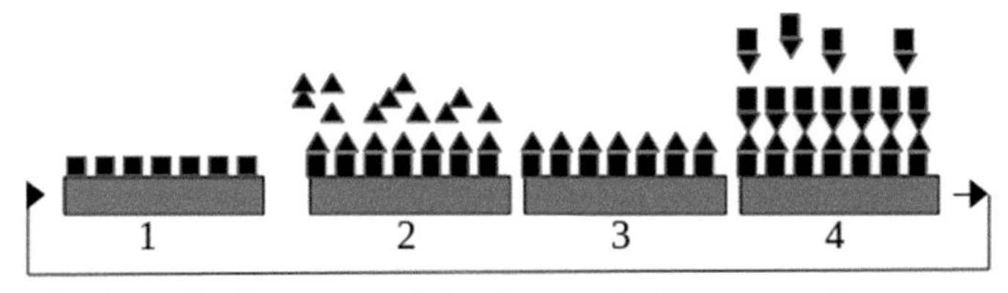

Fig. 11.5 Atomic layer Deposition

11.6.2 Ultra High Vacuum Chemical Vapor Deposition (UHVCVD)

Ultra-high vacuum chemical vapor deposition (UHVCVD) is used for epitaxial layer growth of silicon and related materials as uniformly strained epitaxial layers. Several implementations have been proposed, and of these the multi wafer UHVCVD technique is widely used. *The multi wafer UHVCVD is characterized by the absence of both hydrodynamic boundary layer effects and gas phase chemical reactions, as a result of the ultra high vacuum conditions in the reaction chamber.*

The key advantages of this material deposition technique are:

- Under ultra high vacuum conditions, the molecular mean free path comparable to the chamber dimensions.
- With almost non-existent inter molecule collisions, gas phase reactions are almost eliminated.
- Molecular flow transport eliminates hydrodynamic boundary layer issues.
- Deposited layer growth rate is determined by surface decomposition of the reactant molecules.

Figure 11.5 shows the apparatus used in multi wafer UHVCVD. Wafers are arranged on a wafer boat with inter wafer spacing of the order of a small fraction of the wafer diameter. The wafer boat is placed inside a quartz tube which is heated along its length by a furnace. Gases are injected into the quartz tube at one end and pumped out from the other end using ultra high vacuum pumps. The pumps have *load lock* mechanisms so that low partial pressures of important contaminants e.g., hydrocarbons, water vapor, and oxygen can be maintained. **This key feature is very important for low temperature growth because important contaminants cannot be desorbed at the growth temperature**. The reactants used are hydrides (e.g., silane, germane, diborane, methylsilane, and phosphine diluted in hydrogen or helium to obtain convenient flow rates. The chlorine containing reactants are avoided as chlorine persists in the growth system: undesirable in multi wafer UHVCVD as this results in non-uniform layer deposition. The operating temperatures are in the 500–600 °C range, with reaction chamber pressures in the micro Torr range.

As UHVCVD process is carried out at high temperatures, reducing the thermal budget is very important. The wafer surfaces must be contaminant free, to quickly initiate the epitaxial growth. A common method to create a hydrogen terminated surface is by dipping wafers in dilute hydrofluoric acid with no following rinse (*HF-last clean*). *Consequently, all wafer surface dangling bonds are terminated by hydrogen.* This treated surface does not react with oxygen or water vapor under ambient conditions. Hydrogen terminated wafers are loaded into the furnace and reactant gases are injected quickly to block hydrogen desorption. The wafer is never exposed to any temperature greater than the epitaxial growth temperature. Oxygen and water vapor cannot be allowed to adhere to the wafer surface that is to be processed with UNVCVD. Some other methods have also been implemented to achieve the same goal [12].

11.6.3 Atomic Layer Deposition (ALD)

A variation of CVD is atomic layer deposition, which allows atomic layers of materials to be deposited using chemical reactions. ALD exploits a binary sequence of self-limiting surface chemical reactions which results in films of solid material with Angstrom(s) thickness. It consists of cycles of alternating reactions with one ALD cycle depositing one "atomic layer." The number of deposition cycles provides a tight control on the thickness of the resulting film. The sequence of steps is shown in Fig. 11.6.

11.7 Molecular Beam Epitaxy (MBE)

Molecular Beam Epitaxy(MBE) [13] is used to the deposit thin film compound semiconductors, metals or insulators. MBE enables a tight control of compositional profiles by using a non-equilibrium thermodynamic process. "Epitaxy" is a combination of two Greek roots "epi" and "taxis" which mean to arrange upon. *So epitaxy is the arrangement of one or more thermal particles on top of a heated and ordered crystalline substrate to form a thin layer whose* ***crystalline structure matches that of the substrate,*** *despite their different chemical compositions e.g. SiGe/Si, AlGaAs/GaAs, CdTe/GaAs* etc., *The evaporated atoms and\or molecules do not interact with each other until they reach the substrate because of their long mean free paths—possible only under ultra high vacuum conditions.*

MBE is widely used to produce superlattice structures consisting of many alternate thin (individual layer thickness ~ 10 Angstrom) layers. *Single atom thickness layers (delta doping) can also be fabricated, i.e.,* ***the chemical species above and below the single atom thick impurity atom layer are different.***

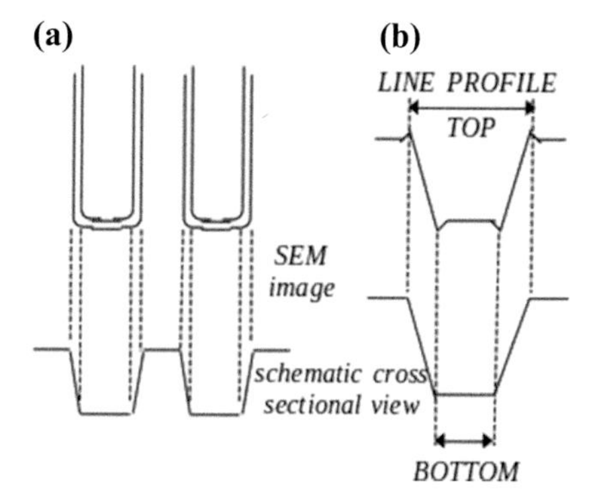

Fig. 11.6 a, b Critical Dimension Scanning Electron Microscopy and line profile

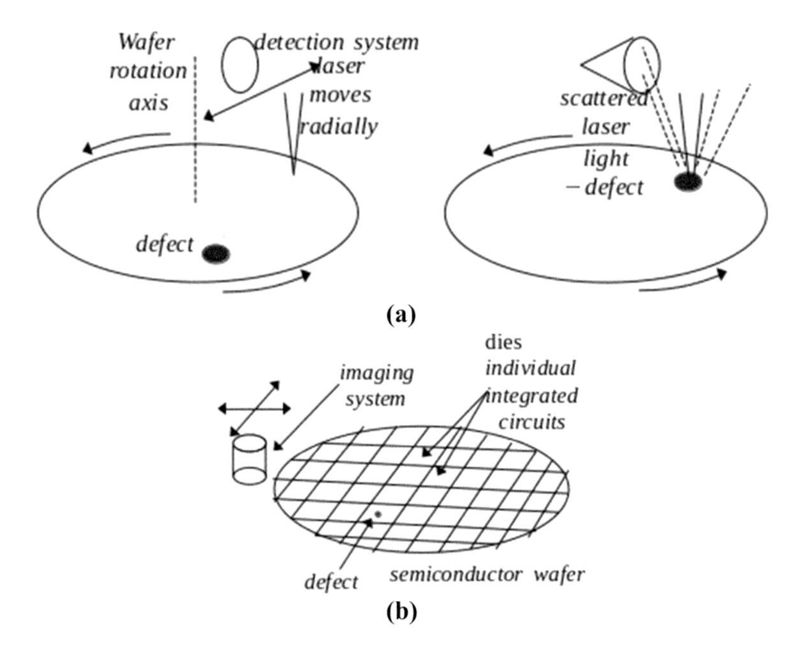

Fig. 11.7 **a** Unpatterned wafer defect detection. **b** Patterned wafer defect detection

MBE enables materials (atoms etc.,) to be deposited at ultra low rates (few Angstrom per second) by evaporating extremely pure solid elements heated in separate pyrolytic boron nitride effusion (Knudsen) cells. Element heating is done either electrically, or using electron beam (for high melting temperature elements).

As the evaporated elements reach the surface they have a probability of 'sticking'. An element atom, after getting attached to the substrate can react with the substrate in three ways.

- If it becomes adsorbed onto the surface through weak, physical, bonds e.g., Van der Waals forces, then it is said to be **physisorbed**.
- If it exchanges electron(s) with the substrate and gets chemically bonded to the surface, then it is said to be **chemisorbed**.
- If it does not desorb then the energy available from the heated substrate will cause it to diffuse about on the surface, promoting growth.

To form a crystallographically oriented material, incoming elements must arrive or diffuse to epitaxial sites and become chemisorbed at that site. Three (Frank van der Merwe, Volmer-Weber and Stranski–Krastanov) layer growth modes can occur depending upon the substrate temperature, the deposition rate and available surface energy. The Frank van der Merwe mode allows precision layer-by-layer growth. Volmer-Weber mode allows growth of three dimensional islands, and the Stranski–Krastanov is an combination|intermediate mode. Unwanted impurity elements in the growing layers are minimized because of the ultra high vacuum $< 10^{-10}$ Torr in the reaction chamber.

MBE is in the fabrication of ultra high frequency|power transistors(HBT, HEMT, SiC etc.,), optoelectronic devices (phototransistors|light emitting diodes|semiconductor lasers|thin film solar cells etc.,). The reaction chamber's interior ultra high vacuum conditions allow efficient diagnostic tools as RHEED (Reflection High Energy Electron Diffraction) to monitor the properties of the deposited layers.

The main MBE high vacuum evaporation chamber is connected to other chambers by gate valves, to avoid contamination of components and materials sources by both external and process gases, and minimize pressure in the main growth chamber. State-of-art MBE systems include a treatment chamber, and a load lock module to insert and remove wafers.

The key pumping system is a combination of Roots, ion, titanium sublimation and liquid nitrogen pumps. The vacuum quality (inside the reaction chamber) metrics are: the mean free path L and the partial pressure of the residual gas molecules. The highest admissible residual gas pressure value depends on L being larger than the distance from the outlet orifice of the beam source to the substrate surface ($L > 0.2$ m).

From the kinetic theory of ideal gases, numerical data for conventional MBE growth of Si, the maximum value of the residual gas pressure is 10^{-5} Torr, i.e. much higher than the typical ultra high vacuum conditions inside the MBE chamber. But the time required for the deposition of 1 mono layer of residual contaminants is 10^5 times the time needed to deposit 1 mono layer of film from the molecular beams. So very low deposition rates should be used in an ultra high vacuum MBE chamber. However, the partial pressure increases during deposition resulting from the increased heat load from the effusion cells and the substrate. Then a titanium filament sublimation pump reduces the residual gas pressure down to minimum permitted values. The vacuum chamber is baked at approximately 250 °C for extended periods of time, to minimize outgassing from the internal walls during the deposition process. A complete bake cycle, the vacuum chamber and the components have an approximate vacuum level of 10^{-10} Torr, using only the main ion pump. In addition, both carbon dioxide and water vapor pressure inside the chamber can be reduced to 10^{-11} Torr with liquid nitrogen cryogenic cooling The substrate holder is located a few centimeters from the effusion cell exit ports, along the center line of the system. The temperature of the substrate can be set during the deposition, starting with the room temperature, up to about 1400 °C, depending on the epitaxial process needed. It can also be heated before deposition for cleaning|surface reconstruction, and afterwards for various heat treatments. An included spectral mass analyzer detects the residual atoms or molecules.

11.8 Metrology in Semiconductor Device Manufacture

Metrology [14] and inspection of the wafers in-between successive processing steps of the complete semiconductor device manufacturing process is essential because of the large number of intermediate (approximately 400 to 600) steps in the overall manufacturing process of semiconductor devices. Completing this entire process for a batch of wafers takes 1- 2 months. If any defects occur early on in the process, these defects will propagate through the subsequent processing steps. *Metrology and inspection steps at critical points of semiconductor manufacturing process ensure that a certain target yield can be confirmed and maintained.*

Some typical inspection|measurement|verification steps include:

- Determining the line width and hole diameter of a circuit pattern at a specified location of a semiconductor wafer.
- Measurement of the thickness of the thin films on the surface of a semiconductor wafer.
- Verify the accuracy of the overlay. Measurement is performed to check the accuracy of the shot overlay of the first and second layer patterns transferred onto a wafer.

Metrology is measurement performed by factoring in errors and accuracy, along with the performance characteristics|tolerances of the measurement equipment. If pattern measurements do not satisfy predefined tolerances, the manufactured device will not operate as designed. Then re-work is essential. The two related concepts behind designating semiconductor metrology equipment as high-performance are:

A: **accuracy and precision**
B: **precision|trueness and repeatability**

Accuracy is a measure of how close the measured quantity is to the "ideal|true value", "precision|repeatability" is a measure of the variations of the measured values of a parameter from multiple measurements.

These two key concepts can be summarized as:

- A small variation (from the ideal value) in a measurement indicates high precision.
- A mean value close to the ideal|true value means high (good) accuracy.

"Small variation" indicates tight tolerance, while "close" depends entirely on the tolerance (e.g., 5%, 2.5%) set to denote the permissible variation from the ideal value.

11.8.1 Critical Dimension Scanning Electron Microscopy

Exploiting the same physics as a scanning electron microscope, the critical dimension scanning electron microscope (CD-SEM) is designed to **measure separation of lines on the sample wafer** with high accuracy and precision. The special features of any of these fully automated measurement machines are:

- Incident electron beam energy is restricted to a maximum value of 1 keV to minimize damage to the wafer surface being inspected.
- Measurement accuracy and repeatability is guaranteed by having very precise high magnification imaging system, such that the measurement repeatability is 1% 3σ of measurement width.

The sequence of processing steps for a CD-SEM line profile measurement are simple.

- Using user input, the CD-SEM system positions the position indicator (often called the cursor) at the required measurement location in the scanning electron microscope (SEM) image.
- The line profile (signal that identifies the changes in the topological profile of measurement feature) of the specified measurement position is measured.
- The line profile data is used to measure the dimensions of the specified location, by counting the number of pixels in the measurement.

Figure 11.7a and b demonstrates these concepts. The two key semiconductor wafer processing steps after which CD-SEM is performed are:

- Examination of photoresist pattern after deposition, before any other wafer processing step.
- Measurement of contact hole and via hole diameters and metal wiring width after etching.

11.8.2 Wafer Defect Detection

A wafer defect inspection system detects both physical and pattern defects (random and systematic) on unprocessed and patterned wafers, as well as determining the x,y coordinates of the defect locations.

- A random defect is a result of events beyond control—damage due to a foreign particle landing on the wafer

- A systematic defect results from events under control—e.g., defects in mask and exposure process, and so will occur at the same position on each wafer of a given batch of wafers being processed.

As a wafer defect detection system identifies faults by comparing the images of circuit patterns on adjacent dies on a wafer under test, systematic defects sometimes cannot be identified easily in conventional wafer defect detection systems. The unpatterned and patterned defect detection schemes are shown in Fig. 11.7a and b. The three common wafer defect detection systems are electron beam, bright and dark field, all using the exploiting the basic physics.

In unpatterned wafer inspection, a finely collimated laser beam is moved back and forth radially across a rotating wafer. When the laser beam is directly incident on a defect|particle of a rotating wafer, the light will be scattered and detected by a detector. From the wafer rotation angle and the radius position of the laser beam, the position coordinates of the particle/defect are calculated and registered. Defects on a mirror wafer include crystal defects and particles. While the bright field inspection system is for detailed examination of pattern defects, the dark-field inspection system is designed for high speed defect inspection of a large number of wafers.

In the electron beam inspection system, electrons incident on the surface of the wafer, and the emitted secondary electrons and backscattered electrons are detected. This defect detection system detects the amount of the secondary electrons as an image contrast (voltage contrast) according to the conductivity of the device's internal wiring. If the conductivity at the bottom of the contact hole of the high aspect ratio is detected, the SiO2 residue of ultra-thin thickness can be detected.

Patterned wafer inspection systems can be electron beam irradiation, or bright|dark field illumination types. A patterned semiconductor wafer consists of adjacent identical integrated circuits, and the possibility of a random defect (e.g., impurity particle dropping on a wafer surface) occurring at the same position in all wafers of a batch is negligible. The pattern on the wafer is captured along the integrated circuit array by electron beam or light. Defects are detected by comparing, (using highly accurate and reliable computer vision systems) between image (1) of the wafer die under inspection and previously captured image (2) of the adjacent die. If there are no defects, the result of the subtraction of Image 2 from Image 1 by digital processing will be zero and no defects are detected. **If the resulting image obtained from subtraction of the image of die 1 from the image of die 2, contains anything, then a defect has been detected**. The defect position coordinates are registered.

11.8.3 Review Scanning Electron Microscopy

Defect Review Scanning Electron Microscopy is the immediate next wafer defect identification step after initial defect detection, using an inspection system, as examined earlier. *Using the output of the initial inspection (as discussed in previous sub-sections) which consists of the identified defect position coordinates on the wafer, the review scanning electron microscope obtains detailed images of these defect sites and classifies each defect.*

The Review Scanning Electron Microscope, similar to the defect inspection system, detects the defect by comparing the image of circuit pattern of a selected die with the image of the circuit pattern of the adjacent dies and obtains the correct position of the defect. The defect is then moved to the center of the field of view and an enlarged image is created. *The defect is then classified.*

Automatic Defect Review (ADR) automatically obtains image of the identified defect using the defect information (coordinates, etc.) obtained in defect inspection. The data is stored and arranged into a database. The image information of the defect stored in the image server is classified according to the cause of the defect by the classification software based on the predetermined rules and is then restored in the classification server. The classified information is sent to Yield Management System (YMS) and the central database of the integrated circuit manufacturer so that it can be used in the failure and defect analysis.

11.9 Thin Film Thickness Measurement Techniques

Common high performance semiconductor device consists of several layers of semiconducting materials each with different electrical properties and thickness. Since the performance characteristics of the device are dependent on the thicknesses of these layers, it is essential that the thickness of each layer be monitored and measured accurately during production. *Selecting a method for a specific measurement is tricky, since while one method may be ideal for measuring the thickness of one type of material, it might be inappropriate for another material.* So thin film thickness measurement technique for a specific case depends on what methods are eligible for the material in question (taking into account film properties as surface roughness, thickness range, properties of film and substrate etc.,) what additional information other than the thickness of the film is to be determined with the analysis, and what the budget is. The key constraints are:

- Thickness measurement must be non-contact and non-destructive.

11.9.1 Oxide Layer Thickness Measurement

Oxide layers are transparent films. If light is irradiated onto the wafer and reflected, various properties of the reflected light wave are changed—easily detected with sensitive metering devices. Identification of vertically deposited multiple layers, each with different optical properties can also be done using optical measurement techniques. To monitor the film thickness across the wafer, several measuring points are quantified (e.g. 5 points on 150 mm, 9 points on 200 mm, 13 on 300 mm wafers). Multiple measurements, each at different locations ensures absolute thickness as well as thickness uniformity across the wafer. If the deposited layer is too thick or too thin material has to be removed (e.g. by etching) or deposited again. The two common techniques are interferometry and ellipsometry.

Interferometry is based on constructive and destructive interference of light—**in phase light waves reinforce, while out of phase light waves cancel eachother.** This phenomenon is used during semiconductor production for measuring translucent films. If light is irradiated onto a wafer some beams of light are reflected on top of it and some penetrate into the film. The latter will be reflected from the bottom of this layer or penetrate into another layer beneath and so on. A polychromatic light beam is shined on the wafer and depending on the film thickness of the radiographed layer, the light waves interfere either constructively or destructively, resulting in a characteristic interference pattern. A photometer can analyze the reflected light and calculate the film's thickness. Interferometry can be used on films whose thicknesses are greater than at least one fourth of the wavelength of incident light.

Ellipsometry is measurement of light polarization. Linearly polarized light is irradiated on the test surface. During reflection of light on top of the wafer or on interfaces of two layers, the light's polarization is changed, easily detected with an analyzer. Combining known optical properties of the film (e.g. angle of refraction, how much of the incident light is absorbed etc.,), of the incident and reflected polarized light, the film thickness can be calculated. Unlike interferometry, ellipsometry measurements can be used for films with thicknesses less than one fourth of the wavelength of the incident light.

11.9.2 Ray Reflectometry, Scanning and Transmission Electron Microscopy

Layered materials can be analyzed with X-ray reflectometry, enabling the estimation of total film thickness, as well as thickness, density and surface roughness of individual layers. The method is suitable for materials with a thickness less than 250 nm, optimally under 100 nm. For accurate measurements, the thickness of the material must be at least one order of magnitude greater than the surface roughness of the film. T*he results obtained with this method are accurate and meaningful only if the estimated composition and structure* *of the sample under test are known beforehand. This is essential since the method relies on fitting the experimental X-ray reflectometry data to corresponding simulated layer model data. As expected, analysis can result in large errors if the composition of the sample is completely unknown.*

Cross-sectional Scanning Electron Moicroscopy (SEM) is ideal for measuring the thickness of semiconductor thin films (single- and multi-layer materials). It also provides information about the surface morphology and elemental composition of the sample. The method is suitable for conductive and semiconductor materials with a thicknesses between 100 nm and 100 μm. Non-conductive materials can also be analyzed with a small modification, i.e., the non-conducting surface has a thin layer of conductive material deposited on the surface. If in addition to thickness information, information on the elemental composition of the film is needed, an SEM equipped with an EDS detector (energy dispersive spectroscopy) is required. The EDS detector measures the x-rays emitted from the sample when the electrons interact with it. As the detector analyses the x-ray spectrum, it recognizes the spectra of individual elements and compounds and allows for their identification and quantification.

Cross-sectional Transmission Electron Beam Microscopy (TEM) is also commonly used to measure thickness of and analyze properties of conductive and semiconductive films (single and multilayer). The thickness range for accurate results varies between few nanometers to 100 nm. In addition, sample thickness can be customized with focussed ion beam. Just like cross-sectional SEM, an additional attached EDS detector enables extraction of information about the elemental composition of the sample. Unlike cross-sectional SEM, the high voltage beam can burn|damage some materials, making it inappropriate for polymeric and organic materials.

11.10 Maintaining Super Clean, Sanitized Semiconductor Fabrication Facilities [15]

As expected, modern semiconductor devices are manufactured in cleanrooms. *With semiconductor device internal physical dimensions getting scaled down low nanometers, it is imperative that ultra clean and sanitized environments exist inside semiconductor device manufacturing facilities to ensure that air particulate levels are within tight tolerances, and de-ionized, distilled water is used for surface cleaning of wafers during intermediate production steps.* Wafer surface contamination has to be eliminated. Process engineers and technicians wear head-to-toe body suits, with their mouth, nose and head completely covered.

To address these issues, real time monitoring systems have been implemented to monitor and eliminate semiconductor wafer contamination during each production step.

Key component of any real time cleanroom cleanliness monitoring system are airborne and liquid particulate monitors. Portable and fixed airborne particle counters (for particle sizes 0.1–0.3 μm), for gas manifold systems are widely used for cleanroom certification and routine monitoring. Equipped with communication network ports, these devices enable real time data logging and eliminate paper use inside the cleanroom.

Wafer cleaning between successive processing steps uses large volumes of de-ionized and distilled water. These water systems need to be monitored both inline and offline using liquid particle counters. State-of-art liquid particulate monitors have resolution in the 50 nm range.

The ISO 14644–1:2015 standard specifies the classification of air cleanliness in terms of concentration of airborne particles in cleanrooms and clean zones. This standard applies only to particle populations having cumulative distributions based on lower limit size between 0.1 to 5 μm. Airborne particle concentration is estimated using light scattering (discrete) airborne particle counters. The airborne particle sizes must be equal to and|or greater than the specified sizes, at designated sampling locations.

The ISO 14698:2003 standard specifies the principles and basic methodology of a formal system of biocontamination control for estimating, monitoring and controlling biocontamination inside a cleanroom. It specifies the methods used for monitoring risk zones in a consistent way and for applying control measures appropriate to the degree of risk involved.

The physical task of removing airborne particulate contaminants is achieved with HEPA filters [16]. **HEPA (High Efficiency Particulate Air) filter** is a type of pleated mechanical air filter, which can theoretically remove at least 99.97% of dust, pollen, mold, bacteria, and any airborne particles with a size of 0.3 microns (μm). The specified 0.3 microns diameter is the worst case—the most penetrating particle size (MPPS). Particles that are larger or smaller are trapped with even higher efficiency. Using the worst case particle size results in the worst case efficiency rating (i.e. 99.97% or better for all particle sizes). All air cleaners require periodic cleaning and filter replacement to function as per specifications.

Minimum Efficiency Reporting Values, or MERVs, report a filter's ability to capture larger particles between 0.3 and 10 microns (μm). This value, developed by by the American Society of Heating, Refrigerating, and Air Conditioning Engineers is helpful in comparing the performance of different filters. A high MERV [16] rating indicates that a filter under test is the ideal one at trapping specific types of particles.

References

1. https://www.pfeiffer-vacuum.com/en/markets/semiconductor/ion-implantation/source-beamline-end-station/.
2. https://www.sciencedirect.com/science/article/abs/pii/0168583X91962083.
3. https://www.leybold.com/en-in/knowledge/vacuum-fundamentals/vacuum-generation/how-does-a-roots-pump-work.

4. https://www.vacuumscienceworld.com/blog/multistage-roots-vacuum-pumps-working-princi ple.
5. https://sens4.com/vacuumunits.html.
6. https://www.mks.com/n/ion-implantation.
7. https://www.microchemicals.com/technical_information/ion_implantation_photoresist.pdf.
8. https://cleanroom.groups.et.byu.net/alignment.parts/Alignment_Tutorial.pdf.
9. https://www.ece.ucdavis.edu/~anayakpr/Papers/Wet%20and%20Dr%20Etching_submitted. pdf.
10. https://www.arrow.com/en/research-and-events/articles/what-is-chemical-vapor-deposition.
11. https://www.azom.com/article.aspx?ArticleID=11585.
12. https://users.ece.cmu.edu/~dwg/research/UHVCVD.pdf.
13. https://www.researchgate.net/publication/256143230_Basics_of_Molecular_Beam_Epitaxy_ MBE_technique/link/57335adf08aea45ee838f482/download.
14. https://www.hitachi-hightech.com/global/en/knowledge/semiconductor/room/manufacturing/ metrology-inspection.html.
15. https://www.iso.org/obp/ui/en/#iso:std:iso:14698:-1:ed-1:v1:en.
16. https://www.epa.gov/indoor-air-quality-iaq/what-hepa-filter.

12 Designing Transistors for Specific Applications

12.1 Transistor Capacitances

Transistor capacitors [1-5] arise because of built-in interfaces between the three regions—base, collector, emitter for a bipolar transistor, and drain, gate, and source for a field effect transistor. The generic structure of a transistor is in Fig. 12.1a Transistor capacitances are defined in terms of current, and time rate of change of voltage as:

$$C = \frac{Q}{V} = \frac{i}{\frac{\partial V}{\partial t}} = \frac{i\partial t}{\partial V} \tag{12.1a}$$

12.1.1 MOSFET Capacitances (SPICE Level 1 Model, Triode Region)

The built-in total(*unlike per unit area*) capacitances inside a metal oxide semiconductor field effect transistor are classified as extrinsic and intrinsic[6–22]. With reference to Fig. 12.1b $C_{DRAIN,GATE}$, $C_{GATE,SOURCE}$, $C_{DRAIN,SOURCE}$, $C_{INTR}(transcapacitance)$ are the intrinsic capacitances, while the rest.

$C_{GATE,DRAIN,OVERLAP}$, $C_{GATE,SOURCE,OVERLAP}$, C_{JDEAIN}, $C_{JSOURCE}$ are labelled extrinsic capacitances.

Using the following definition of transcapacitance, the drain current is re-written as:

$$C_{\text{TRANSCAP}} = C_{\text{GATE, DRAIN}} - C_{\text{DRAIN,GATE}}\, i_{\text{DRAIN}} = \frac{-C_{\text{DRAIN , GATE}}\, \partial C_{\text{GATE, DRAIN}}}{\partial t} \tag{12.1b}$$

A. Banerjee, *Semiconductor Devices*, Synthesis Lectures on Engineering, Science, and Technology, https://doi.org/10.1007/978-3-031-45750-0_12

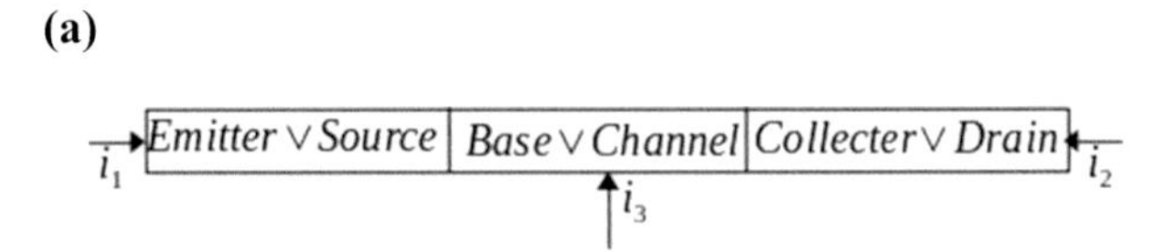

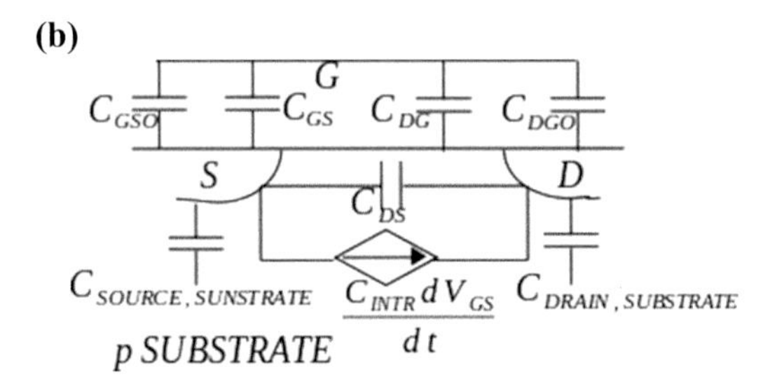

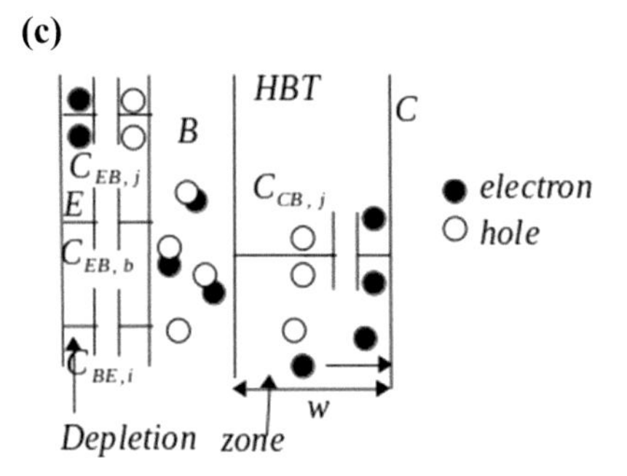

Fig. 12.1 **a** Generic transistor structures. **b** Extrinsic and intrinsic MOSFET capacitances. **c** Electron, hole movement and built-in capacitances in a heterogeneous bijunction transistor (HBT)

The drain source capacitance for a MOSFET in the linear region arises when the drain source voltage increases widening the depletion zone, followed by accumulation of negative substrate charge. Consequently, a positive charge develops in the channel. Because the channel charge is related to both the drain and source charges, the changes in these charges are positive. So the source drain capacitance is negative.

Analytically, it is difficult to derive expressions for the gate-source and gate-drain intrinsic capacitances, but under the simplifying conditions of the triode region operation, the approximate expressions for these two capacitances are:

$$C_{\text{GATE ,SOURCE}} = \frac{2C_{\text{OXIDE}}LW}{3}\left(1 - \frac{\left(V_{\text{GATE, DRAIN}} - V_{\text{THRSH}}\right)^2}{\left(V_{\text{GATE,SOURCE}} + V_{\text{GATE, DRAIN}} - 2V_{\text{THRSH}}\right)^2}\right) \tag{12.1c}$$

$$C_{\text{GATE, DRAIN}} = \frac{2C_{\text{OXIDE}}LW}{3}\left(1 - \frac{\left(V_{\text{GATE, DRAIN}} - V_{\text{THRSH}}\right)^2}{\left(V_{\text{GATE, SOURCE}} + V_{\text{GATE, DRAIN}} - V_{\text{THRSH}}\right)^2}\right) \tag{12.1d}$$

The extrinsic capacitances consist of overlap capacitances, junction capacitances and the gate substrate overlap capacitance. For the gate drain overlap capacitance, the total gate overlap charge, from Gauss's law and the corresponding overlap capacitance is:

$$Q_{GATE,OVERLAP} = \frac{-\varepsilon_{OXIDE} A_{OVERLAP} V_{DRAIN,GATE}}{t_{OXIDE}} \quad C_{OVERLAP,GATE,DRAIN}$$
$$= \frac{-\partial Q_{GATE,OVERLAP}}{V_{DRAIN}} = \frac{\varepsilon_{OXIDE} A_{OVERLAP}}{t_{OXIDE}} \quad (12.1\ e,f)$$

12.2 Heterogeneous|Homogeneous Junction Bipolar Transistor Capacitances

Due to the similar internal layered structure of both heterogeneous|homogeneous bipolar transistors (Fig. 12.1c) the identical internal capacitances are present in both these transistors. As the base controls collector to emitter current flow, all voltages are referenced with respect to the base, e.g., base–collector and base-emitter voltages(V_{BC}, V_{BE}). The cross-sectional areas of both the collector and emitter are larger than that of the gate of FETs, so intrinsic capacitances are very important to the operation of bipolar transistors. The focus is on Npn|npn transistors.

When the base voltage changes, the emitter base potential barrier is lowered, and the depletion zone length region shrinks. The n side shrinks because electrons flow in from the emitter lead and neutralize the charge of some of the ionized donors. The emitter base junction capacitance is:

$$C_{EMITTER,BASE,JUNCTION} = \frac{-\partial Q_{EMITTER,JUNCTION}}{\partial V_{BASE,EMITTER,JUNCTION}} \quad (12.2a)$$

where $Q_{EMITTER,JUNC}$ is the total charge of the electrons that have entered from the emitter contact, and are present at the junction edge. Similarly, holes from the base contact enter to reduce the width of the depletion region in the base side of the junction.

$$C_{BASE,EMITTER,JUNCTION} = \frac{-\partial Q_{BASE,JUNCTION}}{\partial V_{EMITTER,BASEJUNCTION}} \quad (12.2b)$$

As the base emitter junction is planar, the junction capacitance is:

$$C_{EMITTER,BASE,JUNCTION} = \frac{\varepsilon_{SEMICONDUCTOR} A_{EMITTER}}{W} \quad (12.2c)$$

The base storage capacitance is the sum total of the charges of the electrons in the base injected from the emitter, in response to base-emitter voltage change, in order to establish a new electron profile in the base, from which the new collector current is derived.

$$C_{EMITTER,BASE,STORAGE} = \frac{-\partial Q_{EMITTER,STORAGE}}{\partial V_{BASE,EMITTER}} \tag{12.2d}$$

With reference to the hole current, the same capacitance can be re-expressed as:

$$C_{BASE,EMITTER,STORAGE} = \frac{-\partial Q_{BASE,STORAGE}}{\partial V_{EMITTER,BASE}} \tag{12.2e}$$

Using the total charge stored in the quasi neutral base region, and using appropriate boundary conditions, the base storage capacitance is:

$$C_{EMITTER,BASE,STORAGE} = \frac{n_{0p} q^2 A_{EMITTER} e^{\frac{V_{BASE,EMITTER}}{V_{THERMAL}}}}{V_{THERMAL}} \tag{12.2f}$$

The emitter storage capacitance is unimportant in HBTs because injection of holes into the quasi-neutral emitter from the base is blocked by the large potential resulting from the valence band offset.

The capacitance related to the transit of electrons across the wide depletion layer at the reverse biased base collector junction in the active mode of operation is very important. *The electrons come in from the emitter lead, but the holes these electrons draw in from the base are smaller in number* and this effect is very important in estimating the HBT's transition frequency, a key performance characteristic for ultra high frequency transistors. This capacitance is the base emitter transit capacitance:

$$C_{BASE,EMITTER,JUNCTION} = \frac{-\partial Q_{BASE,hole}}{\partial V_{EMITTER,BASE}} \tag{12.2g}$$

The capacitance is in reference to base emitter, as the electrons that trigger its creation are from the emitter. Assuming an uniform electric field in the base collector depletion region of width w, it is maintained by a constant base collector voltage. *Then image charges $+Q$ and $-Q$ form at the edges of the depletion region, resulting from positive charges flowing through the external collector base circuit, and building up at the edges of the depletion zone.* Using the fact that the collector voltage is constant along with Gauss's law, the image charges are expressed as:

$$q_{electron} + q_0 + q_w = 0 \quad (Q - q_0)x + (Q + q_w)(w - x) = Qw \quad q_0 = -q_{electron}\left(1 - \frac{x}{w}\right) \quad q_w = \frac{-q_{electron,x}}{w} \tag{12.2.h,i,j}$$

As the base collector voltage is constant, the electrons have a constant drift velocity, so that the base transit collector transit capacitance is:

$$C_{BASE,EMITTER,JUNCTION} = \frac{\partial Q_{BASE,hole}}{\partial V_{EMITTER,BASE}} = \frac{g_m w}{2 v_{DRIFT,electron}} \tag{12.2k}$$

In the above equation, the term $2v_{DRIFT,electron}$ in the denominator implies that the signal velocity exceeds the drift velocity—a result of image charges at the collector before the electrons reach the collector. *Mobile electrons rearrange the electric field, creating image charges. Image charges are not present in a field effect transistor at the drain—the channel is two dimensional, and is close to the gate.*

In active operation mode, the base collector junction is reverse biased and very few minority carriers are injected across the junction—no charges are stored. The only collector base capacitance is the junction capacitance:

$$C_{COLLECTOR,BASE} = \frac{-\partial Q_{COLLECTOR,JUNCTION}}{\partial V_{BASE,COLLECTOR}} = \frac{\varepsilon_{EMICONDUCTOR} A_{COLLECTOR}}{w} \quad (12.2I)$$

12.3 Transistors for Computer Memory

All computer memory integrated circuits are made from MOSFETs [1, 3, 5–7]. The two computer types dynamic ransom access(DRAM) and flash. While the DRAM uses the standard n channel p substrate(NMOS) as the basic memory cell, flash memory uses the floating gate MOSFET. All memory is organized in rectangular arrays (Fig. 12.1a).

12.3.1 Transistors for Dynamic Random Access Memory (DRAM)

The DRAM uses a n channel p substrate as the memory cell. The MOSFET acts as a switch between the bit line and a parallel plate storage capacitor. The charge stored on the capacitor determines the state (ONE or ZERO) of the cell (Fig. 12.2b). The storage capacitor is connected to the source and has a much larger capacitance than that of the source body np junction. The stored charge connects to the bit line via the inversion layer in the channel when the FET is switched on. The bit line is floating during this part of the operation, so its voltage could change in response to the new charged state of the bit line capacitance. All voltage changes are detected by a sense amplifier attached to the bit line. *In this discussion no voltage change means stored logic value ONE, whereas a slight decrease in bit line voltage is associated with a stored logic ZERO.*

In Fig. 12.2b the *plate* electrode of the capacitor is held at $\frac{V_{DD}}{2}$. Labelling the actual source potential as V_{SRC}, the basic charge-sharing equation is:

$$C_{STORE}\left(V_{SRC} - \frac{V_{DD}}{2}\right) + C_{BIT}V_{BIT} = (C_{BIT} + C_{STORE})V_{BIT}^{P} \quad (12.3a)$$

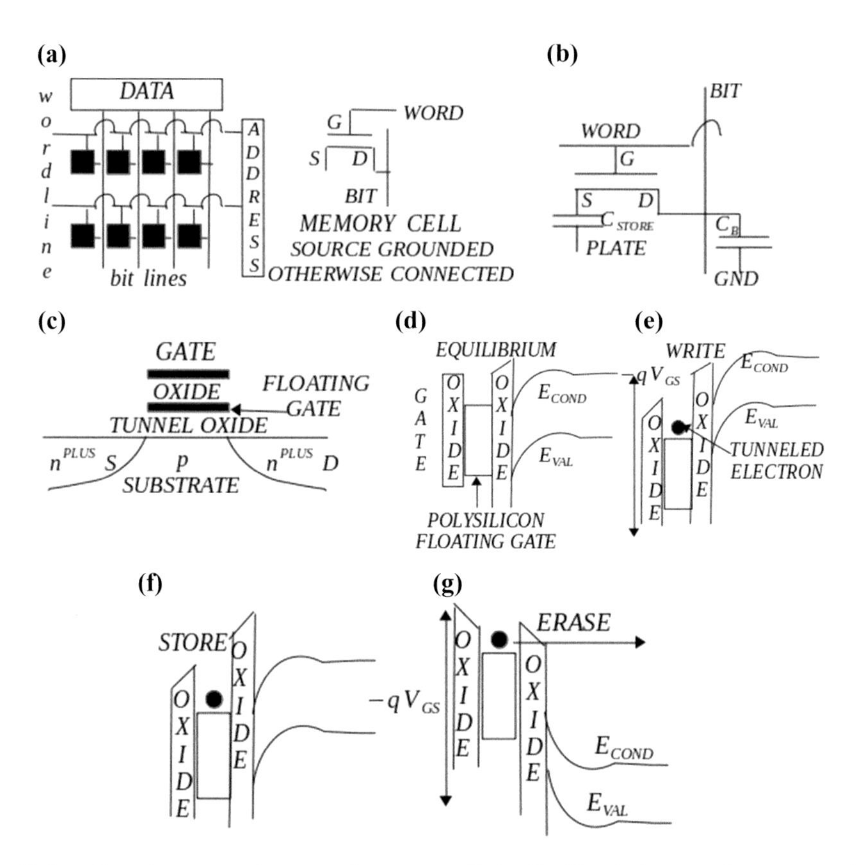

Fig. 12.2 **a** Generic memory layout and memory cell details. The dark rectangles represent individual memory cells. **b** DRAM memory cell. **c** Floating gate NMOS for flash memory. **d**, **e** Energy band diagram for equilibrium and write operation of a floating gate flash memory cell. **f**, **g** Energy band diagrams for store and delete|erase operations of single level floating gate flash memory cell

When writing a ONE, the bit line voltage is raised by internal, charge pumping circuitry to.

$(V_{DD} + V_{THRSH})$—i.e., when the word line is enabled, turning on the n channel pass transistor, the source node rises to V_{DD}. Thus,

$$V_{STORE} = \frac{V_{DD}}{2} \tag{12.3b}$$

To read this ONE, the bit line is precharged to $\frac{V_{DD}}{2}$ and left floating as the transistor is turned on. Charge sharing occurs, and the new bit line voltage is:

$$V_{BIT}^{P} = \frac{V_{DD}}{2}\left(\frac{C_{STORE} + C_{BIT}}{C_{STORE} + C_{BIT}}\right) = \frac{V_{DD}}{2} \tag{12.3c}$$

The bit-line voltage is unchanged, and this is interpreted as a ONE.

To write a ZERO to the cell, V_{BIT} is set to 0, the word line is enabled, turning on the pass transistor, and the source node falls to 0, implying that:

$$V_{STORE} = \frac{-V_{DD}}{2} \tag{12.3d}$$

To read this ZERO, again the bit line is precharged to $\frac{V_{DD}}{2}$ and left floating as the transistor is turned on. Charge sharing results in the bit-line voltage changing to:

$$V_{BIT}^{P} = \frac{V_{DD}}{2}\left(\frac{C_{STORE} - C_{BIT}}{C_{STORE} + C_{BIT}}\right) < \frac{V_{DD}}{2} \tag{12.3e}$$

The physical layout of the DRAM transistors, the word and bit lines have evolved rapidly over the years, especially with sub-micron gate length NMOS transistors replacing the previous micron gate length NMOS transistors.

12.3.2 The Floating Gate NMOS Transistor and Flash Memory

Unlike the DRAM memory, flash memory can retain the logic level(ONE|ZERO) in the memory cell even after the power is switched off, made possible by the floating gate NMOS (Fig. 12.2c).

The MOSFET in a flash memory cell has two polysilicon gates, one of which is completely surrounded by insulating oxide, *floating electrically*. Programming the cell is achieved by applying an appropriate voltage to the top control gate, via the word line. With an appropriate applied voltage, electrons from the channel region punch through the oxide layer below the floating gate and come to rest near the floating gate. The memory cell is now storing a logic ONE. With another appropriate applied voltage, the electrons are pushed out of the oxide layer below the floating gate, and so at the end a logic ZERO is stored on that memory cell. The state of charge on the floating gate determines the threshold voltage of the transistor and the magnitude of the drain current that will pass into the bit line when the cell is probed. A current sensor in the bit line circuitry detects this current, and interprets from it the state of the memory cell.

The floating-gate charge density Q_{FLOAT} controls the channel charge density $Q_{electron}$. Applying Gauss's Law, or equivalently directly from the conservation of charge:

$$Q_{FLOAT} + Q_{electron} + Q_{CONTROLGATE} = 0 \tag{12.3f}$$

Thus, to maintain a given $Q_{electron}$ in the presence of a change in floating gate charge ΔQ_{FLOAT}, the change in control gate charge density is $-\Delta Q_{FLOAT}$. The associated change in control gate voltage follows from Gauss's Law, and is the change in threshold voltage:

$$\Delta V_{THRSH} = \frac{\Delta Q_{FLOAT,GATE}}{C_{OXIDE,TOP}} = \frac{-\Delta Q_{FLOAT} t_{OXID,TOP}}{\varepsilon_{OXIDE}} \quad (12.3g)$$

The word line voltage to address the cell is not sufficient to alter Q_{FLOAT}, so reading the memory cell contents is non-destructive. If the read voltage is less than ΔV_{THRSH}, then the presence of charge on the floating gate puts the transistor in the OFF state. **That is, electrons stored on the floating gate is interpreted as the storage of logic ZERO in the memory cell. The read voltage is such that it is greater than the threshold voltage when** $Q_{FLOAT} = 0$. **The transistor is switched on(ON state) and drain current is received at the bit line: this is interpreted as storage of a logic ONE.** Each memory cell carries one bit of information. This is single level cell operation. *If the floating gate charge could be precisely controlled, then a variable threshold voltage would be obtained. This means that different values of read voltage would be required to turn-on the transistor, depending on the amount of floating charge present. This leads to multi level cell operation and the option of increasing the number of stored bits per cell. If the drain current has a maximum and minimum value.*

($i_{D,MAX}$, $I_{D,MIN}$*), change that can be detected by the sensing circuitry is* ΔI_D, *then the number of possible bits is.*

$$n_{BIT} = \log_2\left(1 + \frac{I_{MAX}}{\Delta I}\right) \quad (12.3h)$$

Obviously, fabrication of a multi level flash memory cell is a challenging task.

The energy band diagrams corresponding to each of the four states(equilibrium, write, store and delete) of a single level floating gate flash memory cell are in (Figs. 12.2d, e, f, g).

Thus for DRAMs, NMOS transistors must be sized so that a very large number can fit on a single die—else state-of-art 8GB etc., *DRAM capacities would be impossible. For flash memories each NMOS transistor must have at least one floating gate, in addition to be being able to be densely packed on the same die.*

12.4 Transistors for Ultra High Frequencies (RF|Microwave)

While the large signal equivalent circuit model is essential for analyzing the DC and switching applications of a transistor, the small signal model is essential for analysing the ultra high frequency performance characteristics [1, 8–16]. The two key ultra high frequency performance metrics of a transistor are the maximum and transition frequencies f_{MAX}, f_T. **A RF|microwave transistor has unity current gain at the transition frequency. At the maximum frequency the RF|microwave transistor has unity power gain.** *The small signal model is the linearized version of the large signal model.*

For a generic transistor, the total current at any of its three terminals has both a DC and a AC component expressed as:

$$i_j(t) = i_{j,DC} + i_{j,AC}(t)$$
$$j = COLLECTOR \vee DRAIN, BASE \vee GATE, EMITTER \vee SOURCE \tag{12.4a}$$

For analysis purposes, j = 2 and the current to be analyzed is the collector current(BJT, HBT) or drain current(FET, HEMT). All voltages are referenced with respect to the the emitter(BJT, HBT) or source(FET, HEMT). Like the current, the voltage at any of the three terminals(base,collector, emitter|gate, drain, source) has both a DC and AC component. These currents are linearized by expanding the corresponding expression in a Taylor's series.

$$\begin{aligned} i_2 &= I_2\left(V_{21,DC}v_{21,AC} + V_{31,DC}v_{AC,31}\right) \\ &= I_2\left(V_{21,DC}, V_{31,DC}\right) + \frac{\partial I_2 v_{21,AC}}{\partial V_{21,DC}} + \frac{\partial I_2 v_{31,AC}}{\partial V_{31,DC}} \end{aligned} \tag{12.4b}$$

Identifying:

$$g_{22} = \frac{\partial I_2}{\partial V_{21,DC}} \quad g_{23} = \frac{\partial I_2}{\partial V_{31,DC}} \quad i_2 = I_2 + g_{22}v_{21} + g_{23}v_{31} \tag{12.4c}$$

where I_2 is the DC current. Using identical analysis,

$$\begin{aligned} g_{33} &= \frac{\partial I_3}{\partial V_{31,DC}} \quad g_{32} = \frac{\partial I_3}{\partial V_{21,DC}} \\ i_3 &= I_3 + g_{33}v_{31} + g_{32}v_{21} \end{aligned} \tag{12.4d}$$

Restricting the Taylor series expansion to produce a linear relationship, means that the second and higher order terms can be neglected. For FETs, this assumption is valid as $i_2 = i_D$ which is already linear in $v_{31} = v_{GATE,SOURCE}$ in the saturation region of operation, and $v_{21} = v_{DRAIN,SOURCE}$ in the linear|resistive region, and the exponent of any relations in the saturation regime never exceeds 2. However, for bipolar devices, $i_2 = i_{COLLECTOR}$ which depends exponentially on v_{31}, so the linearization is possible if the magnitude of the small signal is $v_{31} \stackrel{\text{def}}{=} v_{BE} \ll \frac{2k_BT}{q}$ which is 40 mV at 300 K.

The equivalent circuit representing the small-signal components is shown in Fig. 12.3a. Often the two generator equivalent circuit is converted to a one generator circuit Fig. 12.3b.

The conductor in the top branch of the circuit is real. g_{32} is called the reverse feedback conductance. Its magnitude is much smaller than the other conductances, so it is common practice to name the other conductances as follows:

$$\textit{input conductance } g_{33} \approx g_{33} + g_{32} \quad \textit{transconductance } g_{23} = g_{23} - g_{32} \quad \textit{output conductance } g_{22} = g_{22} + g_{32} \tag{12.4d}$$

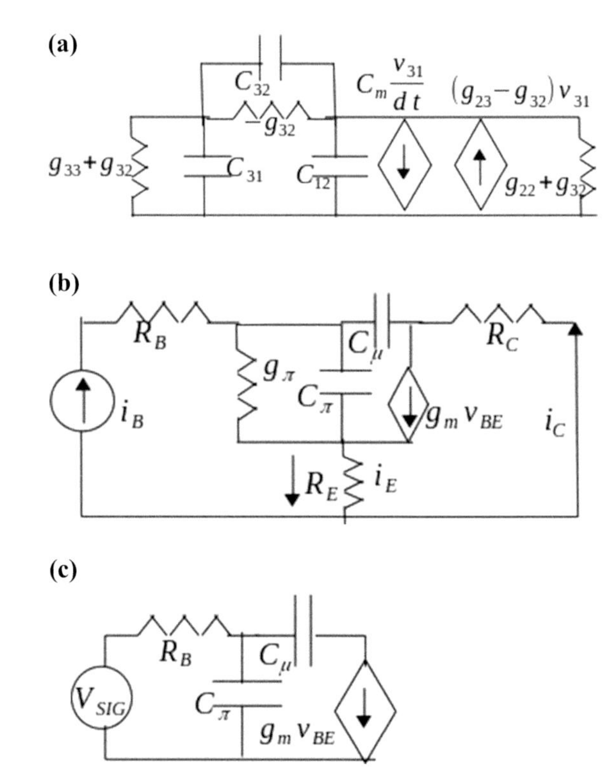

Fig. 12.3 **a** Generic small signal model of bipolar transistor, e.g., HBT. **b** Hybrid pi model of high frequency bipolar transistor. **c** Simplified hybrid pi model of high frequency bipolar transistor, ignoring collector emitter resistances, as being small compared to base resistance

Previously introduced term *transcapacitance(***the effect of terminal 2 on terminal 3 during charging of the capacitor is different from the effect of terminal 3 on terminal 2)**, has its parallel for transport currents—**transconductance**. Transport currents are those currents that involve only conductive components. Transconductance is denoted and defined as:

$$g_m \stackrel{\text{def}}{=} g_{23} - g_{32} \approx \frac{\partial I_2}{\partial V_{31}} \tag{12.4e}$$

where I_2 is the DC collector-emitter|drain-source current and V_{31} is the DC base-emitter|gate-source voltage. In practice, $g_{23} \gg [g_{32}]$. The generic small signal model of a transistor is in Fig. 12.3c. The capacitors (parasitic and real) and resistors in a real heterogeneous junction bipolar transistor are shown in Fig. 12.3d.

The resistors quasi neutral emitter and collector resistors and the path resistances followed by the collector current to the actual collector contact; and the resistance of the base. The base resistor consists of an access resistance ($R_{B,ACCESS}$) due to the path from the base contact to the edge of the base quasi neutral region and base spreading resistance ($R_{B,SPREAD}$) that controls the spreading nature of the resistance in the quasi neutral base. Transferring the resistors and capacitors to the hybrid-π circuit generates

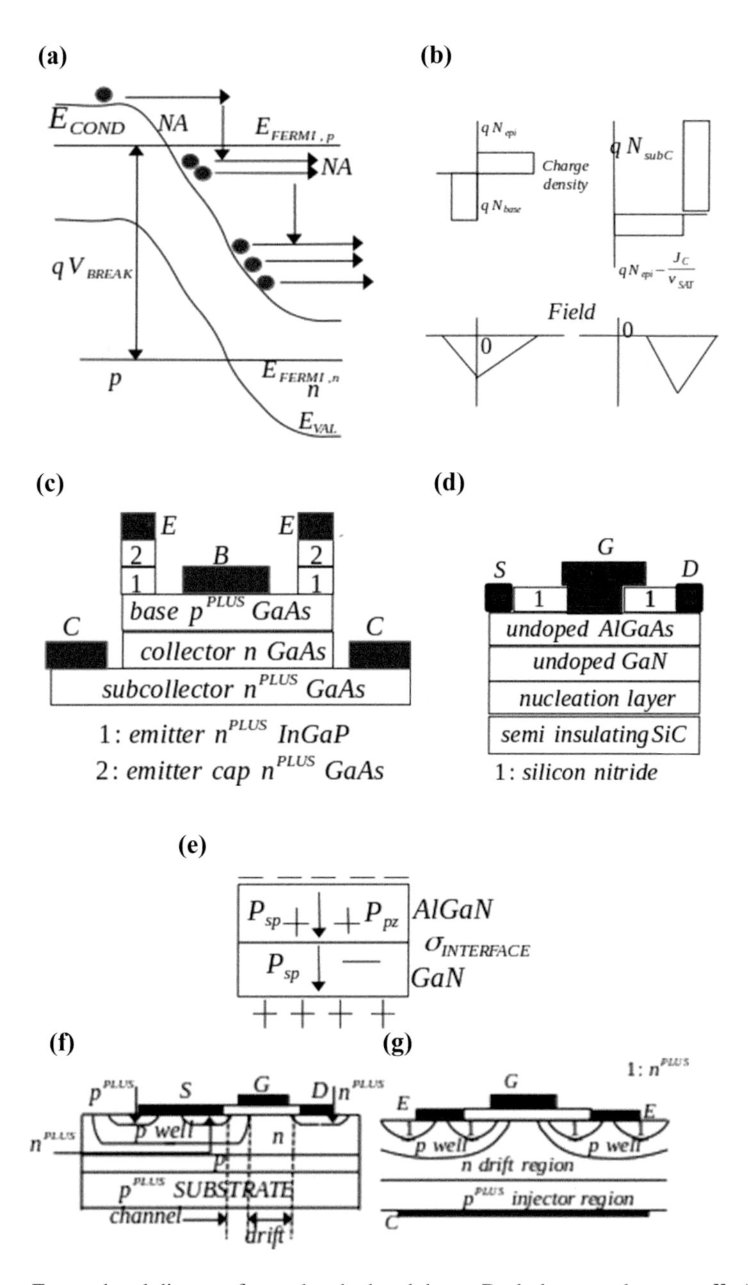

Fig. 12.4 **a** Energy band diagram for avalanche breakdown. Dark dots are electrons. **db** Abrupt np junction charge density and electric field. **c** High frequency and power HBT layered structure. **e** High power and frequency GaN field effect transistor. Charge carrier movement inside GaN field effect transistor. Internal structure of two common high frequency and power MOSFETs. **f**, **g** For the IGBT, the bipolar transistor with base connected to cathode conducts when the MOSFET switches on

the HBT's hybrid model—Fig. 12.3e, which shows a HBT biased in the active mode of operation, So g_{32} ca be ignored.. The output conductance is omitted because it has a near-infinite value in the active mode. *Transcapacitance is not an issue in HBTs, for junction and storage capacitances.*

The input conductance is g_π, the base emitter capacitance is C_π, and the base collector junction capacitance is C_μ. The AC short-circuit at the output is the effective result of holding $V_{COLLECTOR,EMITTER}$ constant. The *transition frequency* f_T, is determined under this condition.

12.4.1 Transition Frequency f_T Definition and Expression

The transition frequency is the extrapolated unity current gain frequency. *The experimental data for the frequency dependency of the square of the magnitude of the current gain* $\left|\frac{i_C}{i_B}\right|^2$ *of an HBT shows that. the measured data ends at a certain frequency that is the upper bandwidth limit of the capabilities of the measuring equipment. But for a decade or so before this frequency limit, the gain rolls off at* $-$ *20 dB/decade, implying a RC circuit with a dominant single pole. The gain can be extrapolated at this slope to higher frequencies, and the frequency at which the gain becomes unity (0 dB), is called the extrapolated transition frequency or transition frequency. The transition frequency is one of two key performance metrics of any high frequency transistor.*

Current gain measurement is performed with the collector and emitter terminals held at constant potentials—for the AC signal the emitter is shorted to the collector. Then.

$$i_B = (g_\pi + j\omega C_{pi})v_{BE} \ + \ j\omega C_{pi} v_{BC} \quad i_C = g_m v_{BE} - j\omega C_\mu v_{BC} \tag{12.4f}$$

To eliminate v_{BR} from (12.4f) it can be re-expressed as below, given that the current gain must be much larger than unity:

$$\begin{aligned} v_{BC} &= v_{BE} + v_{EC} = v_{BE} + i_E R_E + i_C R_C \quad v_{BC} \\ &= v_{BE} + (i_B + i_C)R_E + i_C R_C \approx v_{BE} + i_C(R_C + R_E)\frac{i_C}{i_B} \gg \frac{R_E}{R_C + R_E} \end{aligned} \tag{12.4g}$$

Then the collector current can be re-expressed as follows and the conditions for its validity are:

$$\begin{aligned} i_C &= \frac{(g_m - j\omega C_\mu)v_{BE}}{1 + j\omega C_\mu(R_C + R_E)} |i_C|^2 = \frac{(g_m^2 + \omega^2 C_\mu^2)v_{be}^2}{1 + \omega^2 C_\mu^2 (R_C + R_E)^2} \\ A &: \omega^2 \ll \frac{g_m^2}{C_\mu^2} \quad B : \omega^2 \ll \frac{1}{C_\mu^2 (R_C + R_E)^2} \end{aligned} \tag{12.4h}$$

Similarly, the square of the magnitude of the base current and the condition for its validity is:

$$|i_B|^2 = \left(g_\pi^2 + \omega^2\left(C_\pi + C_\mu(1 + g_m(R_C + R_E))\right)^2\right)$$
$$v_{BE}^2 C : \omega^2 \gg \frac{g_\pi^2}{\left(C_\pi + C_\mu(R_C + R_E)\right)^2} \tag{12.4j}$$

The current gain and consequently the transition frequency are:

$$\left|\frac{i_C}{i_B}\right| = \frac{g_m^2}{\omega^2\left(C_x + C_\mu(1 + g_m(R_C + R_E))\right)^2} 2\pi f_{\text{TRANSTION}}$$
$$= \frac{g_m}{\omega^2\left(C_x + C_\mu(1 + g_m(R_C + R_E))\right)} \tag{12.4k}$$

An ultra high frequency HBT is designed using simple expressions involving the analysis for the transition frequency. The analysis starts with dividing the HBT into separate regions and then estimating the time taken by the input signal to traverse each of the regions. The reciprocal of the transition frequency is called signal delay time τ_{EC}, defined as:

$$\tau_{EMITTER,COLLECTOR} = \frac{1}{2\pi f_{TRANSITION}} = \frac{C_\pi + C_\mu(1 + g_m(R_C + R_E))}{g_m}$$
$$\tau_{EMITTER,COLLECTOR} = \frac{C_{E,B,J}^P}{g_m} + \frac{C_{E,B,j}^P}{g_m} + \frac{C_{E,B,t}^P}{g_m} + C_{C,B}^P\left(R_{E,C} + \frac{1}{g_m}\right)$$
$$\tau_{EMITTER,COLLETOR} = \tau_{BASE} + \tau_{COLLECTOR} + \tau_{EMITTER} + \tau_{CC} \tag{12.4m}$$

The signal delay from the quasi neutral base region is:

$$\tau_{BASE} = \frac{W_{BASE}}{2}\left(\frac{W_{BASE}}{D_{electron}} + \frac{1}{v_R}\right) \tag{12.4n}$$

The signal delay from the change in field in the base—collector depletion region due to the passage of the electrons carrying the signal current is:

$$\tau_{COLLECTOR} = \frac{w_{BASE,COLLECTOR\ DEPLETION}}{2v_{SATURATION}} \tag{12.4p}$$

Charging the base–collector junction capacitance via the dynamic and parasitic resistances introduces delay:

$$\tau_{CC} = \frac{\varepsilon_A A_{\text{COLLECTOR}}}{w}\left(\frac{1}{g_m} + \frac{w_{\text{EMITTER}}}{A_{\text{EMITTER}}\,\sigma_{\text{EMITTER}}} + \frac{w_{\text{COLLECTOR}}}{A_{\text{COLLECTOR}}\,\sigma_{\text{COLLECTOR}}}\right) \tag{12.4q}$$

Finally, the emitter signal delay is the time taken to charge the base-emitter junction capacitance via the dynamic resistance ($\frac{1}{g_m}$) of the transistor is:

$$\tau_{EMITTER} = \frac{\varepsilon_{SEMICONDUCTOR} A_{EMITTER}}{g_m w_{EMITTER}} \tag{12.4r}$$

Ultra high frequency performance depends on device dimension scaling. Lateral scaling reduces collector, emitter areas. Vertical scaling reduces the base width and the base–collector depletion region width w. In high performance InGaP/GaAs HBTs, for example, the base width is so short (about 50 nm) that the major delay in the device is shared by the two delays associated with the collector. Trade-offs must be made regarding w as it affects τ_C, τ_{CC} differently. The delays must become very short, e.g., for transition frequency of 800 GHz, the overall signal delay must be $\approx 2x10^{-12}$ s.

12.4.2 Unity Power Gain Frequency f_{MAX} Definition and Expression

The expression for the transition frequency does not contain the base resistance, because f_T relates to the current gain, and any desired current can be forced through any resistance, provided enough voltage can be applied. This is impossible in real world bipolar transistors there are limits to the available input voltage. **To account for this limitation, an extrapolated frequency related to the power gain -f_{MAX} is used, for the power gain that rolls off with frequency at – 20 dB/decade. Extrapolating the gradient of power gain, to a power gain of 0 dB yields f_{MAX}.**

The simplified hybrid π model of a HBT is in Fig. 12.2f. In this simplified model, the collector and emitter resistances are much smaller than the base resistance, and thus ignored. Also, the input conductance is ignored because the frequency at which the extrapolation is to be made is so high that the following condition holds.

$$\omega^2 \gg \frac{g_\pi^2}{C_\pi^2} \tag{12.4s}$$

From the Thevenin equivalent circuit of the simplified hybrid π circuit, the input impedance and voltages are:

$$v_{THVN} = \frac{v_{INPUT}\left(j\omega C_\mu - g_m\right)}{j\omega C_\mu(1 + g_m R_B) - \omega^2 C_\mu C_\pi R_B} \quad Z_{THVN} = \frac{\frac{1}{R_B} + j\omega\left(C_\mu + C_\pi\right)}{j\omega C_\mu\left(g_m + \frac{1}{R_B}\right) - \omega^2 C_\mu C_\pi} \tag{12.4t}$$

Using conjugate impedance matching conditions yields expressions that can only hold if the corresponding conditions apply:

$$Z_{CIRCL} = Z_{THVN} + Z_{LOAD} \approx 2\Re Z_{THVN} \quad \Re(Z_{THVN})$$

$$= \frac{C_\mu + C_\pi}{C_\mu g_m} \quad C_\pi \approx C_\pi + C_\mu \quad g_m >> \frac{1}{R_B} \quad \omega^2 << \frac{g_m + \frac{1}{R_B}}{C_\pi} \tag{12.4u a,b}$$

The RMS value of the Thevenin voltage holds only if the following conditions apply:

$$|V_{THVN}|^2 \approx \frac{|v_{\text{INPUT}}|^2}{(\omega C_\mu R_B)^2} \quad \omega^2 \ll \frac{g_m^2}{C_\mu^2} \quad \omega^2 \ll \frac{g_m^2}{C_s^2} \quad g_m \gg \frac{1}{R_B} \tag{12.4v}$$

Combining the above expressions, the maximum output power is:

$$P_{\text{OUT},MAX} = \frac{|V_{THVN}|^2}{4\Re Z_{LOAD}} = \frac{g_m |v_{\text{INPUT}}|^2}{4\omega^2 C_\mu (C_\mu + C_x) R_B^2} \tag{12.1w.a}$$

Using the previous assumption that $C_T \approx C_\pi$ the input impedance is.

$$C_\mu \ll C_\pi Z_{INPUT} \approx R_B - \frac{j}{\omega C_\pi} \tag{12.4w b}$$

Thus the input power and maximum available gain(MAG) can be evaluated. The frequency at which MAG = 1 is the extrapolated f_{MAX}.

$$P_{\text{INPUT, MAX}} \approx \frac{|v_{\text{INPUT}}|^2}{R_B} \quad MAG = \frac{P_{\text{OUTPUT, MAX}}}{P_{\text{INPUT, MAX}}}$$

$$= \frac{g_m}{4\omega^2 C_\mu (C_\pi + C_\mu) R_B} f_{\text{MAX}} = \sqrt{\frac{f_{T,INTRINSIC}}{8\pi C_\mu R_B}} \tag{12.4w.c}$$

where the *intrinsic transition frequency is the transition frequency calculated by ignoring parasitic resistances.* So for ultra high frequency operation the transistor must have a good very good intrinsic high frequency performance characteristics. Also, a low base resistance to reduce absorption of the input power, and a high output impedance $\frac{1}{j\omega C_\mu}$ to reduce feedback of power from the output. HBTs are perfectly suited to decreasing the base resistance because the heterogeneous junction at the emitter–base interface can be specifically designed to impede back injection of holes into the emitter. This allows the base doping density to be increased, reducing the base resistance without compromising the current gain. Reducing C_μ involves making the active part of the base–collector junction as thin as possible. Base resistance is a sum of the intrinsic and spreading resistances.

12.4.3 f_T, f_{MAX} For Field Effect Transistors

The small signal, high frequency, equivalent circuit for field effect transistor follows directly from the general hybrid π circuit (Fig. 12.3c, d), by noting that the bipolar transistor's base, collector and emitter become gate, drain and source. Identical arguments hold for the parasitic capacitances. Similarly, the large signal models for the two transistor types can be correlated in a similar fashion. The input conductance is omitted, indicating that there is no tunnelling or leakage through the oxide in the case of a MOSFET, nor any transport current in the Schottky diode of an heterogeneous junction FET. Also, the reverse bias feedback conductance is ignored just like for the HBT. The output conductance.

$\left(g_{22} \overset{\text{def}}{=} g_{DD}\right)$ has been retained because of the drain source channel resistance. There are no circuit elements to represent the substrate, applicable to silicon-on-insulator(SOI) field effect transistors. Using identical analysis as that used for HBTs, the extrapolated, common-source, unity-current-gain frequency of FETs is.

$$2\pi f_T = \frac{g_m}{C_{GD}(1 + g_{DD}(R_D + R_S) + C_{GD}(1 + (g_m + g_{DD})(R_D + R_S)))} \tag{12.5a}$$

A high transconductance is essential for high transistor frequency. Having a high mobility helps—InP and GaAs HEMTs and MESFETs ideal for high frequency applications, specially for HEMTs, where the high mobility o the starting semiconductor material is preserved by the undoped channel material; confinement of the channel charge to a large extent, away from the interface; undoped spacer layer of barrier material for electrons that do spread into the barrier. The 2DEG is confined to a plane, and so there is no scattering in he third direction.

Noting that $g_m \approx \frac{dI_2}{dv_{31}}$, and that for a FET in the saturation mode is $I_2 = I_{DRAIN,SOURCE,SATURATION} \propto \left(V_{GATE,SOURCE} - V_{THRSH}\right)^n$ $1 \leq n \leq 2$. For a HBT, $I_C \propto e^{\frac{V_{BASE,EMITTER}}{V_{THERMAL}}}$.

The corresponding tranconductances per unit current are.

$$\frac{g_m}{I} = \frac{n}{V_{GS} - V_{\text{THRSH}}}(\text{ MOSFET }) = \frac{1}{V_{\text{THFRMAI.}}}(HBT)V_{GS} - V_{\text{THRSH}} = nV_{\text{THERMAL}} \tag{12.5b}$$

At 300 K the operating voltage for the MOSFET is $\approx$ 50 mV, meaning that a very low bias current. An even lower current could be tolerated if it is advantageous to operate in the sub-threshold regime, in which an exponential drain source current vs. gate source voltage holds. Then the transconductance per unit current ratio approximately equalling that of HBTs is possible,

The FET unity gain power frequency is:

$$f_{MAX} = 2\pi F_{T,INTRINSIC}\sqrt{\frac{f_{T,INTRINSIC}}{8\pi C_{DG} R_S}} \tag{12.5c}$$

If the conditions required to derive this equation cannot be met, then it is still possible to arrive at an expression for the power gain that rolls off at $-$ 20 dB/decade, but the equation is much lengthier. The gate capacitance and resistance must be minimized. This is achieved in modern MESFETs and HEMTs by using the 'mushroom' structure for the gate. The small contact region between the gate metal and the underlying semiconductor allows a short gate length to be achieved, and the wider top region keeps the access resistance low.

12.5 Transistors for High Power (High Current and Voltage) and High Frequency

Power amplifiers and switched mode power supplies need transistors [1, 17–21] to conduct currents(10s of A or more), and to withstand high voltages(100s of V). *So the structural details and properties of high power transistors are vastly different from those e.g., for computer memory* etc., These types of high-power transistor include the GaAs HBT and heterogeneous junction GaN FET for power amplification, and the Si MOSFET and hybrid transistor for power supplies. *High currents mean high carrier densities, which can lead to a modification of the space charge region at the output junction (collector/base or drain/body), with consequences for the frequency response and/or the breakdown voltage (avalanche breakdown), and related high current, space charge modifying Kirk Effect.*

Operating transistors at high $V_{COLLECTOR,EMITTER}$, $V_{DRAIN,SOURCE}$ *can lead to electrical breakdown of the base collector or drain substrate junction, respectively, characterized by sudden onset of a large current which, if not interrupted, can overheat the junction and destroy it.* The breakdown process in a reverse biased np junction is shown in Fig. 12.4a. Electrons entering the high field of the junction rapidly gain kinetic energy. If this energy is allowed to exceed the bandgap energy then after collision with a lattice atom, an energy $E \geq E_{GAP}$ can be transferred to another electron, thereby exciting it into the conduction band. Thus, one electron creates another electron (and a hole). This process is the generation equivalent of Auger recombination. The newly generated electron and hole are accelerated by the junction field, creating more electron hole pairs, and a rapidly increasing current—a positive feedback system—**avalanche breakdown**. The corresponding electric field at which avalanche breakdown is initiated, is termed as *break down field strength* E_{BR}. The break down electric field strength and the energy band gap are correlated. The metrics for high power transistors are thermal conductivity, Johnson's figure of merit(JFOM) which characterizes devices for both high frequency and power applications. The large bandgap semiconductor diamond comes out very well in both of the above categories. In a np junction, the maximum electric field occurs at the interface between the two differently doped regions. The left part of Fig. 12.4b shows the case of an abrupt junction with uniform doping on each side of the junction, and under low-current conditions. The right side shows a $pn^M n^P$ where the M, P superscripts denote 'minus', 'plus'. The left hand side diagrams are for a current density sufficiently low for the space charge due to the mobile electrons to be negligible, and for the n^M ("epi") layer to be just fully depleted. The diagrams on the right are for the case when this mobile charge is sufficient to reduce the electric field at the pn^M interface to zero. The depletion region penetrates the n^P layer, the *sub-collector*. The applied bias is the same in both sets of figures. The electric field and the break down voltage value(with the Depletion Approximation) are:

$$E_x = \frac{2(V_{BI} + V_{APP})}{W} \quad V_{BR} = \frac{\varepsilon E_{BR}^2}{2q}\left(\frac{1}{N_{ACCEPTOR}} + \frac{1}{N_{DONOR}}\right) \tag{12.6a}$$

This $pn^M n^P$ structure arises at the base–collector junction in Npn HBTs, and in the drain-substrate region of lateral diffused MOSFETs. The lightly doped region is usually deposited by vapor phase epitaxy—the “epi” layer. The electric field profile in|around the epitaxial layer, and how it responds to an increasing electron current is very important, and is analyzed as a one dimensional problem. A HBT in the forward active operation mode has electrons injected from the emitter which pass through the p type base into the lightly n doped epitaxial layer. Then these electrons are collected in the heavily n doped sub-collector region. Ignoring holes in the epitaxial layer, Poisson’s equation in 1 dimension gives:

$$\frac{dE}{dx} = \frac{\left(qN_{n,epitaxial} - \frac{J_C}{v_{SATURATION}}\right)}{\varepsilon_{EMICONDUCTOR}} \tag{12.6b}$$

where the electrons are moving in the epitaxial layer at their saturation velocity. For low collector current density J_C the field gradient is positive for x > 0. At $J_C \stackrel{\text{def}}{=} J_{\text{CRITICAL}} = qN_{epi}v_{SAT}$ the field becomes constant. At higher collector current densities the field gradient in the epitaxial layer becomes negative for x > 0. Then the effect of the positive space charge of the donor ions in the epitaxial layer is overcome by the negative space charge of the electrons carrying the current. At a critical current density the field goes to zero at the base|epitaxial layer boundary. *The current density at which the field disappears at the p—epitaxial layer junction is known as the Kirk current* J_{KIRK}. The onset of the Kirk Effect can be delayed by using a semiconductor for which saturation velocity is high, and by choosing a high doping density for the epitaxial layer. Kirk effect is unavoidable: if the epitaxial layer thickness is reduced to zero, then the breakdown becomes very severe.

With ever increasing carrier frequencies for both wired and wireless telecommunication networks, power transistors must be able to handle both high current|voltages and ultra high frequencies—e.g., transistors used in cellphone repeater amplifiers. InGaP/GaAs HBTs, and AlGaN/GaN heterogeneous junction FETs are best. HBTs operating at modest voltages and power, are used in the transmitter stage of cellphones, and final stages of power amplifiers In radio base stations, where operation is at tens of volts and hundreds of watts. GaN based heterogeneous junction FETs are also used for these applications. These FETs are used for electronic circuits operating in harsh environments due to the high bandgap of the transistor material.

The structure of a high power, high frequency HBT is shown in Fig. 12.4c. The semiconducting material are chosen for their high charge carrier mobility. Device fabrication tricks as interdigitated emitter and base contacts facilitate a large current flow via the large total emitter area. Simultaneously, reducing base access, controlling base spreading resistance and a thick, lightly doped collector ensures a high breakdown voltage. Splitting the base current between three contacts reduces the power dissipation in the base spreading resistance to 1/12 of its value in the single base case. A tall emitter separates the emitter and base metallizations. The thick top part of the emitter reduces the parasitic emitter resistance. The thick, lightly doped collector allows a high break down voltage to be achieved to over a high J_{CRIT}. At $J_C = J_{CRIT}$ there is no field at the base–collector junction. There is no potential barrier to stop the holes from leaving the base. If the collector current density increases more, holes flood into the epitaxial layer, widening the quasi neutral base region. This increases the base transit time. The reduced width of the base—collector space charge region could enhance the collector signal delay time, but this is counter balanced by the increase in the collector charging time. So τ_{EC} increases and the transition frequency f_T decreases at high currents. Before the onset of Kirk Effect, the transition frequency increases as the transconductance improves.

$$g_m \approx \frac{\partial I_{COLLECTOR}}{\partial V_{BASE,EMITTER}} = \frac{I_{COLLECTOR}}{\gamma(I_{COLLECTOR})V_{THERMAL}} \tag{12.6c}$$

where γ is the **junction ideality factor**. At emitter current densities(that induce recombination in the emitter–base space charge region) can be neglected, $\gamma = 1$, while low level injection conditions apply. In high injection conditions, assuming that the injected minority carrier concentration at the edge of the depletion region in the base of an Npn transistor is such that $n(x_{dp}) = p_{p0}$.

$$n(x_{dp}p_{p0}) = n_i^2 e^{\frac{V_{APP}}{V_{THERMAL}}} \quad n(x_{dp}) = n_i e^{\frac{V_{APP}}{2V_{THERMAL}}} \tag{12.6d}$$

Minority carrier concentration increase is matched by a corresponding increase in majority carrier concentration to maintain charge neutrality—**conductivity modulation**. At the collector end of the base the minority carriers are extracted, so that there are large concentration gradients of both electrons and holes. *Thus both electrons and holes will diffuse at different rates*—**ambipolar diffusion**. *As a result an induced electric field retards the diffusion of the faster carrier. Diffusion dominates, with an high effective effectively diffusivity*—**Webster Effect**. Under appropriate operating conditions, a diffusive electron current with the boundary condition of (12.6d) results in an ideality factor of $\gamma = 2$, and reduces the transition frequency at high currents. As the channel ideality factor increases the current the same channel decreases. **This is a very interesting but complicated charge carrier transport mechanism involving isothermal|non-isothermal drift diffusion and isothermal|non-isothermal energy balance**.

Electrons injected into the high field region of the reverse biased base collector space charge region are accelerated to velocities above the saturation velocity, before thermalizing collisions occur—reducing the signal delay time in the collector space charge region and consequently improving the transition frequency. The increased velocity also increases the current.

Gallium nitride(GaN) based high power heterogeneous junction FETs have two material properties that are superior to those of GaAs—breakdown field strength and thermal conductivity. The high electron saturation velocity is also key to high frequency applications in devices where the field is high enough for this velocity to be attained.

The layered structure of a sample GaN heterogeneous junction FET is in Fig. 12.4d. The gate length and the separations between gate and source|drain electrodes are about 100–500 nm. The operating voltages are about 28 V for wireless base station applications, so velocity saturation is likely. An AlGaN surface is electrically active, so that a passivation layer such as silicon nitride is essential. The gate electrode extends over the passivation layer towards the drain, forming a field plate. This reduces the field in the channel at the edge of the gate, thereby improving the breakdown voltage. But the field plate adds capacitance to the device, undesirable for high frequency performance. This can be curbed by recessing the gate, as a result of increasing the transconductance through the closer coupling of the gate to the channel a two dimensional electron gas at the AlGaN|GaN interface. A common substrate material is silicon carbide(SiC), which has excellent thermal conductivity. A buffer layer of GaN is deposited before epitaxial growth of the actual layers of the device, which crystallize in the wurtzite(hexagonal close packed lattice) structure. Crystal growth is along the c-axis, from the {0001} basal plane in the [0001] direction. The atoms are arranged in two, repeating, closely packed bilayers, each layer of which is an hexagonal arrangement of either gallium or nitrogen ions. The GaN bond is strongly ionic, with nitrogen more electronegative than gallium. Thus the material is spontaneously polarized, with the polarization vector pointing towards the substrate. For the ternary material $Al_xGa_{1-x}N$ the lattice constant is approximately $a = (0.3189 - 0.0077x)nm$. Therefore, AlGaN grown pseudomorphically on GaN is under tensile strain, adding a piezoelectric polarization $\vec{P}_{Pz}$ to the spontaneous polarization that is inherent to the material. The situation for AlGaN on Ga-face GaN is in Fig. 12.4e. The surface polarization charge densities are negative on the top of the AlGaN and positive on the bottom of the GaN. At the interface between the two materials the surface polarization density is:

$$\sigma_{INTERFACE} = \left(\vec{P}_{SPONTANEOUS,2} + \vec{P}_{PIEZOELECTRIC,2}\right) \cdot \hat{n}_2 + \vec{P}_{SPONTANEOUS,1} \cdot \hat{n}_1 \tag{12.6e}$$

The sheet polarization charges are bound charges fixed in space. Th electrons are drawn to the interface during the period of cooling after the growth of the layers or from the ohmic contacts in the fabricated heterogeneous junction FET. **So a two dimensional sheet**

of electrons is created at the interface without doping either of the layers. The two dimensional gas consists of electrons confined in a potential notch, one side of which is due to the electron affinity mismatch between the two materials. The relation expressed in terms of the mole fraction x for Al, is χ (x) $= 4.1 - 1.87 \times$ eV. A typical value is x $= 0.15$, gives a barrier of about 0.28 eV, while also maintaining the lattice constant mismatch to be within appropriate tolerances. The spontaneous and piezoelectric polarizations in AlGaN are so large that the surface concentration of electrons at the interface is approximately $\frac{10^{13}}{cm^2}$ giving AlGaN|GaN heterogeneous junction FETs a high current carrying capability. Also, the large bandgap and breakdown voltage allow operation at typical base station voltage levels of 28 V. Thus additional voltage conversion circuitry is unnecessary. These properties in combination with high electron velocity and high thermal conductivity produce a transistor with for power amplifiers operating at high frequencies.

Power transistors used in high current|voltage DC-DC converters are used as a switch. When the switch is closed the transistor needs to pass a large current. As this current is derived from the input voltage source, the ON resistance of the transistor must be low, to curb unwanted power dissipation within the transistor. When the switch is open, the transistor must be able to withstand a voltage at least equal to that of the output voltage—specified by the **forward blocking voltage**. The transistor must be able to switch quickly between the ON and OFF states. As bipolar transistors are current controlled devices, the pulse width modulated control circuitry for duty cycle adjustment becomes very complex. Also, to get the required low ON resistance, bipolar transistors must be operated in the saturation mode.

However MOSFETs, which are voltage controlled devices need simpler pulse width modulation control circuitry and are unipolar, so switching times are not determined by slow recombination processes. So field effect transistors are now the choice for power supply designers.

The two MOSFETs specifically designed for high power and high frequency switching applications are the Laterally Diffused MOSFET(LDMOS) and the Insulated Gate Bipolar Transistor(IGBT)—Fig. 12.4f, g For the LDMOS**,** the source is embedded in a p well, and these two regions are formed in a sequential diffusion process—hence 'diffused' in the transistor title. 'Lateral' comes from the lateral layout of the device, allowing three terminal contacts on the top of the structure. The lateral arrangement enables the integration of the power transistor with standard CMOS FETs, which can be used for the control circuitry. The drain consists of the usual n + contact region, and a lightly doped drain extension in the form of an n epitaxial layer grown on the p substrate. The junction between the p well and the epitaxial layer forms a pn diode, and the junction between the epitaxial layer and the p substrate forms a second diode.

The breakdown voltage of the first pn diode is given by 12.6a, with the carrier concentrations N_{epi}, N_{pwell}. The low value for epitaxial layer doping concentration ensures a low ON resistance. The ON resistance is determined by the lateral resistance of the epitaxial layer. A thin epitaxial layer is advantageous because of a two dimensional *reduced surface field* (RESURF) effect. At the second diode junction, the space charge region extends into the n epitaxial layer by a distance approximately given as:

$$y_{sn} = \sqrt{\frac{2\varepsilon_{SEMICONDUCTOR} N_{SUBSTRATE}\left(V_{BR,JUNCTION1} + V_{DRAIN,SOURCE}\right)}{qN_{epitaxial}\left(N_{epitaxial} + N_{SUBSTRATE}\right)}} \tag{12.6f}$$

The key is to choose the thickness of the epitaxial layer to be less than y_{dn} at the desired forward blocking voltage. Then the depletion region from the second pn junction reaches through to the surface of the FET and augments the depletion region surrounding the first depletion region—effectively extending it in the x-direction. The voltage drop across the first pn junction is spread over a longer region, and so reduces the electric field E_x below E_{BREAK}, shifting the region of likely breakdown to the second pn junction.

By making $N_{SUBSTRATE} < N_{epi}$, *the breakdown voltage at the second pn junction can extend the breakdown voltage of the first pn junction, increasing the forward blocking voltage. But a parasitic* $n^{PLUS}pn^{MINUS}$ *BJT formed by the source-substrate-epitaxial regions, and if the current is high enough, the Kirk Effect will come into play and influence the voltage distribution in the structure.* In the LDMOS the point of highest field moves to the $n^{PLUS}n^{MINUS}$ junction between the drain and epitaxial layer regions, and this determines the breakdown voltage. *The LDMOS is an interesting design problem involving the choice of epitaxial layer thickness and doping concentrations and the interaction between blocking voltage and operating current.*

12.6 Low Noise Transistors

The reader is requested to refer to a previous chapter dedicated to semiconductor device noise mechanisms.

12.7 Figures of Merit for High Power and High Frequency Transistors

As power diodes and transistors have to operate reliably with very high currents and voltages, the power semiconductor device industry has adapted a set of figures of merit to characterize the performance characteristics of these devices. These are:

- BFM: Baliga's figure of merit—a measure of the specific on-resistance of the drift region of a vertical field effect transistor. A related figure of merit is for high frequency transistors
- BSFM: Bipolar transistor switching speed figure of merit

- BPFM: Bipolar transistor power handling capacity figure of merit
- BTFM: Bipolar power transistor switching product
- FPFM: Field effect transistor power handling capacity figure of merit
- FSFM: Field effect transistor switching speed figure of merit
- FTFM: Field effect transistor power switching product
- JFM: Johnson's figure of merit denotes the power frequency product for low power transistors
- KFOM: Keyes' figure of merit denotes the thermal limitation to switching transistors inside integrated circuits

The Baliga figure of merit identifies material parameters that minimize conduction losses in low frequency unipolar transistors. The Baliga high frequency figure of merit indicates that using silicon carbide(SiC) devices for high frequency applications can significantly reduce power loss.

$$BFOM = \varepsilon_r \mu E_{GAP}^3 \quad BHFFOM = \mu E_{BREAK}^2 \sqrt{\frac{V_{GATE}}{4V_{BREAK}^3}} \tag{12.7a}$$

$$JFOM = \left(\frac{E_{BREAK} v_{electron,saturation}}{2\pi} \right)^2 \quad KFOM = \kappa \sqrt{\frac{c v_{electron,saturation}}{4\pi \varepsilon_{STATIC}}} \tag{12.7b}$$

The other listed figures of merit are semiconductor device manufacturer specific, i.e., some figures of merit are quoted by some vendors, but not by others.

References

1. Sze, S. M. (2008, January). Physics of semiconductor devices, 3rd Edition. Retrieved form https://www.amazon.in/Physics-Semiconductor-Devices-3ed-S-M/dp/8126517026/ref=sr_1_1?qid=1670906380&refinements=p_27%3AS.M.+Sze&s=books&sr=1-1
2. Tsividis, Y. (1999). Operation and modeling of the MOS transistor, 2nd Edn., Sec. 9.2.1, Oxford University Press.
3. Pulfrey, D. L., & Tarr, N. G. (1989). Introduction to Microelectronic Devices, Prentice-Hall.
4. John, D. L. (2008). Limits to the signal delay in ballistic nanoscale transistors. *IEEE Transactions on. Nanotechnolgy, 7,* 48–55.
5. Streetman, B., & Banerjee, S. Solid State Electronic Devices 7th Edition https://www.amazon.com/Solid-State-Electronic-Devices-7th/dp/0133356035/ref=sr_1_1?qid=1670906725&refinements=p_27%3ASanjay+Banerjee&s=books&sr=1-1
6. Trinh, C. et al. (2009). A 5.6MB/s 64Gb 4b/cell NAND flash memory in 43 nm CMOS. *ISSCC Digest Technology Papers,* 246–248.
7. Schloesser, F. et al. (2008). A 6F 2 Buried Wordline DRAM Cell for 40 nm and beyond. *IEEE IEDM Technology Digest,* p. 33.4.

8. Hafez, W., Snodgrass, W., & Feng, M. (2005). 12.5 nm base pseudomorphic heterojunction bipolar transistors achieving fT = 710 GHz and fmax = 340 GHz. *Applied Physics Letters, 87,* 252109.
9. Hafez, W., & Feng, M. (2005). Experimental demonstration of pseudomorphic heterojunction bipolar transistors with cutoff frequencies above 600 GHz. *Applied Physics Letters, 86,* 152101.
10. Liu, W. (1999). Fundamentals of III-V Devices: HBTs, MESFETs, and HFETs/HEMTs, pp. 226–231, John Wiley & Sons Inc.
11. Vaidyanathan, M., Pulfrey, D. L. (1999). Extrapolated fmax for Heterojunction Bipolar Transistors. *IEEE Transactions on Electron Devices, 46,* 301–309.
12. Lee, S., Wagner, L., Jagannathan, B., Csutak, S., Pekarik, J., Zamdmer, N., Breitwisch, M., Ramachandran, R., & Freeman, G. (2005). Record RF performance of Sub-46 nm L gate NFETs in microprocessor SOI CMOS technologies. *IEEE IEDM Technology Digest,* 241–244.
13. Yamashita, Y., Endoh, A., Shinohara, K., Hikosaka, K., Matsui, T., Hiyamizu, S., Nimura, T. (2002). In 0.52 Al 0.48 As/In 0.7 Ga 0.3 As HEMTs with an Ultrahigh f T of 562 GHz. *IEEE Electron Device Letters, 23,* 573–575.
14. Castro, L. C., & Pulfrey, D. L. (2006) Extrapolated fmax for CNFETs. *Nanotechnology, 17,* 300–304.
15. Gonzalez, G. (1984). Microwave transistor amplifiers: analysis and design, (2nd Edn.) Chapter 1, Prentice-Hall.
16. Mason, S. J. (1954). Power gain in feedback amplifier. *IRE Trans. Circuit Theory, 1,* 20–25.
17. Ho, S. C. M., & Pulfrey, D. L. (1989). The effect of base grading on the gain and high-frequency performance of AlGaAs/GaAs heterojunction bipolar transistors. *IEEE Transactions on Electron Devices, 36,* 2173–2182.
18. Baliga, B. J. Fundamentals of Power Semiconductor Devices. Retrieved from https://www.amazon.in/Fundamentals-Semiconductor-Devices-Jayant-Baliga/dp/3319939874.
19. DiSanto, D. (2005). Aluminum Gallium Nitride/Gallium Nitride High Electron Mobility Transistor Fabrication and Characterization, Table 1-1, Ph.D. Thesis, Simon Fraser University.
20. Johnson, E. O. (1965). Physical limitations on frequency and power parameters of transistors. *RCA Review, 26,* 163–177.
21. Apanovich, Y., Blakey, P., Cottle, R., Lyumkis, E., Polsky, B., Shur, M., & Tcherniaev, A. (1995). Numerical simulation of submicrometer devices including coupled nonlocal transport and nonisothermal effects. *IEEE Transactions on Electron Devices, 42,* 890–898, 298.
22. Ambacher, O., Smart, J., Shealy, J. R., Weimann, N. G., Chu, K., Murphy, M., Schaff, W. J., Eastman, L. F., Dimitrov, R., Wittmer, L., Stutzmann, M., Rieger, W., & Hilsenbeck, J. (1999). Two-dimensional electron gases induced by spontaneous and piezoelectric polarization charges in N- and Ga-face AlGaN/GaN heterostructures. *Journal of Applied Physics, 85,* 3222–3233.
23. Ludikhuize, A. (2000). A review of RESURF technology, proceedings of. 12th IEEE International Symposium on Power Semiconductor Devices and ICs, pp. 11–18.

Performance Characteristics of Selected Commercial Heterogeneous Transistors and TCAD (Technology Computer Aided Design) Tools

13

13.1 Test Bench

The generic test bench used for the both AC|DC characteristic measurement of the high performance transistors is in Fig. 13.1. All current measurements were done with a Keithley 6485 picoammeter [5]. A Keithley 2230G-30–1 programmable triple output DC power supply was used to provide the collector|drain and base|gate voltages [6, 7]. The collector|drain and base|gate resistances were selected using the corresponding transistor's datasheet "Maximum Absolute Ratings" table information provided by the device manufacturer [1–4]. **All DC sweep measurements(Ice|Ids vs Vbe|Vgs or Ice|Ids vs. Vce|Vds were performed with continious and pulsed DC power modes**. I*n the Ice vs. Vbe continious Vce mode, the Vbe voltage was swept over the specified range(as per device datasheet) at different Vce levels(as per device datasheet). For the corresponding pulsed Vce, Ice vs. Vbe measurement, the Vce was applied as pulses, with the pulse amplitudes the same as in the continious DC measurement case.* **As each RF transistor has parasitic capacitors and inductors, pulsed DC measurement allows these parasitic components to be discharged when the applied DC pulse amplitude is zero.**

A. Banerjee, *Semiconductor Devices*, Synthesis Lectures on Engineering, Science, and Technology, https://doi.org/10.1007/978-3-031-45750-0_13

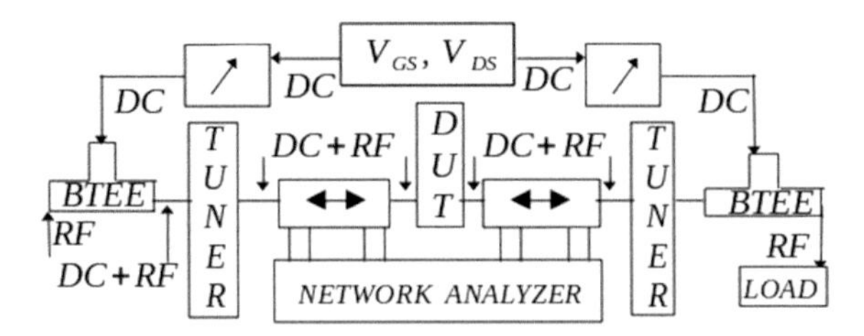

Fig. 13.1 Generic test bench for RF|microwave transistors. The test bench components for RF only(network analyzer, sampler, tuner, BTEE) are not essential for DC sweeps (Ids—Vgs @ constant Vds and Ids – Vds @ constant Vgs

13.2 BFQ790 Heterogeneous Junction (SiGe) Medium Power NPN RF Transistor

The BFQ790 SiGe transistor is a heterogeneous junction silicon germanium (SiGe) medium power RF NPN device from Infinieon Technologies AG [1]. The results of the DC sweep measurements are shown in Fig. 13.2a, b, c, d, e and f.

13.3 TP65H150G4LSG Heterogeneous Junction Gallium Nitride (GaN) Power FET

The TP65H150G4LSG is a gallium nitride(GaN) heterogeneous junction power field effect transistor[3] with an upper voltage limit of 650 V. The DC performance characteristics are shown in Fig. 13.3a, b, c, d and e.

13.4 EPC2216 GaN Power FET DC Performance Characteristics

The EPC2216 is another GaN power FET whose DC performance characteristics are shown in Fig. 13.4a, b, c, d and e.

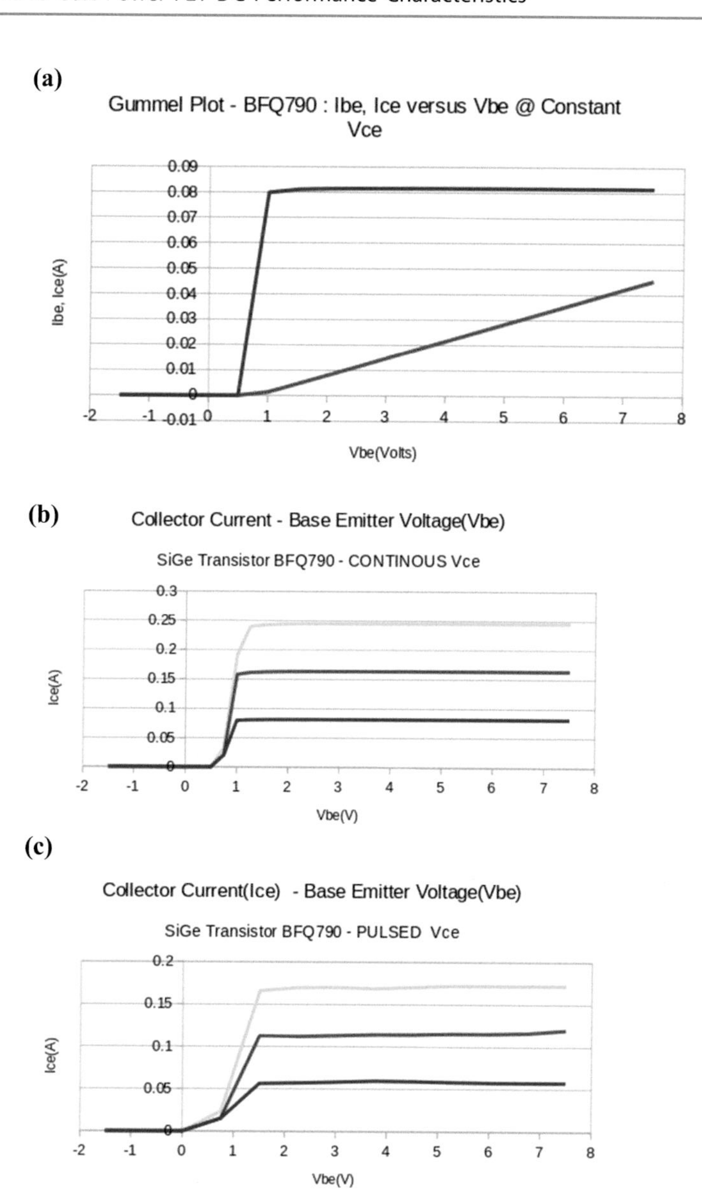

Fig. 13.2 **a** Gummel plot(Ibe, Ice vs. Vbe) for BFQ790, **b** Collector current(Ice)—base emitter voltage(Vbe) at continuous collector emitter voltage(Vce) for BFQ790. **c** Collector current(Ice)—base emitter voltage(Vbe) at pulsed collector emitter voltage(Vce) for BFQ790. Current collapse, compared to continuous collector emitter voltage case is evident. **d** Collector current(Ice)—collector emitter voltage(Vce) at continuous base emitter voltage(Vbe) for BFQ790. **e** Collector current(Ice)—collector emitter voltage(Vce) at pulsed base emitter voltage(Vbe) for BFQ790. Current collapse compared to continuous base emitter voltage case is evident. **f** Transconductance -collector current(Ice) for BFQ790

Fig. 13.2 (continued)

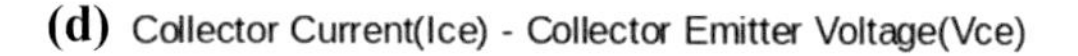

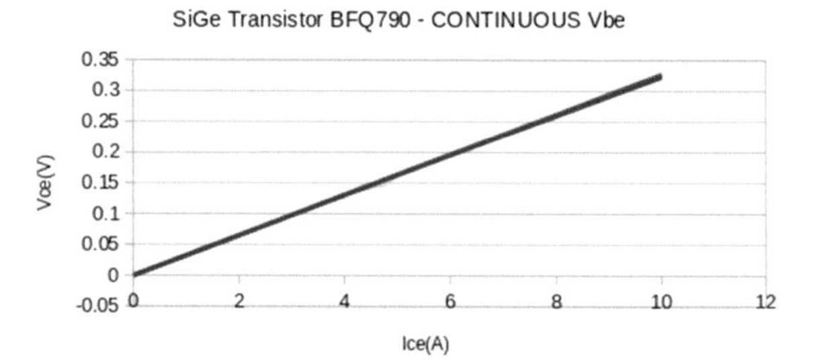

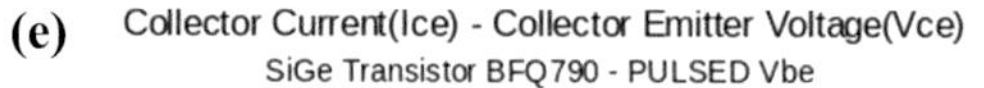

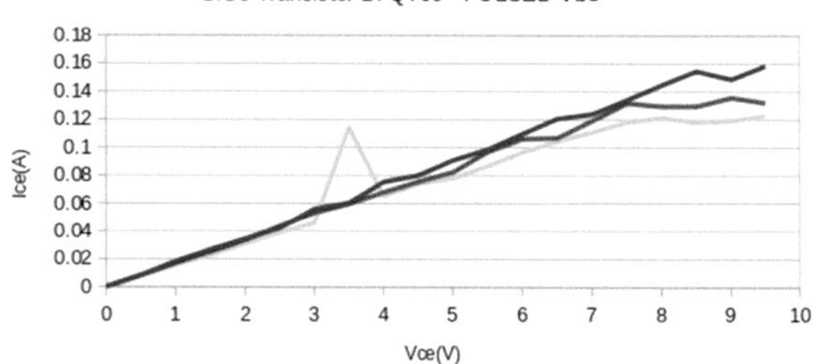

(f) Transconductance(gm) - Ice

BFQ790 SiGe NPN Transistor - CONTINIOUS Vce

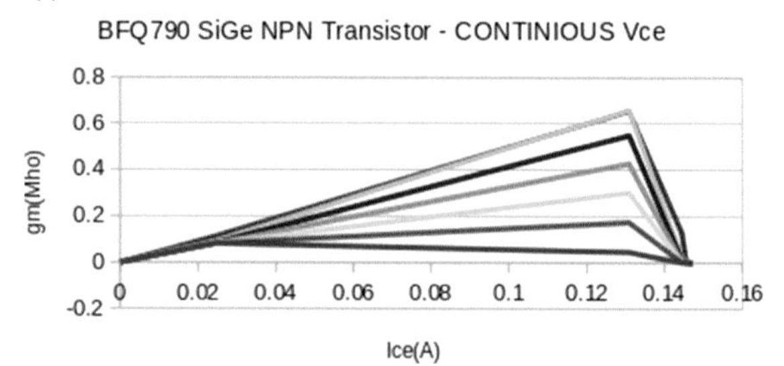

13.5 NE34018 Gallium Arsenide (GaAs) Field Effect Transistor

This GaAs FET operates in the RF|microwave frequency range [4], whose typical DC performance characteristics are in Fig. 13.5a, b, c and d.

The transistors used in the author's laboratory test bench for generating the DC sweep curves were obtained as test samples from the respective manufacturers.

13.6 Technology Computer Aided Design (TCAD) Tools for Analyzing, Estimating and Visualizing Internal Electrical Properties of Semiconductor Devices

In the recent past, a number of open source [8–10] and commercial [11, 12] Technology Computer Aided Design (TCAD) tools have become sophisticated enough to **analyze, estimate and visualize key internal electrical properties of semiconductor devices** (e.g., charge carrier density, charge carrier energy, electrical potential, electric field (x,y components), charge carrier velocity etc.,) Notable among the open source versions are

(a)

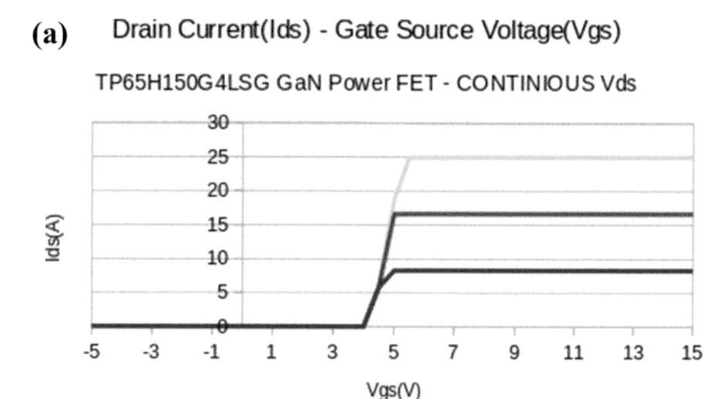

(b)

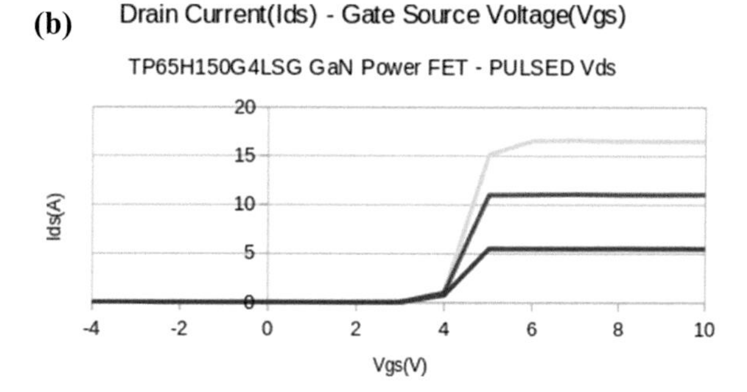

(c)

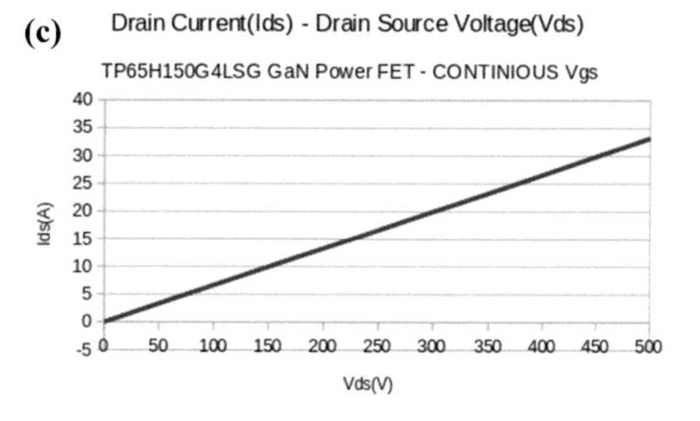

Fig. 13.3 **a** Drain current(Ids)—gate source voltage(Vgs) continuous drain source voltage(Vds) for TP65H150G4LSG. **b** Drain source current(Ids)—gate source voltage(Vgs)—pulsed drain source voltage(Vds) for TP65H150G4LSG. Current collapse compared to continuous drain source voltage case is evident. **c** Drain current(Ids)—drain source voltage(Vds) continuous gate source voltage(Vgs) for TP65H150G4LSG. **d** Drain current(Ids)—drain source voltage(Vds) pulsed gate source voltage(Vgs) for TP65H150G4LSG. Current collapse compared to continuous gate source voltage case is evident. **e** Transconductance(gm)—drain current(Ids) continuous drain source voltage(Vds) for TP65H150G4LSG

Fig. 13.3 (continued)

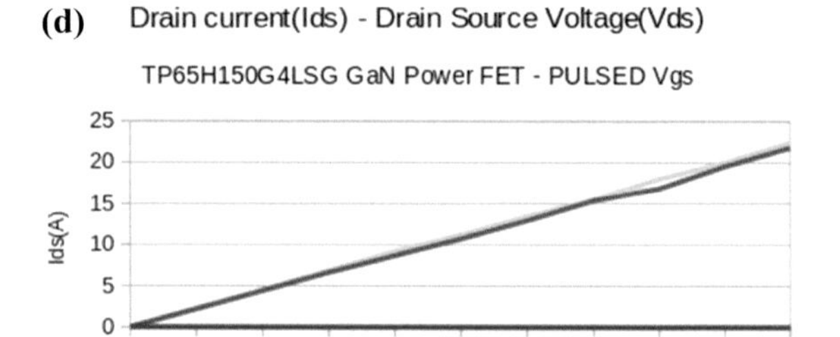

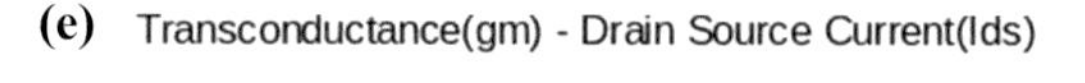

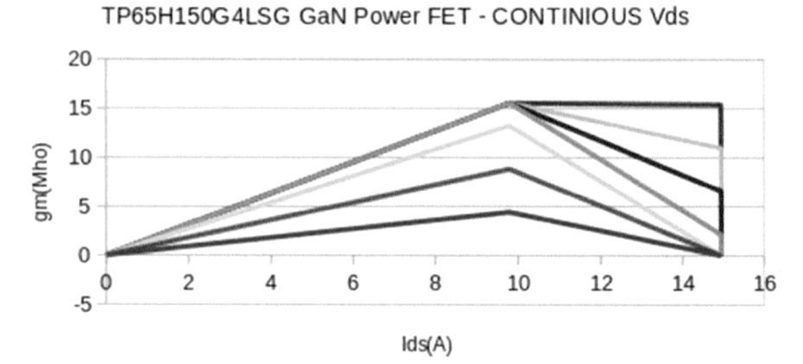

Archimedes [8], Charon [9] (from Sandia National Laboratories). The commercial varieties [11, 12] even allow semiconductor device processing steps to be simulated and analyzed to optimize semiconductor device manufacturing process flow.

TCAD is a software package|tool that **models both semiconductor device fabrication and semiconductor device operation.** *Semiconductor device fabrication modeling is Process TCAD, semiconductor device internal electrical property analysis and estimation is termed Device TCAD.* Semiconductor device fabrication steps include ion implantation, photolithography, molecular beam epitaxy, chemical vapor deposition, etc., Analysis and estimation of key internal semiconductor device properties (e.g., charge carrier density, charge carrier energy, electrical potential, electric field (x,y components), charge carrier velocity etc.,) involves solving key underlying physics equations as Maxwell Boltzmann Transport equation, drift–diffusion equation, hydrodynamic equation, Schrodinger's Wave equation etc., including real semiconductor device issues as charge carrier scattering from acoustic|optical phonons, impurities, crystal lattice defects etc., Sophisticated numerical analysis techniques as Monte Carlo [13] and two|three dimensional finite element [14] analysis.

Device TCAD programs perform two|three dimensional analysis. For a typical two dimensional TCAD program [8], the user supplied input must include:

- Physical dimensions in appropriate units

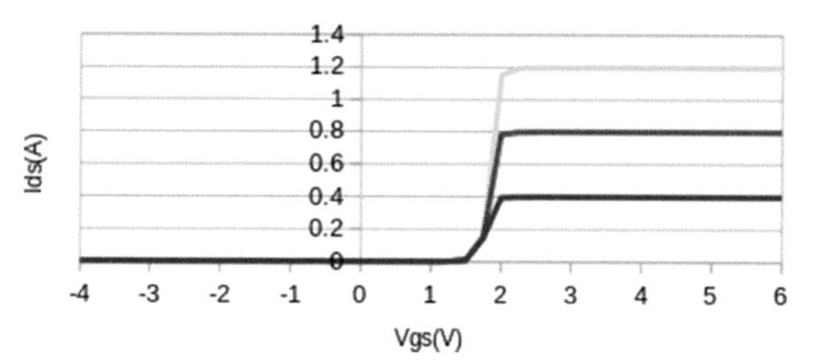

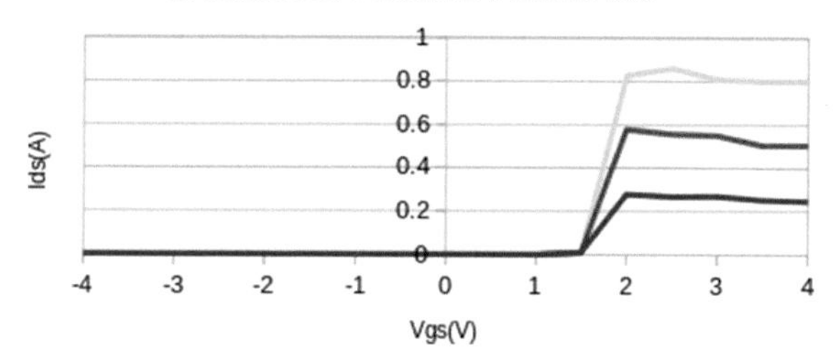

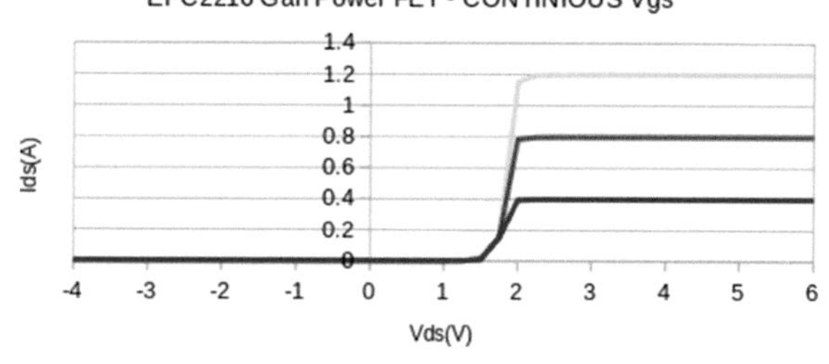

Fig. 13.4 **a** Drain current(Ids)—gate source voltage(Vgs) continuous drain source voltage(Vds) for EPC2216. **b** Drain current(Ids)—gate source voltage(Vgs) pulsed drain source voltage(Vds) for EPC2216. Current collapse compared to continuous drain source voltage case is evident. **c** Drain current(Ids)—drain source voltage(Vds) continuous gate source voltage(Vgs) for EPC2216. **d** Drain current(Ids)—drain source voltage(Vds) pulsed gate source voltage(Vgs) for EPC2216. Current collapse compared to continuous gate source voltage case is evident. **e** Transconductance(gm)—drain current(Ids) for EPC2216

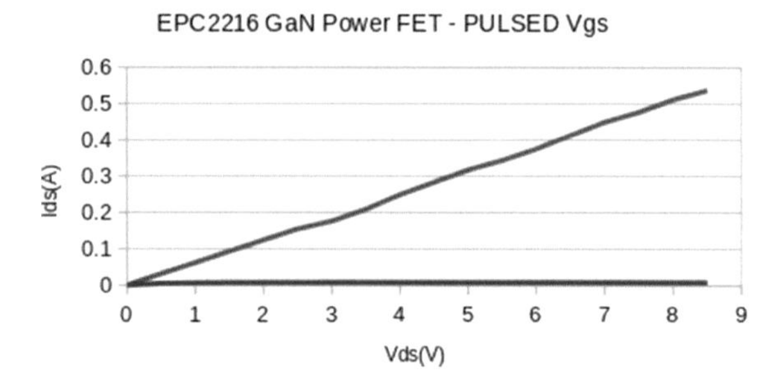

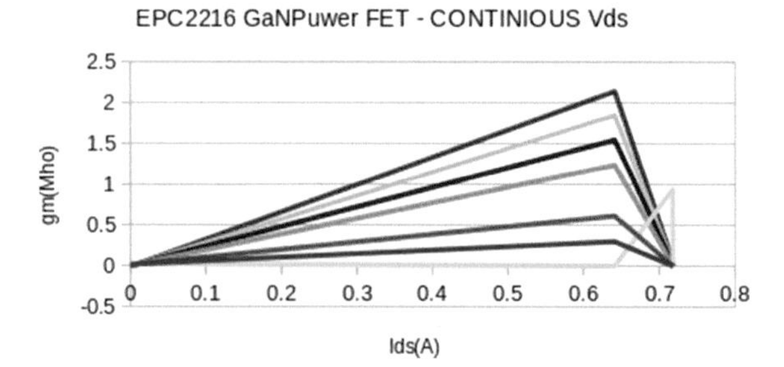

Fig. 13.4 (continued)

- Type of semiconductor material (gallium arsenide|germanium|indium phosphide|silicon) used in the device
- Type of numerical analysis—Monte Carlo|finite element, including the option of whether movement of both charges carriers (electron|holes) or one of the pair are to be analyzed
- Type of internal charge carrier scattering process (acoustic phonon, optical phonon, impurity,lattice imperfections scattering etc.,) to be included in the calculations
- Type(acceptor|donor) and concentration in different regions of the device
- Type(insulator|Ohmic|Schottky) of contacts to the various regions of the device
- Lattice temperature and result output format (popular Gnuplot [15] or 'mesh' [16]) format

The following three dimensional surface plots (Fig. 13.6a, b, c, d, e, f, g, h, I, j) show the key internal electrical properties of an abrupt junction np rectifying|signal diode obtained with the Archimedes 2.0.1 tool [8]. *Both acoustic and optical scattering are included in the calculations.* Both charge carriers (electron, hole) properties (density, energy, velocity etc.,) are measured. The three dimensional surface plots were generated with the excellent

Fig. 13.5 **a** Drian current(Ids)—gate source voltage(Vgs) at continuous drain source voltage(Vds) for NE34018. **b** Drain current(Ids)—gate source voltage(Vgs) pulsed drain source voltage(Vds) for NE34018. **c** Drain source current(Ids)—drain source voltage(Vds) continuous gate source voltage(Vgs) for NE34018. d Drain source current(Ids) drain source voltage(Vds) pulsed gate source voltage(Vgs) for NE34018

(a)

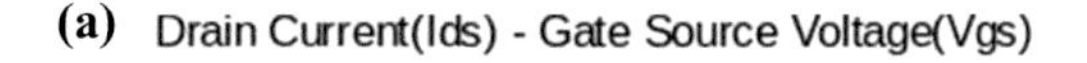

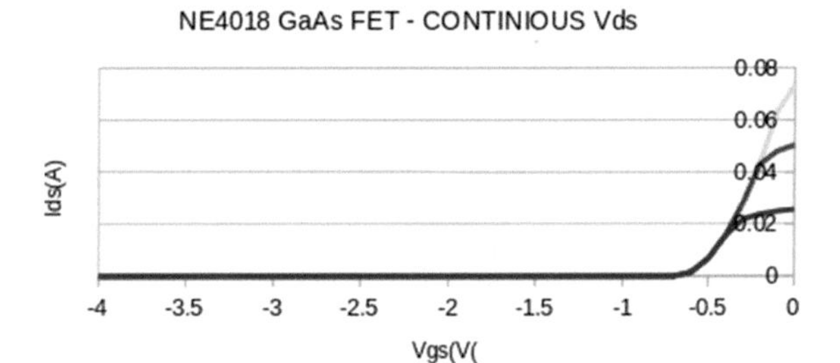

(b) Drain Current(Ids) - Gate Source Voltage(Vgs)

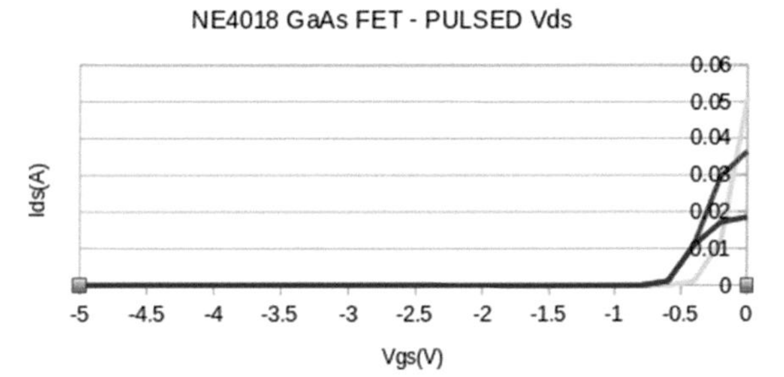

(c) Drain Current(Ids) - Drain Source Voltage(Vds)

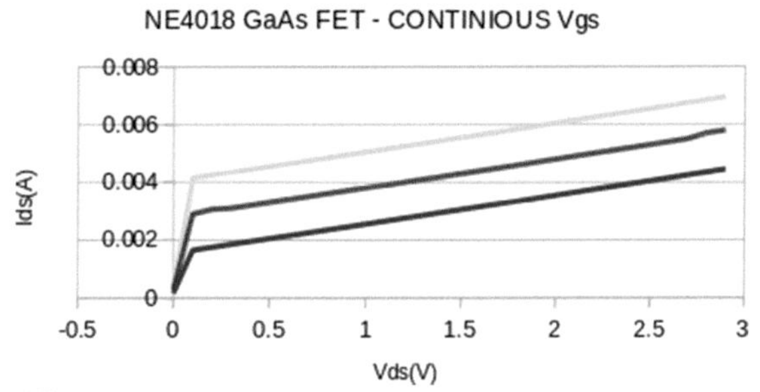

(d)

Drain Current(Ids) - Drain Source Voltage(Vds)

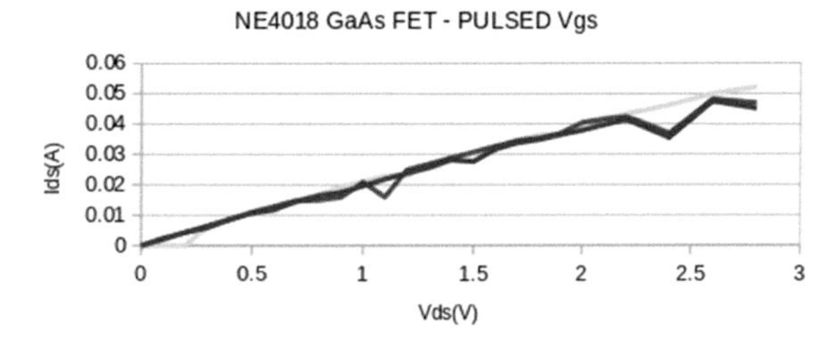

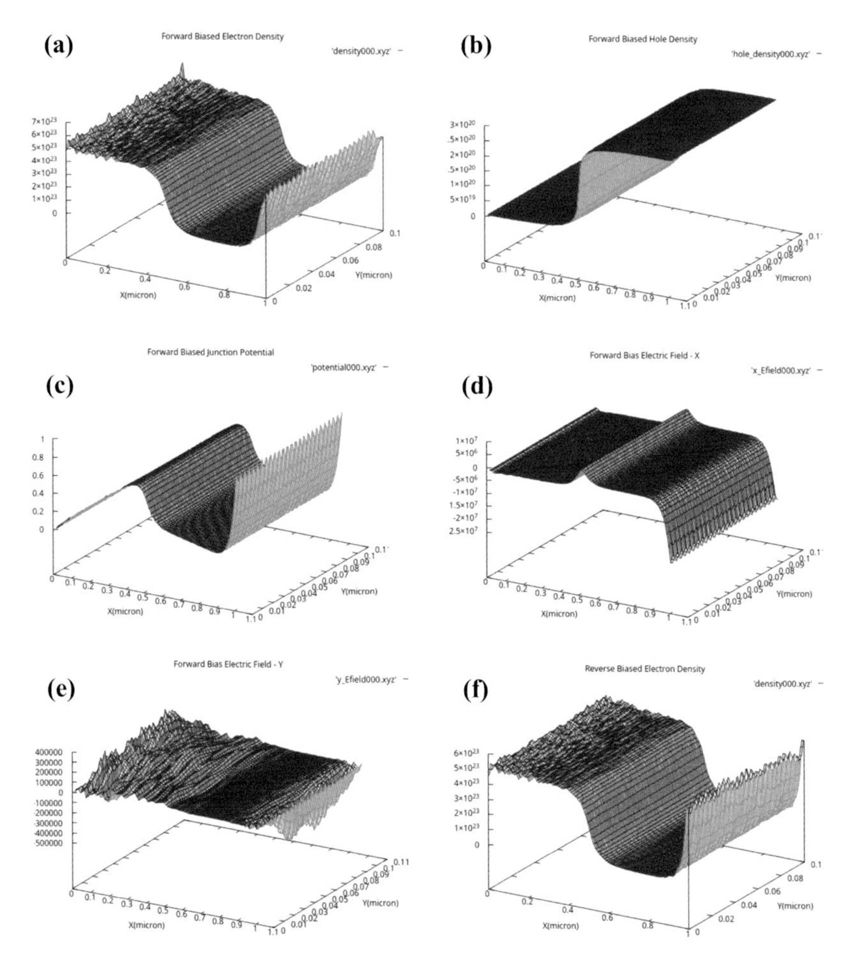

Fig. 13.6 **a** Forward bias abrupt junction np silicon diode electron density (unit $\frac{n}{m^3}$), **b** Forward bias abrupt junction np silicon diode hole density (unit $\frac{n}{m^3}$), **c** Forward bias abrupt junction np silicon junction potential (unit Volt), **d** Forward bias abrupt junction np silicon electric field x component (unit $\frac{V}{m}$), **e** Forward bias abrupt junction np silicon diode electric field y component (unit $\frac{V}{m}$), **f** Reverse bias abrupt junction np silicon diode electron density (unit $\frac{n}{m^3}$), **g** Reverse bias abrupt junction np silicon diode hole density (unit $\frac{n}{m^3}$), **h** Reverse bias abrupt junction np silicon junction potential (unit Volt), **i** Reverse bias abrupt junction np silicon electric field x component (unit $\frac{V}{m}$), **j** Reverse bias abrupt junction np silicon diode electric field y component (unit $\frac{V}{m}$)

freely available and downloadable graphics plotting software package Gnuplot [15] from MIT.

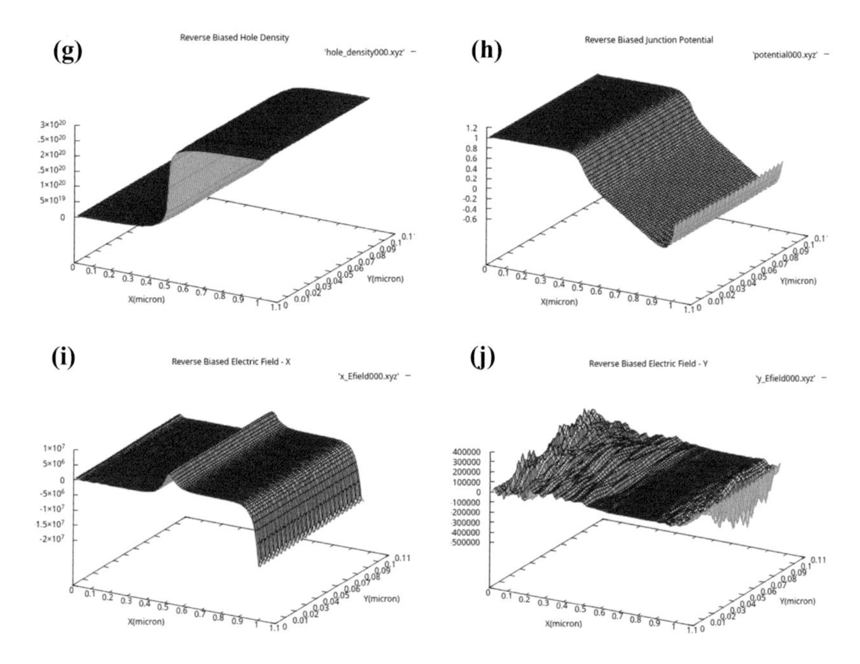

Fig. 13.6 (continued)

References

1. https://www.infineon.com/dgdl/Infineon-BFQ790-DS-v02_00-EN.pdfiieId=5546d4624cb7f111014d237c2a8f54fc
2. https://epc-co.com/epc/Portals/0/epc/documents/datasheets/EPC2216_datasheet.pdf
3. https://www.transphormusa.com/en/document/datasheet-tp65h150g4lsg-650v-gan-fet/
4. https://pdf1.alldatasheet.com/datasheet-pdf/view/6373/NEC/NE34018.html
5. https://download.tek.com/manual/6487-900-01(C-Mar2011)(User).pdf
6. https://www.tek.com/en/products/keithley/dc-power-supplies/2220-2230-2231-series
7. https://download.tek.com/datasheet/2220_2230DataSheet_0.pdf
8. https://www.gnu.org/software/archimedes/
9. https://charon.sandia.gov/
10. https://devsim.org/
11. https://silvaco.com/tcad/
12. https://www.synopsys.com/silicon/tcad.html
13. https://www.ibm.com/topics/monte-carlo-simulation
14. https://www.simscale.com/docs/simwiki/fea-finite-element-analysis/what-is-fea-finite-element-analysis/
15. http://www.gnuplot.info/
16. https://mech.fsv.cvut.cz/~dr/software/T3d/guide/node13.html

14 Semiconductor Optoelectronic Devices

14.1 Basic Solar Cell Structure (Photo Diode)

A solar cell [1–16] is a large area np junction diode, that converts sunlight to electricity via the *photovoltaic effect*. Absorbed photons generate electron–hole pairs, which are separated within the diode resulting in photocurrent. This current flows into an external load circuit, which converts the input current to a voltage. Thus optical energy|power is converted to electrical energy|power—Fig. 14.1a.

In Fig. 14.1a the top metal contact covers a miniscule fraction of the front surface area to allow maximum possible semiconductor surface area to be exposed to the incident light. *The conversion of incident light energy to electrical energy consists of four steps—light absorption, electric current generation, internal transport of photo generated charges, conversion of incoming photocurrent to photovoltaic power in an external circuit.* This analysis uses a single junction diode made from a homogeneous single crystal semiconductor. Real world solar cells are multi-crystalline silicon homogeneous junction cells, thin film compound semiconductor heterogeneous junction cells or state-of-art multi junction, multi semiconductor tandem cells.

14.1.1 Light Absorption by Solar Cells

Some of the incident solar light is reflected, and the remainder is transmitted into the device. The spatial part of the electric field component of the incident sunlight's transverse electromagnetic wave is:

$$E_y(x) = E_0 e^{\frac{j2\pi x}{\lambda_{SEMICONDUCTOR}}} \quad k = \frac{\omega n_{COMP}}{c} = \frac{2\pi(n_R + jk_R)}{c} \tag{14.1}$$

A. Banerjee, *Semiconductor Devices*, Synthesis Lectures on Engineering, Science, and Technology, https://doi.org/10.1007/978-3-031-45750-0_14

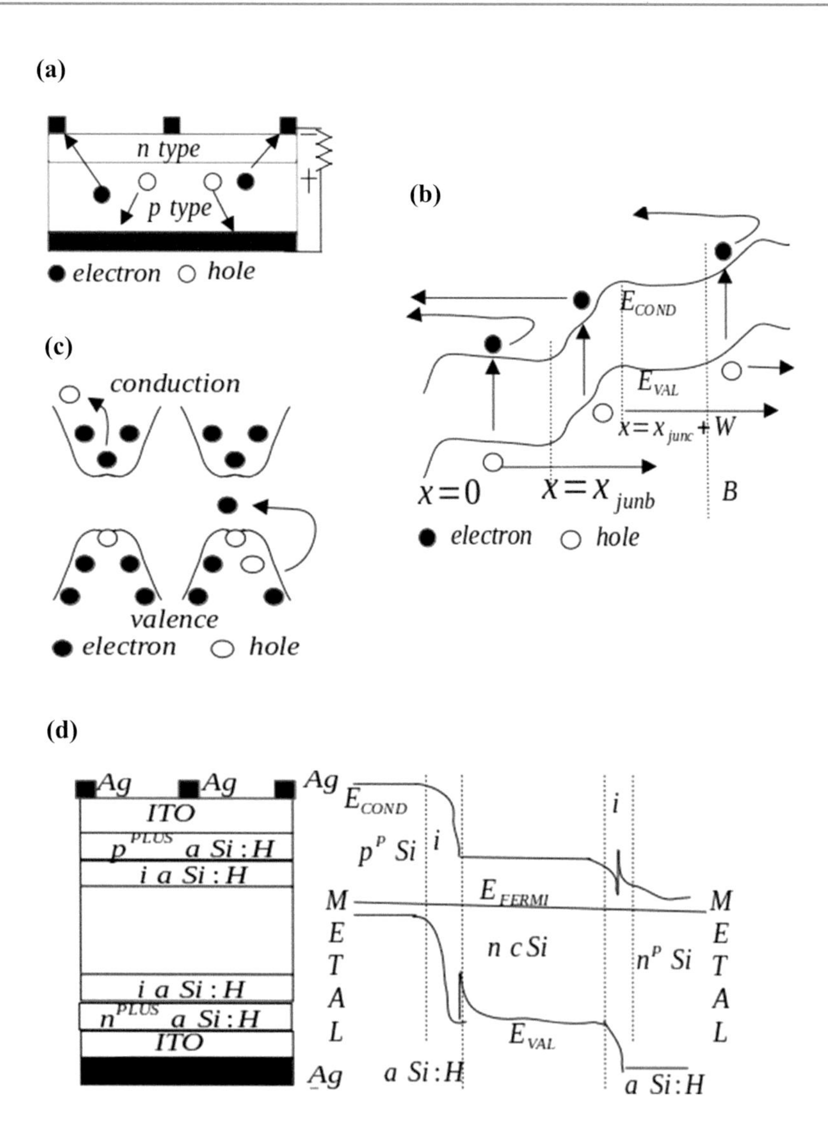

Fig. 14.1 **a** Layered solar cell structure. Light is incident on the top surface. **b** Energy band diagram for a solar cell. B represents the blocking barrier that selectively allows one type of charge carrier to pass through and blocks the other type. **c** Processes that do not generate any photocurrent in the external circuit. **d**, **e** Structure and energy band diagram of heterogeneous junction solar cell where a Si:H denotes hydrogenated amorphous silicon. C Si crystalline silicon

where n_R, k_R are the real and imaginary parts of the complex refractive index of the semiconductor material, and c, λ are the incident light's velocity and wavelength in free space. After substituting for the wave number in (14.1), the electric field is:

$$E_y(x) = E_0 e^{\frac{2\pi(jn_R - k_R)x}{\lambda}} \tag{14.2}$$

The Poynting vector corresponding to an electromagnetic wave embodies energy carried by it and its optical power density. For an uniform plane wave, travelling in the x direction, the magnitude of the power density (as related to its Poynting vector) is:

$$\vec{S} = \vec{E} \times \vec{H} \quad S_x = E_x H_z = \sqrt{\frac{\varepsilon}{\mu}} |E_y|^2 = \sqrt{\frac{\varepsilon}{\mu}} E_y E_y^p = \sqrt{\frac{\varepsilon}{\mu}} E_0^2 e^{-\alpha x} \quad \alpha = \frac{4\pi k_R}{\lambda} \tag{14.3}$$

where a, ε, μ are respectively the absorption coefficient, the electric permittivity and magnetic permeability of the semiconductor material.

14.1.2 Generation of Photovoltage and Photocurrent

The power density S at some free-space wavelength λ can be converted into the photon flux $\phi(\lambda)$ (units $\frac{photons\ second}{m^2}$) at that wavelength by dividing by the photon energy hc/λ. The balance equation for monochromatic photons within the semiconductor is:

$$\frac{\partial P(\lambda)}{\partial t} + \nabla \cdot \phi(\lambda) = R_{RAD}(\lambda) - A(\lambda) \tag{14.4}$$

where P is the volumetric photon density, R_{RAD} is the generation rate of photons within the semiconductor due to the radiative recombination of previously generated electron–hole pairs, and A is the rate of loss of photons due to absorption of photons within the volume. Ideally, for a solar cell all of the absorbed photons must create separable electron–hole pairs, $A(\lambda) = G_{OP}(\lambda)$, the generation rate. But unwanted excitation of free carriers G_{fc} and bound electron–hole pairs (*exciton*) G_{ex} do not contribute to the solar cell's output current—as it moves through the device as a neutral entity. So,

$$A = G_{OP} + G_{fc} + G_{ex} = A\eta_{INTERNAL} + G_{fc} + G_{ex} \tag{14.5}$$

$G_{OP} = \eta_{INTERNAL} A$ is the internal quantum efficiency of the solar cell. In steady state, for a sunlight beam, in one dimension, the internal quantum efficiency is:

$$G_{IO} = \eta_{INTERNAL}(\lambda)\alpha(\lambda)\phi_0(\lambda)\phi_0 = \sqrt{\frac{\eta}{\mu}\frac{E_0^2\lambda}{\hbar c}} \tag{14.6}$$

where ϕ_0 is the photon flux at the solar cell surface. Most silicon solar cells have a high generation rate near the surface of the cell. The roll-off is exponential for all wavelengths, but is pronounced at shorter wavelengths where α is high. So, the peak photon density occurs at approximately 500 nm.

Current generated by photon absorption is the photocurrent. It is created by separating the photo generated electron–hole pairs the electrons and holes to flow in opposite directions. This is achieved by making the solar cell a np diode. The key considerations to fabricate such a np junction are:

- How deep into the device should the metallurgical junction be placed
- The type (n, p) of the top section of the diode
- Appropriate acceptorldonor densities for the two regions (n,p)

As the generation rate decays exponentially with increasing depth into the solar cell, the np junction should be placed very close to the front surface. Thus, the top part of the semiconductor, which is called the **emitter**, is thin. Since the current must move in a lateral path to the metallic contacts on the front surface, the emitter region must be heavily doped to avoid unwanted, parasitic series resistance. The lower part of the diode is called the **base**. The base must be thick (compared to the emitter layer) to collect the long wavelength photons, and provide mechanical strength for the solar cell. Minority carriers generated in this region would travel long distances to the junction, so the doping type of the base must provide the more mobile minority carrier. For solar cell materials as Si and GaAs, this means that the base should be p type. *So a solar cell is a shallow junction, np diode, with a heavily n doped emitter.*

The energy band diagram for a standard solar cell is in Fig. 14.1b. As the generation rate decreases with distance into the cell, both electrons and holes in both of the quasi-neutral regions diffuse to the right. In the emitter, charge carrier separation occurs at the junction, where the electrons are reflected by the built-in potential barrier. To achieve the same effect for carriers generated in the base, a back-surface field is created by heavily doping the end of the p type region. There is also a field at the front surface of the cell due to the donor density gradient, which occurs during the diffusion doping process. In the remaining region of the cell, the space-charge region, carriers generated (in the space charge region) are separated by the built-in junction field.

The expression for the component of the total photocurrent due to generation in the emitter quasi neutral region is obtained by analyzing the diffusion of minority carrier holes, under the assumption that no large gradients in the kinetic energy per carrier exist Then at steady state,

$$\frac{-dJ_{hole}^{E}}{qdx} + \frac{p_0 - p}{\tau_{hole}} + G_{OP} = 0$$

$$J_{hole}^{E}(x) = \frac{-k_B \mu_{hole} T_{lattice} dP}{dx} \stackrel{\text{def}}{=} \frac{-qD_{hole} dP}{dx} 0 < x < x_{junction} \tag{14.7}$$

with the boundary conditions:

$$D_{hole} dP \frac{\iota}{dx} = S_F(p(0) - p_0) \quad x=0 \quad p(x_{junction}) = p_0 \tag{14.8}$$

where S_F is the front surface recombination velocity. The second boundary condition implies that all excess minority-carrier holes reaching the edge of the space charge region are swept across the junction by the built-in field. Using (14.6) for G_{OP} and *assuming an extremely idealized internal quantum efficiency of 100%, the solution for monochromatic light of wavelength λ is*:

$$J_{\text{hole}}^{E}\left(\lambda, x_{\text{junction}}\right) = \frac{\alpha q \phi(0) L_{\text{hole}}}{\alpha^2 L_{\text{hole}}^2 - 1} \tag{14.9}$$

Similar expression for the base generated minority carrier electron photocurrent can be derived using similar technique as for the hole current in the emitter.

In the space charge layer all carriers generated in that region are separated by the built-in field, and the expression for the photo generation current in this region is obtained by analyzing the photocurrent due to either electrons or holes. Assuming no recombination occurs as the carriers are swept out of the space-charge region very fast, and considering the photocurrent for electrons in combination with the Depletion Approximation (to calculate the width W of the space charge region) gives:

$$\frac{dJ_{electron}^{DEPLETION}}{qdx} + G_{OP} = 0 \tag{14.10}$$

Integrating over $x_{junc} \leq x \leq x_{junc} + W$ and *assuming perfect internal quantum efficiency*, gives:

$$j_{electron}^{\text{DEPLETION}} = q\mathrm{e}^{-\alpha x_{junction}}\left(1 - \mathrm{e}^{-\alpha W}\right) \quad x_{\text{junction}} \leqslant x \leqslant x_{\text{junction}} + W \tag{14.11}$$

Then the total photocurrent is the sum of the above three components:

$$J_{PHOTO}(\lambda) = J_{hole}^{E}\left(\lambda, x_{junction}\right) + J_{electron}\left(\lambda, x_{junction} + W\right) + J_{electron}^{DEPLETION} \tag{14.12}$$

The effect of surface fields on minority carriers is controlled by *surface recombination velocity.* The rate of recombination of electrons at the surface is the same as if a flux of electrons of density $(n - n_0)$ were drifting with an average velocity S into the surface and being removed. Thus, in the quasi neutral base at x = B, the electron current is diffusive:

$$\frac{-D_{electron}\, d(n - n_0)}{dx} = \frac{-D_{electron}\, dn}{dx}$$

$$\stackrel{\text{def}}{=} S(n - n_0) S = \infty \quad n(B) = n_0 J_{electron} \; \textit{finite Ohmic}$$

$$S = 0 \quad n(B) > n_0 J_{electron} = 0 \quad \textit{Blocking} \tag{14.13}$$

The electrons and holes constituting the photocurrent flow in the direction of a reverse bias current. When this current flows through an external load, it generates a voltage across that load to forward bias the diode. A current that opposes the photocurrent is developed, so some net current loss occurs. The above expressions for the photocurrent are valid only if the illumination is not so intense so that:

- there are very small deviations in quasi-Fermi levels in the space-charge region from constant values.
- minority carrier lifetime does not change from its value 'in the dark'.

The *diode dark current* is computed using the boundary conditions at the edges of the space charge region, and a boundary condition at the contact ends of the quasi neutral regions. For electrons injected into the base:

$$n\left(x_{junction} + W\right) = n_{0p}e^{\frac{V_{lattice,junction}}{V_{THERMAL}}} \quad \left(\frac{-D_{electron}dn}{dx}\right)_{x=B} = S_B(n(B) - n_0) \qquad (14.14)$$

where S_B is the back-surface recombination velocity, i.e., at x = B, and $V_{L,junc}$ is that part of the load voltage that is dropped across the junction. In absence of series resistance this is the load voltage. Neglecting any current change from recombination in the space charge region, the hole dark current (injected across the forward biased junction into the emitter), the total dark current is the sum of the two injected currents. The load current is the difference of the photo and dark currents:

$$J_{DARK} = J_0\left(e^{\frac{V_{lattice}}{V_{THERMAL}}} - 1\right) J_{LOAD} = J_{PHOTO} - J_{DARK} \qquad (14.15)$$

When the dark current exactly negates the photocurrent, an open circuit exists, and the load voltage is the open circuit voltage. J_0 can be re-written in terms of the minority carrier properties of diffusion length and surface recombination velocity for the emitter and base, the appropriate acceptor, donor concentrations N_A, N_D. The intrinsic carrier concentration depends exponentially on the bandgap so that J_0 can be re-written in terms of the bandgap, the minority carrier properties of diffusion length and surface recombination velocity for the emitter and base and the acceptor, donor concentrations (constant C in 14.16)

$$\begin{aligned} V_{OPEN\,CIRCUIT} &= V_{THERMAL}\ln\left(\frac{J_{PHOTO} + J_0}{J_0}\right) \\ J_{OPEN\,CIRCUIT} &= \frac{E_{GAP}}{q} - V_{THERMAL}\ln\left(\frac{C}{J_{PHOTO}}\right) \end{aligned} \qquad (14.16)$$

For a typical solar cell, when $0 < V_{LOAD} < V_{OPEN\,CIRCUIT}$, the solar cell's I-V curve is in the 4th quadrant, where the current is negative and the voltage is positive—the power is negative, indicating generation, not dissipation. The power at the point of maximum power generation is expressed as a density in $\frac{\text{Watts}}{\text{m}^2}$, given by:

$$\begin{aligned} P_{MAX\,POWER} &= J_{MAX\,POWER} V_{MAX\,POWER} \\ &= FILL_{FACTOR} J_{SHORT\,CIRCUIT} V_{OPEN\,CIRCUIT} \end{aligned} \tag{14.17}$$

The presence of parasitic resistances in the solar cell forces $|J_{SHORT\,CIRCUIT}| < |J_{PHOTO}|$. *Series resistance arises as the vertical path through the p doped base region is much larger than the corresponding vertical path through the very thin highly n doped region.* As a low voltage device, a solar cell is sensitive to series resistance. Shunt resistance arises mainly from internal imperfections at the junction, as well as any conductivity along the vertical edges of the device.

The $FILL_{FACTOR}$ is typically less than unity because the exponential form of the solar cell's I-V characteristic is an approximation to the perfect characteristic that would have $J_{LOAD} = J_{SHORT\,CIRCUIT} = J_{PHOTO}$ $V_{LOAD} < V_{OPRNCIRCUIT}$. The $FILL_{FACTOR}$ ensures that a real cell makes the transition from $J_{LOAD} = J_{SHORT\,CIRCUIT}$ to $J_{LOAD} = 0$ in a smooth manner. This 'softness' of the diode characteristic becomes less important as the bounding area $J_{PHOTO} V_{OPEN\,CIRCUIT}$ is increased. Using a high bandgap material does not improve the $V_{OPEN\,CIRCUIT}$, as the photocurrent density is reduced by the increased transparency of the material to sunlight. Consequently, the photovoltaic conversion efficiency $\eta_{PHOTOVOLTAIC}$ peaks at some value of the bandgap energy.

Thin film solar cells have been designed and built to fulfil the need for solar cells be deposited in situ onto large-area surfaces. A candidate is copper indium gallium diselenide, CIGS. Depending on the ratio of indium to gallium in the material, the bandgap lies 1.04 eV $(gallium\,free) < E_{GAP} <$ 1.7 eV $(selenium\,free)$. Inserting dopants in this ternary compound crystal is difficult, so excess carriers are created by deliberately creating vacancies in the deposited film. For example, if the growth conditions are adjusted so that there is a deficiency of copper in the CIGS layer, then the selenium atoms surrounding the copper vacancies will lack electrons. Selenium atoms accept electrons from elsewhere, and the material becomes p-type. **CIGS solar cells have a p type base and use another semiconductor for the n type emitter—heterogeneous junction diode**. In CIGS solar cells the n-type semiconductor has a large bandgap and does not absorb much sunlight—incident solar energy penetrates the absorbing CIGS film quickly.

14.1.3 Heterogeneous Junction Solar Cell

The structure and energy band diagram of a common silicon based heterogeneous(np junction is between amorphous and crystalline silicon) junction solar cell [16] is in Fig. 14.1 d, e.

Silicon based heterogeneous junction solar cells exploit the fact that the high recombination active contacts are moved away from the crystalline surface by insertion of a film with wide bandgap. To reach the full device potential, the heterogeneous|homogeneous interface state density should be minimum—achieved by having a nanometer thick hydrogenated amorphous silicon (a-Si:H) film, whose bandgap is wider than that of crystalline silicon (c-Si). Amorphous silicon films reduce the crystalline silicon (c-Si) surface state density by hydrogenation. Such films can be doped easily (n|p) type, without lithography fabrication of contacts with very low values for the saturation current density. T*he outcome is impressive large-area (>100 cm*^{2}*) energy-conversion efficiencies (~25%).*

The basic device features on the illumination side are successively an intrinsic hydrogenated amorphous silicon (a-Si:H) passivation layer and a p doped amorphous silicon emitter both deposited by plasma enhanced chemical vapor deposition (PECVD). On top of the silicon layers, an antireflective *transparent conductive oxide* (TCO) is deposited by physical vapor deposition (PVD) and the charge collection is made by a screen-printed metallic contacting grid. On the back side, an electron collecting stack is used, and it is composed of an intrinsic hydrogenated amorphous silicon (a-Si:H) passivation layer, a doped n type amorphous silicon (both deposited by PECVD), a TCO layer and a metallic contacting layer (deposited by PVD).

Low defect density in high purity silicon wafers enables heterogeneous junction silicon based solar cells to have efficiencies over 25%. To achieve this high efficiency device, the surface of the wafer must be free of electronically active defects. High quality surface passivation is achieved by using plasma deposited hydrogenated amorphous silicon (a-Si:H), resulting in large carrier lifetimes and high efficiencies.

A key challenge in fabricating a high efficiency solar cell from a high quality silicon wafer is to ensure that the electrons and holes are collected at two spatially separated terminals. *Selective collection exploits properties of semi-permeable electronic membranes. Such membranes provide a low-resistance electrical connection for one type of charge (e.g. electrons) while blocking with minimal leakage the other type (holes).* Doped amorphous silicon layers (n|p type a-Si:H) is the ideal way to provide the desired charge carrier selectivity. However, a-Si:H films introduce limitations as parasitic absorption of light and non-ideal selectivity (non-negligible resistance to charge extraction and low lateral conduction).

A number of interesting solar cell structure are being actively investigated at present.

Dopant-free solar cells: Although it was believed for long that an efficient photovoltaic device needs doped contacts of opposite polarities, recent understanding of the physics of solar cells does not support this. Experimental demonstration of a high-efficiency *entirely dopant-free crystalline silicon cell*, using slightly sub-stoichiometric molybdenum trioxide (MoO3) and lithium fluoride (LiF) as hole|electron selective contacts—promises entirely new device architecture, with much simplified processes and extremely simple designs.

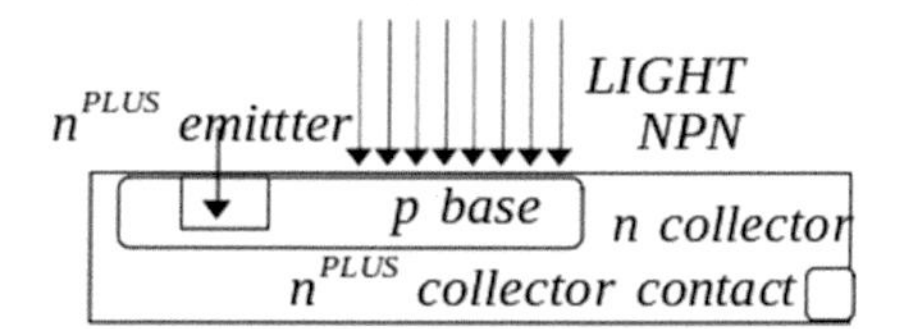

Fig. 14.2 Layered structure of a NPN phototransistor

Interdigitated back contact (IBC) solar cells: Metal contacts extract the electrical charges from a silicon solar cell. In the traditional architecture solar cells negative (electrons) and positive (holes) charges are collected on each side of the wafer, the IDC design collects both charge types on the rear of the wafer. All metal contacts are placed at the rear of the wafer, preventing shadowing and allowing a higher current to be generated. Conceptually simple but difficult to implement in a real world solar cell.

Small-area devices: So long photovoltaic cells were small area devices (1 cm^2 or less). Recent record efficiencies for wafer based silicon devices were obtained with areas>100 cm^2. The large diffusion length of photogenerated carriers in silicon (typically of millimeter scale) makes edge recombination a particular issue, and the fabrication of small devices challenging.

A phototransistor is a simple extension of a photodiode. Figure 14.2 shows a NPN phototransistor.

14.2 Light Emitting Diodes (LEDs)

14.2.1 Heterogeneous Junction Light Emitting Diodes

A light emitting diode (LED) is a pn-junction diode in which generates light (in a selected frequency band of the electromagnetic spectrum, e.g., infra red, visible etc.,) based on radiative recombination, under forward bias operating conditions. There is conversion of electrical energy to optical energy, *opposite to what happens in a solar cell.* In most LED designs, the internally generated photons escape through the top surface which cannot be covered by a top metallic contact. A number of performance metrics are associated with a LED as listed below [17].

- Voltage efficiency. This relates the applied voltage to the bandgap of the semiconductor. The bandgap determines the desired color of emitted light.
- Current efficiency. This relates the current due to recombination in the desired part of the device to the current due to recombination elsewhere. Estimating this efficiency is very important for heterogeneous junction LED design. Modern high intensity LEDs are heterogeneous junction devices.

- Radiative efficiency. Correlates amounts of radiative recombination and unwanted, non-radiative radiation. Estimating this efficiency is key to selecting the LED material (direct bandgap), and its properties as doping level and purity.
- Extraction efficiency. Quantifies hiw efficiently the photons can come out of the semiconductor, and determines the substrate, contacts, and device shape.
- Wall-plug efficiency. Measures the overall efficiency of the electrical to optical conversion process.

Voltage Efficiency: At equilibrium, electrons occupy states near the bottom of the conduction band, and holes reside near the top of the valence band. An applied forward bias perturbs the distribution of both electrons and holes, but the energy of any photons resulting from radiative recombination will be close to that of the bandgap. The voltage efficiency is defined as:

$$\eta_{VOLTAGE} \stackrel{\text{def}}{=} \frac{\hbar\omega}{qV_{APP}} \approx \frac{E_{GAP}}{qV_{APP}} \tag{14.18}$$

This equation defines the applied voltage required to get a voltage efficiency of unity when using the semiconductor that has been chosen for a particular frequency of light output—$V_{APP} < \frac{\hbar\omega}{q}$ would lead to higher voltage efficiencies. As the carrier concentrations depend exponentially on potential, low biases do not result in a high overall efficiency.

Current efficiency: The rate of radiative recombination is proportional to the local electron hole concentration product. In a forward biased diode, the minority carrier concentrations are elevated above their equilibrium values over distances that extend from the space charge layer by several minority carrier diffusion lengths. Photon generation will occur over a length of the order of microns.

To obtain intense photon generation, the minority carriers must be concentrated into a smaller region. Modern high brightness LEDs use dissimilar materials for the p- and n-regions of the diode (Fig. 14.3). *This structure uses a low bandgap material sandwiched between two higher bandgap materials to form a potential well in which the minority carriers are trapped, thereby increasing their concentrations and the intensity of the generated light. The region of carrier confinement is called the active layer.*

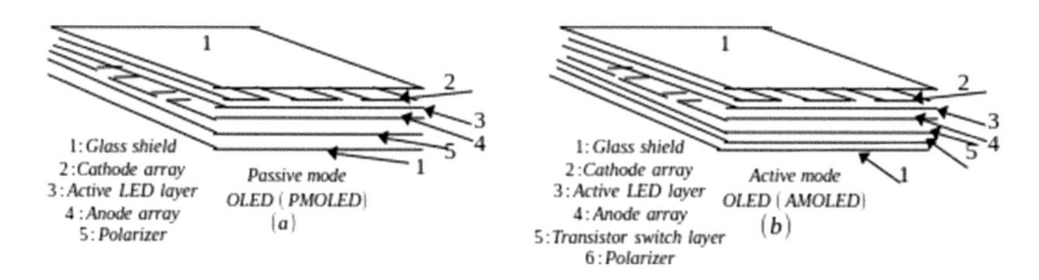

Fig. 14.3 **a**, **b** Passive mode organic light emitting diode (PMOLED - **a**) and active mode organic light emitting diode (AMOLED - **b**)

In a heterogeneous junction, the current due to electrons injected from the n region recombine in the active region with holes that are injected from the p region. The electron current density, in steady state and with no non thermal generation of electron hole pairs, is:

$$\frac{d\ J_{electron}}{q\ d\ x} - \frac{\Delta n}{\tau_{electron}} = 0 \tag{14.19a}$$

Integrating over the active region bounded by $x = 0$ and $x = H$, while noting that the minority carrier lifetime is uniform over the active region,

$$\int \left(\frac{\Delta n dx}{\tau_{RADIATION}} \right) = \frac{\tau_{electron}}{q\tau_{RADAAION}} \int \left(\frac{d J_{electron}}{dx} \right)$$
$$= \frac{\tau_{electron}}{q\tau_{RADIATON}} (J_{electron}(H) - J_{electron}(0))\ 0 \leqslant x \leqslant H \tag{14.19b}$$

where $\tau_{RADIATION}$ the radiative recombination lifetime. Now the current efficiency can be defined as:

$$\eta_{CURRENT} = \frac{J_{electron}(H) - j_{electron}(0)}{J_{TOTAL}} \tag{14.20}$$

under the assumption that the electron current density is zero in the p region.

$J_{electron}(0)$ is reduced to prevent electrons from escaping into the p region, achieved by making a large energy barrier for electrons at the interface between the active region and the p region. Likewise, a barrier at the active—n region junction prevents the holes from escaping from the active region. These properties are included in the LED by implementing Type I heterogeneous junctions. *LEDs are p-on-n diodes, unlike solar cells, which are n-on-p diodes.* The junction region in the LED does not have to be close to the surface, as the photons are all of energy very close to the bandgap, and so the relevant absorption coefficient is low. To provide mechanical rigidity the bottom layer of the LED is thick, and n type material is used to minimize series resistance. 'Vertical' resistance dominates in this device because there is little lateral current, and the LED has a much smaller cross-sectional area than a solar cell. As a LED operates in forward bias, the current is large and series resistance is avoided to obtain the desired junction.

14.2.2 Radiative Recombination Efficiency

A LED operates if recombination is confined to a well-defined active region, and to ensure that maximum light is generated, the recombination is radiative. So the active material should be a direct bandgap semiconductor. For a high rate of radiative recombination, high concentrations of both types of carrier are needed. As a high level of electron and hole injection into the active layer must be established,—the LED must be forward biased.

The background doping density of the active layer is usually chosen to make the region p type. So electrons are the minority carriers, and their superior minority carrier mobility ensures a more uniform carrier distribution in the active layer. Therefore a more spatially uniform generation of photons is ensured. The net rate of recombination for electrons is $\frac{\Delta n}{\tau_{electron}}$. The radiative recombination efficiency is:

$$\eta_{RADIATION} = \frac{\tau_{electron}}{\tau_{radiation}} \tag{14.21a}$$

The principal non-radiative contributors to $\tau_{electron}$ are generation-recombination center recombination and Auger recombination. Recombination generation recombination in the active layer is reduced by ensuring that the active region material is of high crystalline purity|perfection. The active-layer material must be lattice matched to the substrate material on which it is epitaxially grown.

14.2.3 Extraction and Wall Plug Efficiency

The *extraction efficiency* is the ratio of the optical power that actually escapes from the structure to the optical power that is generated within the diode.

$$\eta_{EXTRACTED} = \frac{S_{OUT}}{S_{GEN}} \tag{14.21b}$$

Extracting the light is inherently difficult because the relatively large value of refractive index for most semiconductors means that the critical angle beyond which total internal reflection occurs, is small. For example, GaAs has a refractive index of about 3.5, so, at the semiconductor/air interface, from Snell's Law:

$$\theta_{CRITICAL} = \arcsin\left(\frac{1}{3.5}\right) \approx 1.7^{o} \tag{14.21c}$$

The spontaneous light emission from an LED is omnidirectional, and if the light is assumed to emanate from a point source, it is easy to estimate the optical power that exits through a segment of a spherical surface defined by a polar angle equal to the critical angle. Then the corresponding output optical power is approximately given by:

$$\eta_{EXTRACTION} = \frac{S_{OUT}}{S_{GEN}} = \frac{1 - \cos(\theta_{CRITICAL})}{2} \tag{14.21d}$$

For a point source in GaAs, this means that only about 2% of the optical power would escape into the surrounding air. However, semiconductor layers in an LED are usually deposited epitaxially onto a substrate, and the resulting structure is planar. Thus, the light source is planar rather than a point, so a significant fraction of light generated inside the LED can escape. The semiconductor die can be sawed and wedge-shaped diodes, in which total internal reflection is reduced by effectively increasing the critical angle.

Wall plug Efficiency: The output power density of a LED is:

$$S_{OUT} = \eta_{CURRENT}\, \eta_{VOLTAGE}\, \eta_{EXTRACTION}\, \eta_{RADIATION}\, J_{DIODE}\, V_{DIODE} \tag{14.21e}$$

where V_{DIODE}, J_{DIODE} are the applied bias and diode current density, respectively. Rearrangement of this equation leads to an expression for the overall efficiency of the electrical to optical conversion process, LED wall-plug efficiency:

$$\frac{S_{OUT}}{P_{INPUT}} = \eta_{CURRENT}\, \eta_{VOLTAGE}\, \eta_{EXTRACTION}\, \eta_{RADIATION} \tag{14.21f}$$

The wall-plug efficiency is a radiometric measurement based on measurable physical properties of power, that can be measured by a calibrated photodetector and a wattmeter. Common figures-of-merit for LED optical performance are based on photometric units, i.e., optical power that is perceived by the human eye. The eye is most sensitive to the color green. So, the optical power perceived by the eye can be expressed in watts as $\int \gamma S^P_{OUT}(\lambda)d\lambda$ where the prime signifies power, not power density, and $S_{OUT}(\lambda)$ is the spectral power density in W/m. The perceived optical power is expressed in units of lumens (lm), by applying a conversion factor of 683 lm/W. This factor is based on the old light intensity unit of candle power, originally defined in terms of the light output of a standard candle. Thus, S watts of optical power are perceived by the eye as lumens of luminous flux according to:

$$\Phi = 683.0 \int \gamma S^P_{OUT}(\lambda)d\lambda. \tag{14.21g}$$

14.2.4 Luminous Efficiency and Efficacy

The luminous efficacy measures the effectiveness of the eye in perceiving optical power:

$$Luminous\ Efficacy = \frac{\Phi}{S^P_{OUT}} \frac{lumens}{\text{Watt}} \tag{14.21h}$$

The luminous efficiency is also expressed in lumens per Watt and relates the perceived optical power to the electrical power supplied to the LED:

$$Luminous\ Efficacy = \frac{\Phi}{I_{DIODE}V_{DIODE}} \frac{lumens}{\text{Watt}} \tag{14.21i}$$

Luminous efficiency is the product of the wall plug efficiency and the luminous efficacy.

The light from blue green InGaN LED, mixed with output with red light from AlGaInP LED generates white light. The human eye senses color through rod and cone cells in the retina that are specifically sensitive to red, green, or blue light. Combinations of certain intensities of these colors are perceived by the eye as white light. For standardization and colorimetric reasons, three color matching functions have been developed. Each is centered on a particular wavelength (red, green, or blue) and has some spectral content such that specific combinations of the intensities of the three sources can produce any desired color. The chromaticity coordinates corresponding to the red, green, and blue color-matching functions are labeled x, y, z, respectively. The x-coordinate, e.g., measures the ratio of the stimulus of the red cells in the eye to the total stimulus of all the color cells to the entire visible spectrum. Thus, $x + y + z = 1$, and it is only necessary to stipulate two chromaticity coordinates when specifying a particular color. This is the basis of the chromaticity diagram.

14.2.5 Organic Light Emitting Diodes (OLED)

An *Organic Light Emitting Diode* (OLED) [21] has an emissive electroluminescent layer of film made up of ***organic*** (carbon based) molecules. Light is emitted when electrical current travels through the organic molecules. The advantages of an OLED is that it is thinner than a conventional LED, and generates brighter, higher contrast display with faster response times, wider viewing angles, and less power consumption (all properties in reference to a conventional LED).

The OLED has a simple planar layered structure going from top to bottom.

- Topmost glass shielding layer.
- Cathode: a negatively charged cation attracting layer.
- Emissive layer: consisting of organic molecules or polymers that performs the key task of transporting electrons from the cathode to the anode.
- Conductive layer: also consisting of organic molecules or polymers that transport holes from the anode layer
- Anode: Positively charged electron or anion attracting layer
- Substrate: Bottom glass shielding layer.

Light generation mechanism of an OLED is straightforward. Electrical current is applied to both the anode and cathode. **Current traverses both the organic material emissive and conductive cathode and anode layers, to the anode layer**. *The current provides electrons to the emissive layer and removes electrons from the conductive layer. So holes are generated in the conductive layer. The generated holes in the conductive layer need to be re-filled with electrons.* **To recombine with electrons, the holes jump from the conductive layer to the electron filled emissive layer. Electron hole recombination generates energy in the visible part of the electromagnetic spectrum—the bright, electroluminescent light that is visible through the outermost layer of glass (substrate and seal)**. Instead of having a backlight, like in LEDs, the OLED's display is self-illuminating because of its organic material. Consequently, OLEDs are significantly thinner than standard liquid crystal displays.

OLED displays are of three varieties, passive mode (PMOLED), active mode (AMOLED) and transparent OLEDS.

The physical structure of the passive mode (PMOLED) is simple. The anode and cathode layers consist of nanometer wide strips at tight angles to eachother, with the emissive and conductive layers sandwiched in between. Light is generated at the intersection points of the cathode and anode layers (Fig. 14.3a).

The active mode OLED (AMOLED) Active-matrix OLEDs (AMOLED) have a thin film transistor layer (TFT) beneath the anode layer which forms a matrix (Fig. 14.3b). The TFT array determines which pixels get switched on in order to form an image. The TFT arrays require less power than external circuitry, so AMOLEDs are more energy efficient than PMOLEDs. With faster refresh rates than PMOLEDs, AMOLEDs are ideal for large displays.

Transparent OLEDs are used specifically for head up displays in military aircraft. When the OLED is switched off it is almost 85% transparent. When powered on, light can pass through in both directions. Transparent OLEDs can also be either a passive- or active-matrix. The military aircraft pilot can look straight ahead, without having to move his|her head up|down|sideways to check key aircraft performance parameters.

14.3 Phototransistor Structure and Performance Characteristics

A phototransistor is an extension of the photodiode, whose equivalent electrical circuit consists of a photodiode whose output photocurrent drives the base of a small signal transistor. Phototransistors performance characteristics are a combination of the performance characteristics of both these devices, as listed below.

14.3.1 Spectral Response

The output of a phototransistor is dependent upon the wavelength of incident light: starting from near UV wavelengths, through the visible and into the near infra red part of the spectrum. Without optical filters, the peak spectral response is in the near infra red at approximately 840 nm. The peak response is at a shorter wavelength than that of a typical photodiode, since the diffused junctions of a phototransistor are formed in epitaxial rather than crystal grown silicon wafers. Phototransistors will respond to fluorescent or incandescent light sources but display better optical coupling efficiencies when matched with infra red LEDs as GaAs (940 nm) and GaAlAs (880 nm).

14.3.2 Sensitivity

For a given light source illumination level, the output of a phototransistor is defined by the area of the exposed collector base junction and the DC current gain of the transistor. The collector base junction of the phototransistor functions as a photodiode generating a photocurrent which is fed into the base of the transistor section. Thus, just like the standard photodiode, doubling the size of the base region doubles the amount of generated base photocurrent (IP), which then gets amplified by the DC current gain of the transistor. Similarly, like signal transistors, hFE (forward DC current gain) varies with base drive, bias voltage and temperature. At low light levels the gain starts out small but increases with increasing light (or base drive) until a peak is reached. As the light level is further increased the gain of the phototransistor starts to decrease. The DC current gain also increases with increasing the collector emitter voltage.

14.3.3 Linearity

Unlike a photodiode whose output is linear with respect to incident light over 7–9 decades of light intensity, the collector current (IC) of a phototransistor is linear for only 3–4 decades of illumination, because the DC gain (hFE) of the phototransistor is a function of collector current (IC) which in turn is determined by the base drive. The base drive may be in the form of a base drive current of incident light. While photodiodes operate best when linear output versus light intensity is extremely important, as in light intensity measuring equipment, the phototransistor in contrast performs best as a switch. When light is present, a phototransistor or photodarlington can be considered "on", a condition during which they are capable of sinking a large amount of current. When the light is removed these photodetectors enter an "off" state and function electrically as open switches.

14.3.4 Collector-Emitter Saturation Voltage—$V_{CE,SATURATION}$

Saturation condition is when both the emitter base and the collector base junctions of a phototransistor become forward biased. From a practical standpoint the collector emitter saturation voltage, $V_{CE(SAT)}$ indicates how closely the photodetector approximates a closed switch. This is because $V_{CE(SAT)}$ is the voltage dropped across the detector when it is in its "on" state. $V_{CE(SAT)}$ *is specified as the maximum collector emitter voltage allowed at a given light intensity and for a specified value of collector current.*

14.3.5 Dark Current—I_{DARK}

When the phototransistor is placed in the dark and a voltage is applied from collector to emitter, a certain amount of photocurrent will flow—the *dark current (ID)*. This current consists of the leakage current of the collector base junction multiplied by the DC current gain of the transistor. Therefore the phototransistor is never completely "off", i.e., it is not and ideal switch with "close-open" transitions. The dark current is specified as the maximum collector current permitted to flow at a given collector emitter test voltage. The dark current is a function of the value of the applied collector emitter voltage and ambient temperature.

14.3.6 Speed of Response

The speed of response of a phototransistor is controlled totally by the capacitance of the collector base junction and the value of the load resistance. These dominate due to the Miller Effect which multiplies the value of the RC time constant by the current gain of the phototransistor. Consequently, for devices with the same active area, the higher the gain of the photodetector, the slower will be its speed of response. A phototransistor takes a certain amount of time to respond to sudden changes in light intensity. This response time is usually expressed by the rise time (tR) and fall time (tF) of the detector where:

tR–The time required for the output to rise
from 10 to 90% of its on state value.
tF–The time required for the output to fall
from 90 to 10% of its on state value.

As long as the light source driving the phototransistor is not intense enough to cause optical saturation (characterized by the storage of excessive amounts of charge carriers in the base region), risetime equals falltime. If optical saturation occurs, tF can become much larger than tR. Phototransistors must be properly biased in order to operate. Specifically, applied voltages must not exceed the collector emitter breakdown voltage (VBRCEO) or the emitter–collector breakdown voltage (V_{BRECO}). Else permanent damage to the phototransistor results.

The generic layered structure of a phototransistor is in Fig. 14.3. Although at start phototransistors were fabricated with germanium and silicon, currently gallium arsenide is a popular choice for phototransistor fabrication. Although phototransistor normally implies a bipolar transistor, photo field effect transistors are also available, but less commonly used. Also, phototransistors are not as widely used as photodiodes, as they have narrower operating wavelength ranges, lower quantum efficiency, smaller active detection area and detection bandwidth. Although the phototransistor has higher responsivity than a photodiode, it does not translate to reduced noise equivalent power, as photocurrent and dark current are both amplified by the device. Moreover the large base collector capacitance reduces detection bandwidth, so that rise-fall times are in the microsecond range. Although avalanche phototransistors have higher responsivity than ordinary phototransistors, the collector emitter current generation mechanism is different, i.e., avalanche breakdown.

Field effect phototransistors (photoFET) are voltage controlled devices. Light incident on the gate region generates a voltage that switches the transistor on. Also, the incident electromagnetic radiation does not have to be in the visible wavelength range—infra red radiation can also trigger a photoFET.

14.4 Charge Coupled Devices

An optical instrument, e.g., a telescope integrates the incident photons, over long time exposures, enabling the user can see faraway objects. The human eye responds to the rate at which photons reach the retina. *At optical frequencies|wavelengths, photoelectric effect ejects electrons from the surface the photons are incident on—the charge coupled device (CCD).*

The quantum efficiency of a light detector is the ratio of the number of photons detected to the number of incident photons. Charge coupled devices (CCDs) have quantum efficiencies of 10–90% over the entire optical frequency range, and this can be extended using clever techniques. A CCD consists of a thin (approximately 10μ) n type silicon layer, on top of a thick (about 250–300μ) p type silicon layer. The n doped region creates *fixed positive* charges, and the p doped region contains *fixed negative* charges. Before a CCD is exposed to radiation, the n type substrate is purged of electrons. Consequently, the positive free charges in the p-type silicon, repelled by the overlying positively charged n type substrate, move to the bottom of the p type substrate and form an "undepleted region" (Fig. 14.4a). Conduction is possible because of free holes. *However, above this undepleted region, and in both the n type and p type silicon layers, the lack of free charges prevents conduction.* In the p depleted layer an electric field in the downward direction (motion of a positive test charge) occurs as a result of the high positive charge in the n type region, which has a greater concentration of fixed positive charges than in the p type region. **The device is now ready for operation—capturing the photoelectrons, which are forced upward into the upper n layer, whose positive charge traps them**.

Incident photons generate photoelectrons and for standard CCDs, one electron is produced for each detected photon. Silicon with band gap energy of 1.14 eV, absorbs photons of energies of 1.1–1.4 eV, energizing valence electrons to gain enough energy to move into the conduction band. The wavelength sensitivity|quantum efficiency of CCDs can be deduced from the path length of light as a function of wavelength in silicon. The quantum efficiency also depends on the width of the CCD. Photons of incident light with

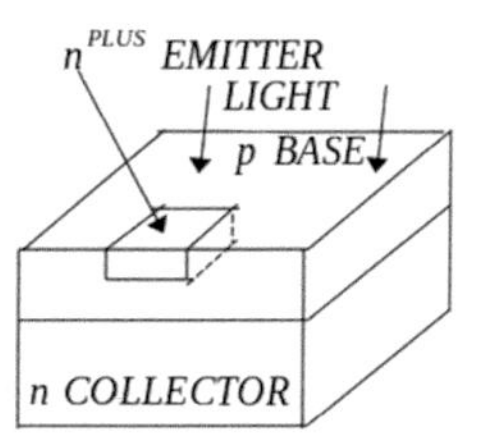

Fig. 14.4 Generic phototransistor physical layered structure. Individual devices could be heterogeneous|homogeneous junction type

wavelength >0.35μ and <0.8μ are absorbed by the backside illuminated CCDs, which are only ~15μ thick, as well as by the front side illuminated CCDs of ~300μ width. At wavelengths shorter than ~0.35μ, 70% of the incident light is reflected. An anti-reflection coating can partly remedy this and extend the sensitivity to 0.3 μm, but at wavelengths larger than 1μ, the CCD is transparent.

The photoelectrons released into the conduction band do not recombine with the silicon, which they would normally do. These generated photoelectrons are quickly forced into the n type layer as a result of the electric field, where they are trapped in the potential well. At the end of the observation, the number of stored electrons depends on the integrated intensity of light. The amount of charge that a pixel can hold after normal integration is its **well capacity**. Integration times must be set so as not to exceed the well capacity of the detector.

Picture elements, or pixels are generated by applying potential across the top n type layer. Each pixel generally has 3 electrodes of different positive voltages, so that the electrons gather at the spots within the array (and within the pixel) where the potential is highest. In Fig. 14.4a, electrode "1" induces the highest potential, ~+5 V, and the photons gather at this place. CCDs consist of an array of ~15μ wide pixels.

The electrodes' (generally 3), applied voltage can be increased|reduced within a upper and a lower threshold value. To read out the CCD, the potentials are sequenced simultaneously, to move the charges, along a column, which is connected in parallel, towards an "output register"—an unexposed row of pixels at the end of the exposed chip. This shifting of electrons to the output registrar is achieved by altering the voltages in the 3 gates, simultaneously in all pixels. The voltage of electrode "1" is dropped to ~2.5 V, while"2" is raised to a higher voltage. In the next step "1" is dropped to zero, and 2 raised to 5 V so that the electrons migrate to electrode "2". In the next two steps, the electrons migrate to electrode "3". In the final 2 steps the electrons move to electrode "1" in the adjacent pixel, which has been vacated by the same process, conducted simultaneously. Therefore, in 6 voltage adjustment steps, the electrons migrate to the adjacent pixel. Once an entire row is shifted into this output register row, the charges on the "output register" row pixels are shifted, similarly using variable voltages, towards the output electronics. The sequence of operations are partially shown in Fig. 14.4b. The number of photons that can be detected depends on: well capacity, digitization rate at which raw pixel data is converted to usable data and the electron capacity beyond which the output of the CCD becomes non-linear (*readout noise*). It is very difficult to detect the onset of readout noise.

The readout time is typically kept to 50 microsecond/pixel, to prevent amplifier heating (with the higher current). This minimizes readout noise—approximately 10 electrons/pixel/readout. Read noise occurs in part during the analog to digital conversion, as well as semiconductor device specific noise currents, as thermal, shot noise(s) etc., Readout noise can be controlled by binning the pixels before readout. Then if four pixels are binned, the signal is affected by only one read out noise, rather that four times the value. The output register pixels (non-illuminated) are normally 5–10 times the number of illuminated pixels so that binning is possible even for high signals.

When CCD images are converted from analog to digital, a DC bias is added to prevent the signal from being negative at any point—analog to digital converters have unable to tackle negative numbers. The signal is kept positive to use all the bits in the A/D converter on the signal rather than the minus sign. The value of the bias varies with position across the CCD and with the read out time. The bias can be measured with *overscan* strips or the measurement of a *bias frame*. Overscan strips are added columns to the CCD, which are recorded and read out with no signal, after the entire CCD image is read out. They are averaged for each row, thereby giving a measure of the zero signal point, and subtracted from each image pixel in that row. Another scheme is to eliminate the bias is to record bias frames, i.e. images taken with no integration time and a closed shutter. These images represent the zero point in the data.

A very fast and efficient analog to digital conversion scheme is delta sigma conversion. This scheme trades amplitude for temporal accuracy—a zero crossing resulting from input signal amplitude transition from positive to negative results in zero value in the output signal amplitude. During the interval when the input signal amplitude is negative, the output signal amplitude is zero.

14.5 Semiconductor Lasers

14.5.1 Laser Fundamentals [18]

For any material immersed in electromagnetic radiation, two things can happen.

An atom with an initial energy E_i absorbs a photon of energy $h\nu$ and transitions to a new state with energy E_f $E_f > E_i$—stimulated absorption. Albert Einstein proposed that the probability per unit time that this happens is proportional, through a constant, to the energy density of the radiation, at the frequency ν, that is $f(\nu)$ (14.22a). The frequency of the stimulated events with transition $i \rightarrow f$ referred to the unit of volume of the material is (14.22a).

$$R_{if} = B_{if} f(v) \quad N_i = n_i R_{if} \quad N_f = n_i B_{if} f(v) \tag{14.22a}$$

where n_i is the number of atoms per unit volume, in the initial energy state and B is Einstein's B coefficient. From the higher energy state (excited), the atom can return to its initial energy state (ground) via spontaneous and stimulated emission.

In *spontaneous emission*, each excited atom can return to the ground state any time it wants to, so that the **emission of radiation is *out of phase***. Upon reaching the ground state, the relaxed atom emits the extra energy as radiation—basis for non laser light. In contrast, for *stimulated emission*, ***all* the excited atoms return to the ground state at the same time, emitting the extra energy at the same time—*in phase***. This is light emission by stimulated emission of radiation, LASER.

The probability R_{fi} per unit time that the atom returns to its ground state is (14.22b), where A_{fi} is Einstein's A coefficient, and represents the spontaneous emission term, while $B_{fi}f(\nu)$ is the stimulated emission term. Spontaneous emission does not depend on the radiation $f(\nu)$, while the stimulated emission does, and is proportional to the stimulating radiation density, that is, $f(\nu)$. The emission frequency with transition $f \rightarrow i$, just like in the $i \rightarrow f$ case is (14.22b):

$$R_{\beta} = A_{\beta} + B_{\beta} f(v) \quad N_i = N_f = n_f R_{\beta} = n_f \left(A_{\beta} + B_{\beta} f(v)\right) \tag{14.22b}$$

where n_f is the population of the final state.

At equilibrium, the number of photons emitted per time must be equal to the number of absorbed photons, and this expression after some manipulation gives the radiation density as:

$$f(v) = \frac{\frac{A_{\beta}}{B_{\beta}}}{\frac{n_i B_{if}}{n_f B_{fi}} - 1} \quad n_i = n_0 e^{\frac{-E_i}{k_B T_{lattice}}} \quad n_f = n_0 e^{\frac{-E_f}{k_B T_{lattice}}} \quad f(v) = \frac{\frac{A_{\beta}}{B_{\beta}}}{e^{\frac{\hbar \nu}{k_B T_{lattice}}} \frac{B_{if}}{B_{fi}} - 1} \tag{14.22c}$$

From Planck's blackbody radiation law (14.22c), the expressions for the ratios of Einstein's A, B coefficients are (14.22d):

$$f(v) = \frac{8\pi \hbar \nu^3}{c^3 \left(e^{\frac{\hbar \nu}{k_B T_{lattice}}} - 1\right)} \quad B_{if} = B_{fi} \quad \frac{A_{fi}}{B_{fi}} = \frac{8\pi \hbar \nu^3}{c^3} \tag{14.22d}$$

The B coefficient for the stimulated emission is equal to coefficient of stimulated absorption. Also, higher ν implies higher level of spontaneous emission. At thermal equilibrium the ratio of the probability of a spontaneous decay and that of a stimulated decay is:

$$\frac{A_{fi}}{B_{fi} f(v)} = e^{\frac{\hbar \nu}{k_B T_{lattice}}} - 1. \tag{14.22e}$$

A laser is a device in which, through external methods, the populations of the initial and final states are brought to values different from the thermal equilibrium values.

In a typical laser, the atoms in the excited state are allowed to fall back to a state which has an energy value intermediate between the excited state energy and the final ground state energy. *The key property of the intermediate state is that the atoms in this state have long mean life times, e.g., of the order of milliseconds.* **Therefore, a large number of atoms collect in this intermediate (metastable) state and after that they all return to the ground state *at the same time—i.e., in phase* emitting stimulated emission.**

14.5.2 Semiconductor Lasers

Semiconductor lasers [19–22] are *direct bandgap* semiconductor material based solid state lasers to exploit stimulated emission from interband transitions. These transitions result from high carrier density in the conduction band. Figure 14.5a illustrates how gain is achieved in an optically pumped semiconductor material. In absence of an external stimulus (the optical pump) all the electrons are in the valence band. An external optical pump beam with photon energy slightly above the semiconductor's band gap energy can push some of these electrons high into the conduction band. From these high energy excited states, the electrons fall back to energy levels at the bottom of the conduction band—metastable state. Simultaneously, the holes generated in the valence band move to the top of the valence band. Electrons near the bottom of the conduction band can then recombine with these holes, emitting photons with an energy near the bandgap energy. This process is enhanced by external pump providing photons with appropriate energy values. *In indirect band gap semiconductors (e.g. silicon), the conduction band electrons in the holes acquire substantially different wavenumbers. Consequently, optical transitions cannot take place as momentum is not conserved.*

Although most semiconductor lasers are optical diode lasers, optically pumped semiconductor lasers and quantum cascade lasers are also available. Diode lasers are pumped with electrical pulses at the np junction. Optically pumped lasers use an external optical pump and quantum cascade lasers exploit *intraband* transitions. Direct bandgap semiconductor laser materials include:

- Aluminum arsenide (AlAs)
- Aluminum gallium arsenide (AlGaAs)
- Gallium phosphide (GaP)
- Indium gallium phosphide (InGaP)
- Gallium nitride (GaN)
- Indium gallium arsenide (InGaAs)
- Indium gallium nirtride arsenide (InGaNAs)
- Indium phosphide (inP)

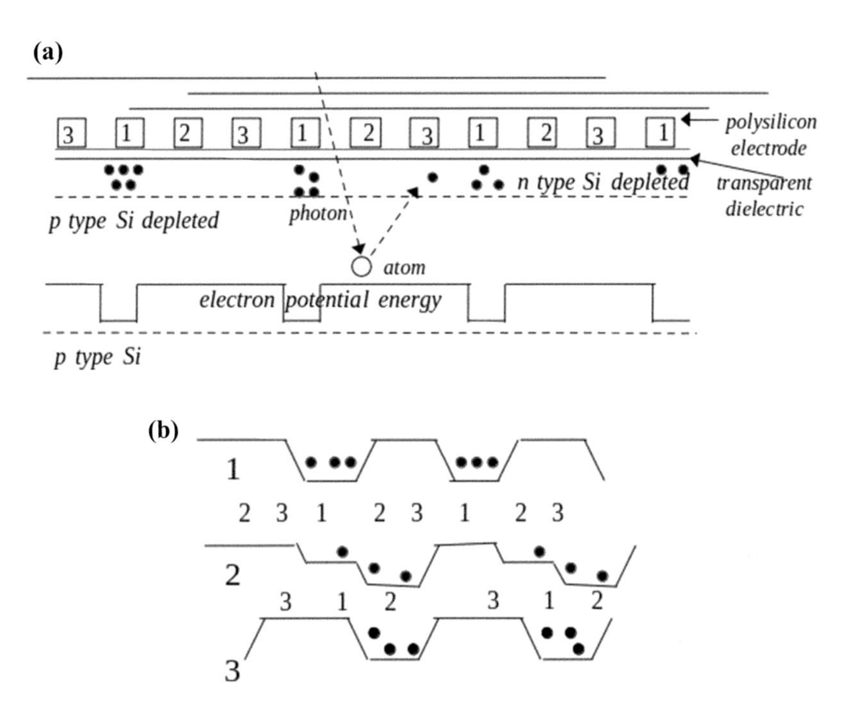

Fig. 14.5 **a** General charge coupled device structure. Black dots are ejected photoelectrons. **b** Key signalling steps to read out data from a charge coupled device

- Indium gallium phosphide (InGaP)

As the photon energy of a laser diode is close to the bandgap energy, compositions with different bandgap energies allow for different emission wavelengths. For the ternary and quaternary semiconductor compounds, the bandgap energy can be continuously varied in some substantial range. In AlGaAs (equivalently AlxGa1 − xAs) an increased aluminum content (increased x) causes an increase in the bandgap energy. Most common semiconductor lasers operate in the near infra red spectral region, some others generate red light (e.g. laser pointers using GaInP) or blue or violet light (with gallium nitrides). For mid infra red emission, there are e.g. lead selenide (PbSe) lasers and quantum cascade lasers. Organic semiconductor compounds are being researched upon for use in semiconductor lasers. Only optically pumped organic semiconductor lasers have been demonstrated so far—for various reasons it is difficult to achieve a high efficiency with electrical pumping.

A wide variety of semiconductor lasers, each for a specific end application are available.

- Small edge emitting laser diodes generate high quality low milliwatt (approximately 0.5 W), used in laser pointers, in CD players and for optical fiber communications.

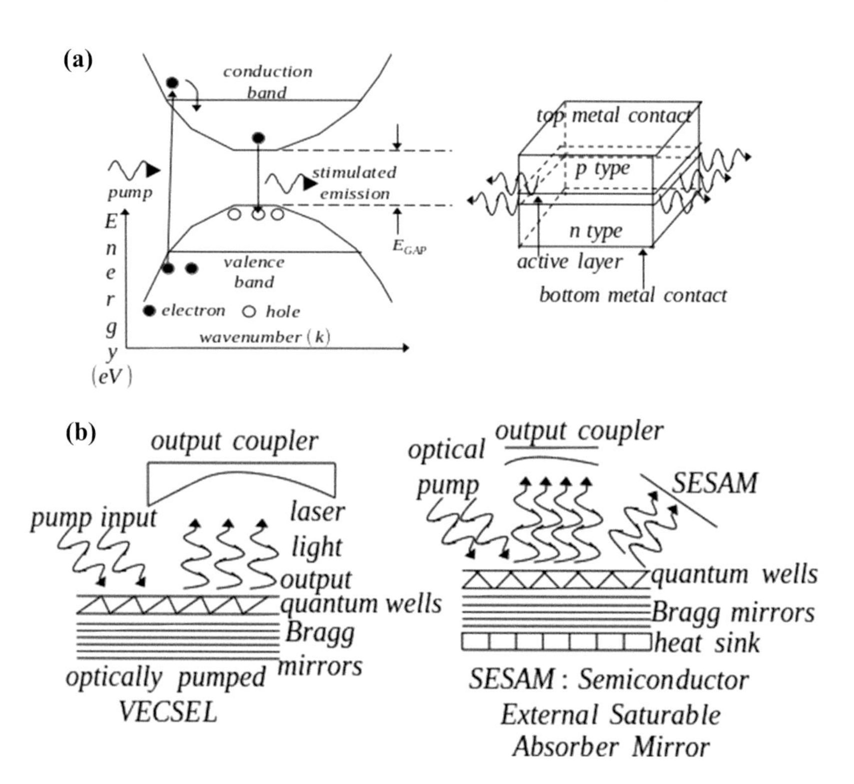

Fig. 14.6 **a**, **b** Optical pumping in a semiconductor laser and electrically pumped semiconductor laser. **c**, **d** Optically pumped VECSEL

- External cavity diode lasers contain a laser diode as the gain medium of a longer laser cavity. They are often wavelength tunable and exhibit a small emission linewidth.
- Both monolithic and external-cavity low-power levels can also be mode locked for ultra short pulse generation. Broad area laser diodes generate low beam quality low output power beams.
- High power diode bars are an array of broad area emitters, generating high power (tens of watts) poor quality output.
- High power *stacked* diode bars can generate extremely high power (hundreds or thousands of watts). Surface emitting lasers (VCSELs) emit low power, very high beam quality laser radiation perpendicularly to the wafer.
- Optically pumped vertical cavity surface emitting lasers (VECSELs) generate high beam quality multi -watt output, even in mode locked operation. Electrically pumped photonic crystal surface emitting lasers promise to reach similar performance. Quantum cascade lasers operate on intraband transitions (as compared to interband transitions) and usually emit teraHertz frequency mid infra red beams.

When electrons acquire sufficient energy that they are raised to the conduction band through external pumping (electrical|optical), the empty states left behind in the valence band are holes. Light is produced when an electron near the bottom of the conduction band recombines with a hole in the valence band. The photon emitted during this recombination process carries an energy $fh = E_{GAP}$. As $f = \frac{c}{\lambda}$, a semiconductor laser can operate only in a specified wavelength band near $\lambda = \frac{ch}{E_{GAP}}$.

Semiconductor lasers operating in the wavelength range 1.3–1.6 microns are used for optical fiber communications. These use the quaternary compound In1$-x$GaxAsyP1$-y$ that is grown in a layer form on InP substrates using molecular beam epitaxy. The lattice constant of each layer must match the lattice constant of InP to maintain a well defined lattice structure so that defects are not formed at the interfaces between any two layers with different bandgaps. The fraction x and y cannot be chosen arbitrarily but are related as x/y = 0.45 to ensure matching of the lattice constant. The bandgap of the quaternary compound can be expressed in terms of y only by the empirical relation:

$$E_{GAP}(y) = 0.12y^2 - 0.72y + 1.35 \tag{14.23a}$$

where $0 <= y <= 1$. The smallest bandgap occurs for $y = 1$. The corresponding ternary compound $Ind_{0.55}Ga_{0.45}As$ emits light near 1.65μ. By a suitable choice of the mixing fractions x and y, In1$-x$GaxAsyP1$-y$ lasers can be designed to operate in the wide wavelength range 1.0–1.65μ that includes the region 1.3–1.6μ important for optical fiber communication systems.

Electrically pumped semiconductor lasers use a *pn* junction [19]. The three dimensional physical layered structure of such a laser is in Fig. 14.5b. The central core layer("active layer") is sandwiched between the *p* type and *n* type cladding layers, both of which are doped so heavily that the Fermi-level separation Efc − Efv exceeds the bandgap energy Eg under forward biasing of the p–n junction.

The central core layer("active layer") is the light emitting semiconductor material. The cladding layers have bandgaps much larger than that of the active layer. **The bandgap difference between the two semiconductors confines the electrons and holes to the active layer. The active layer has a slightly larger refractive index than the surrounding cladding layers that acts a planar waveguide whose number of modes can be controlled by changing the active-layer thickness. This heterostructure design confines both the injected carriers (electrons and holes) and the light generated within the active layer through electron–hole recombination**. *Also the two cladding layers are transparent to the emitted light owing to their higher bandgap, resulting in a low-loss structure. These features enable semiconductor lasers to be used for a wide variety of applications.*

When the injected carrier density in the active layer exceeds a threshold, *population inversion* occurs and the active region exhibits optical gain. An input signal propagating inside the active layer is then amplified by a factor of e^{gL}, where g is the gain coefficient and L is the active-layer length. g *is* calculated numerically based on the rates at which photons are absorbed and emitted through stimulated emission and these parameters depend on details of the band structure of the active material. The plot of optical gain (g) versus carrier density (N) for a 1.3μ wavelength InGaAsP has an interesting structure. Initially when population inversion has not yet occurred, $g < 0$. As N increases, g becomes positive over a spectral range that increases with N. The peak value of the gain, gp, also increases with N, together with a shift of the peak toward higher photon energies, varying almost linearly for carrier densities above a threshold value. The optical gain in a semiconductor laser increases rapidly after population inversion.

Based on empirical data, the nearly linear dependence of gp on N is approximated by

$$gp(\mathrm{N}) = \sigma g(N - N\mathrm{T})$$

where $N\mathrm{T}$ is the *transparency value of the carrier density* and σg is the *gain cross section* or *differential gain.* Typical values of $N\mathrm{T}$ and σg for InGaAsP lasers are in the range $1.0\text{–}1.5 \times 10^{18}$ cm^{-3} and $2\text{–}3 \times 10^{-16}$ cm^{-2}, respectively. This approximation is valid in the high gain region where $gp > 100$ cm^{-1}.

Semiconductor lasers with a larger value of σg perform better, since the same amount of gain can be realized at a lower carrier density or, equivalently, at a lower injected current. In quantum-well semiconductor lasers, σg is typically larger by about a factor of 2. The linear approximation in the equation above for the peak gain can still be used in a limited range. A better approximation replaces it with $gp(N) = g0[1 + \ln(N/N0)]$, where $gp = g_0$ at $N = N_0$ and $N0 = eN\mathrm{T} \approx 2.718\ N\mathrm{T}$ since $gp = 0$ at $N = N\mathrm{T}$.

A laser can work only when *opti*cal *gain* is combined with *optical feedback*, which converts any amplifier into an oscillator. In most lasers the feedback is provided by placing the gain medium inside a Fabry–Perot (FP) cavity formed by using two mirrors. *Semiconductor lasers do not require external mirrors since the two cleaved facets act as mirrors owing to the large refractive index difference across the air semiconductor interface.* The facet reflectivity normal to this interface is:

$$R_m = \frac{n - 1}{n + 1} \tag{14.23b}$$

where n is the refractive index of the gain medium. Typically, $n = 3.5$, resulting in 30% facet reflectivity. Even though FP cavity formed by two cleaved facets is lossy, the gain in a semiconductor laser is large enough that high losses can be tolerated.

The threshold condition that triggers the laser operation is obtained by analysis of how the amplitude of an optical mode changes during one round trip inside the FP cavity. Assume that the mode has initially an amplitude $A0$, frequency f, and propagation constant $\beta = \frac{2\pi f \bar{n}}{c}$ where $\bar{n}$ is the mode index. After one round trip, its amplitude increases by $\exp[2(g/2)L]$ because of gain (g is the power gain) and its phase changes by $2\beta L$, where L is the length of the laser cavity. Simultaneously, its amplitude decreases by $\sqrt{R_1 R_2 e^{-\alpha_{INTL} L}}$ cause of reflection at the laser facets and because of internal losses resulting from free carrier absorption and interface scattering. The facet reflectivities R1 and R2 can be different if facets are coated to change their natural reflectivity. In the steady state, the mode should remain unchanged after one round trip:

$$A_0\sqrt{R_1R_2}e^{((g-\alpha_{INTERNAL})+2j\beta)L} = A_0$$

$$g = \alpha_{INTERNAL} + \frac{1}{2L}\ln\left(\frac{1}{R_1R_2}\right) = \alpha_{INTERNAL} + \alpha_{MIRROR} = \alpha_{CAVITY} \quad (14.23c)$$

$$\beta L = m\pi f = f_m = \frac{mc}{2\overline{n}L}$$

where m is an integer. The first equation shows that the gain g equals total cavity loss α_{CAVITY} at the threshold and beyond. *g is not the same as the material gain gm.* The optical mode extends beyond the active layer while the gain exists only inside it. So $g = \Gamma g_m$, where Γ is the confinement factor of the active region with typical values <0.4.

From the phase condition above, it is clear that the laser frequency f must match one of the frequencies in the set f_m, where m is an integer. These frequencies correspond to the longitudinal modes and are determined by the optical length $\overline{n}L$. The spacing Δf_L between the longitudinal modes is the free spectral range associated with any FP resonator and is given by $\Delta f_L = \frac{c}{2n_{GROUP}L}$ which includes material dispersion. Typically, $\Delta f_L = 150$ GHz for $L = 250\mu$.

A semiconductor laser generally emits light in several longitudinal modes of the cavity simultaneously. The gain spectrum g(ω) of semiconductor lasers is wide enough (bandwidth ~10 THz) so that many longitudinal modes of the FP cavity experience gain simultaneously. The mode closest to the gain peak becomes the dominant mode. Under ideal conditions, the other modes should not reach threshold since their gain always remains less than that of the main mode. In reality, the difference is so small (~0.1 cm^{-1}) such that one|two neighboring modes on each side of the main mode share a sizeable fraction of the laser power together with the main mode. Since each mode propagates inside the fiber at a slightly different speed because of group velocity dispersion, the multimode nature of a semiconductor laser often limits the bit rate of lightwave systems operating near 1.55μ. To circumvent this issue, lasers can be designed to oscillate in a single longitudinal mode.

The *vertical cavity surface emitting laser* (VECSEL) [20] is another semiconductor laser consisting of a surface light emitting semiconductor integrated circuit and a laser resonator. Unlike other types of semiconductor lasers, a VECSEL can generate high beam quality, very high power, *diffraction limited* optical output. Compared to both doped insulator solid sate and gas lasers, the output wavelength of a VECSEL is tunable.

Physically, a VECSEL (Fig. 14.5c, d) consists of one|more Bragg mirrors and the active light generation region with several quantum wells, with a total thickness of a few microns. The laser is fabricated on a semiconductor substrate and the whole structure is mounted on a heat sink. The laser resonator is created with an external mirror at a distance varying between a few millimeters and a few centimeters. *The constraint is that for high beam quality output, the length of the resonator should not be much smaller than the Rayleigh length of the intracavity beam.* This external resonator arrangement controls the laser mode. The external resonator may be folded with additional flat|curved mirrors and may contain additional optical elements. These additional optical elements include:

- An optical filter for single frequency operation, i.e., wavelength tuning
- A nonlinear crystal for crystal for intracavity frequency tuning
- A saturable absorber for passive mode locking

A VECSEL may be pumped either electrically or optically. For electrically pumped VECSEL, a ring electrode is used to contact the active area. *This technique limits the usable active area and thus the output power, since it is difficult to pump large areas uniformly, avoiding a weakly pumped region at the center of the active area. Thus electrically pumped VECSELs can generate a maximum of 1 W of output power.*

The optical pumping method circumvents the shortcomings of electrical pumping. Optical pumping enables large active areas to be illuminated uniformly. No specially doped regions|structures for carrying current is needed. The pump light is provided by a high brightness broad area laser diode or diode bar. Due to the very short absorption length of the semiconductor gain structure (at least for spacer pumping), the beam quality of the pump light is unimportant, a poor beam quality only requires working with a strongly converging pump beam, which requires more space and may make it more difficult to arrange the intracavity elements. A diode bar addresses this problem, enabling tens of watts of output power. *An external resonator is necessary for diffraction limited output, when the mode area is large. Therefore, VCSELs (having a monolithic resonator) are not suitable for high powers with perfect beam quality, even with optical pumping.*

As the quantum wells are very thin, input optical power absorption is low, if only the quantum wells are present. To circumvent this issue, a gain structure consisting of *spacer layers* between the quantum wells are added to absorb input optical pump radiation. *The carriers generated in these layers can be efficiently transferred to the quantum wells, as these have a smaller band gap than the spacer layers.* **Efficient carrier transfer means that the bandgaps of the spacer and quantum well materials must be widely separated, and this means that the pump wavelength be much shorter than the laser wavelength**. *This increases the quantum defect and thus the dissipated power.*

Although *in well pumping*, i.e. directly pumping the quantum wells is a proposed alternative. Efficient input pump power absorption can be achieved by using a multipass pumping scheme, as in a solid state thin disk laser. However, fabricating this VECSEL is difficult and the optical spectrum the pump radiation must satisfy tight tolerances.

The quantum wells in a VECSEL are fabricated in **resonant periodic gain** configuration—i.e., *each well is in an anti node of the electric field distribution for the lasing wavelength.* As different wavelengths have a different standing wave periods, their field distributions will overlap less perfectly with the quantum wells. This results in lower confinement factor and consequently reduced effective gain.

The number of quantum wells used in a VECSEL can vary. A larger number of quantum wells result in a higher gain, but the gain structure becomes thicker with a higher sensitivity to fabrication errors, strain and temperature effects. Also inhomogeneous gain saturation can occur as a result of internal temperature gradients and different excitation levels of the quantum wells. This can be a problem in narrow linewidth or mode locked VECSELs.

Temperature changes affect both the wavelength of maximum intrinsic gain of the quantum wells and the field distribution. Heating is unavoidable, so gain structures are designed so that an optimum match of all parameters is achieved at the expected operating temperature (not room temperature)—non ideal designs result in significant power reduction (roll-over) for high pump powers.

Lattice constant mismatch between quantum well material and the material used for the Bragg mirrors introduces physical strain. Lattice strain induced issues crop up in the VECSEL output e.g., dark lines from photoluminesence quenching. The problem is complicated by lattice strain propagating along certain crystal lattice directions. Lattice strain issues are curbed with additional layers with a intermediate lattice constant and appropriate thickness. This *strain compensation* balances compressive and tensile strain. This similar to the graded doping in metamorphic HEMT examined previously.

References

1. Honsberg, C. B, Corkish, R., & Bremner, S. P. (2001). A new generalized detailed balance formulation to calculate solar cell efficiency limits. In *17th European Photovoltaic Solar Energy Conference* (pp. 22–26).
2. Swanson, R. M. (2005). Approaching the 29% limit efficiency of silicon solar cells. In *Thirty-First IEEE Photovoltaic Specialists Conference*, January 2005, Lake Buena Vista, FL, USA, (pp. 889–94).
3. https://www.energy.gov/eere/solar/photovoltaic-cell-and-module-design.
4. https://www.sciencedirect.com/science/article/pii/B9780128129593000010.
5. ASTM G173-03e1. Standard Tables for Reference Solar Spectral Irradiance at Air Mass 1.5: Direct Normal and Hemispherical for a 37° Tilted Surface. Available from ASTM International, West Conshohocken, PA. http://www.astm.org/Standards/G173.htm.
6. Markvart, T., & Castener, L. (2005). *Solar cells: Materials*. Elsevier.
7. Green, M. A., & Keevers, M. J. (1995). Optical properties of intrinsic Silicon at 300 K. *Progress in Photovoltaics: Research and Applications, 3*, 189–192.
8. Jhao, J,, Wang, A., Campbell, P., & Green, M.A. (1997). 22.7% Efficient PERL Silicon solar cell module with textured front surface. In *Proceedings of IEEE 26th Photovoltaic Specialists Conference* (pp. 1133–1136).
9. Zhao, J., Wang, A., & Green M.A. (1999). 24.5% efficiency Silicon PERT cells on MCZ substrates and 24.7% efficiency PERL cells on FZ substrates. *Progress in Photovoltaics: Research and Applications, 7*, 471–474.
10. Tarr, N. G., & Pulfrey, D. L. (1979). An investigation of dark current and photocurrent superposition in solar cells. *Solid-State Electronics, 22*, 265–270.

11. Green, M. A. (1982). *Solar cells: operating principles, technology and system applications* (p. 89). Prentice-Hall.
12. Repins, I., Contreras, M.A., Egaas, B., DeHart, C., Scharf, J., Perkins, C.L., To, B., & Noufi, R. (2008). 19.9% efficient ZnO/CdS/CuInGaSe^2 solar cell with 81.2% fill factor. *Progress in Photovoltaics: Research and Applications, 16*, 235–239.
13. King, R. R., Law, D. C., Edmondson, K. M., Fetzer, C. M., Kinsey, G. N., Yoon, H., Sherif, R. A., & Karam, N. H. (2007). 40% Efficient metamorphic GaInP/GaInAs/Ge multijunction solar cells. *Applied Physics Letters, 90*, 183516.
14. Noufi, R., & Zweibel, K. (2006). High efficiency CdTe and CIGS thin-film solar cells: highlights and challenges. In *IEEE 4th World Conference on Photovoltaic Energy Conversion* (pp. 317–320).
15. International Energy Agency, World Energy Outlook (2006). ISBN: 92 64 10989 7
16. Weber, K.J., Blakers, A.W., Deenapanray, P.N.K., Everett, V., & Franklin, E. (2013). Sliver cells. In *Proceedings IEEE 31st Photovoltaic Specialists Conference* (pp. 991–994), 2005.17,16. Pullfrey, D. *Understanding modern transistors and diodes*. Cambridge University Press, 2013. ISBN 13 978-0521514606.
17. https://www.epfl.ch/labs/pvlab/research/heterojunction_solar_cells/.
18. https://johnloomis.org/ece445/topics/egginc/pt_char.html.
19. https://www.rp-photonics.com/semiconductor_lasers.html.
20. https://www.fiberoptics4sale.com/blogs/wave-optics/semiconductor-laser-physics.
21. https://www.rp-photonics.com/vertical_external_cavity_surface_emitting_lasers.html.
22. https://www.jameco.com/Jameco/workshop/Howitworks/how-organic-light-emitting-diodes-work.html.

Gallium Nitride—The Reigning King of Ultra High Frequency|Power Transistors

15

15.1 Face Centered Cubic Crystal (Zinc Blende) Gallium Nitride

Gallium Nitride appears in two basic crystalline forms (*phases*)—the f*ace centered cubic* (FCC–Zinc Blende structure) [1–16] (Fig. 15.1a) and the *hexagonal close packed* (HCP—Wurtzite structure). **The unit cell (for any crystal—e.g., GaN) consists of that unique combination of constituent atoms (e.g., gallium and nitrogen for GaN) in the three dimensional space, which translated in appropriate directions, can re-create the entire crystal**. The properties of the face centered cubic (FCC) and hexagonal close packed (HCP) crystal structures are different. So while FCC GaN is ideal for some applications, it is inappropriate for others for which HCP is ideal. FCC crystalline structure is a metastable, β- phase (space group F 43 m) form of gallium nitride, which can be stabilized in epitaxial films. The stacking sequence for the (111) close-packed planes in this structure is ABCABC. **Since the α-(hexagonal close packed) and β-(face centered cubic) phases of group III-nitrides only differ in the stacking sequence of nitrogen and metal atom planes, these hexagonal and cubic phases can coexist in epitaxial layers, e.g., due to stacking faults.** *Both α, β phases of GaN lack inversion symmetry.*

Although the HCP lattice structure GaN is the material of choice for semiconductor device industry, as it can be deposited ***without lattice constant mismatch*** on sapphire substrates, the FCC crystalline structured GaN is also used in special applications. FCC lattice structure GaN has higher saturated electron drift velocity and a slightly lower bandgap than HCP lattice structure GaN. The key crystalline and electronic electronic properties essential for semiconductor device design and fabrication with both FCC and HCP crystal structures gallium nitride (GaN) are:

- Energy gap
- Electron affinity

A. Banerjee, *Semiconductor Devices*, Synthesis Lectures on Engineering, Science, and Technology, https://doi.org/10.1007/978-3-031-45750-0_15

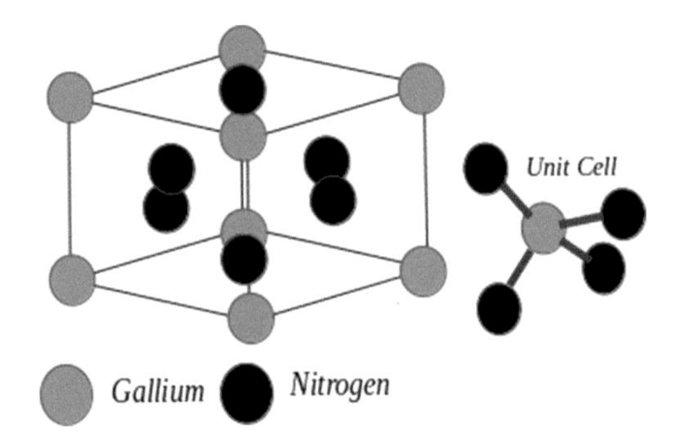

Fig. 15.1 Three dimensional face centered cubic lattice structure of GaN. The nitrogen atoms are located at the center of each square face, while the gallium atoms are at the vertices of the cube

- Conduction band Γ, X valley separation
- Conduction band Γ, L valley separation
- Conduction band effective density of states
- Valence band spin orbital splitting
- Valence band effective density of states
- Effective electron mass
- Effective mass of density of states
- Effective heavy hole mass
- Effective light hole mass
- Effective hole mass split-off band

The numerical values for these properties for both crystal structures are freely available and downloadable [1–16].

The intrinsic, conduction and valence band carrier concentrations are:

$$n_i = \sqrt{N_{COND} N_{VAL}} e^{\frac{-E_{GAP}}{k_B T_{lattice}}} \; N_{COND} = 2.3x10^{14} T^{\frac{3}{2}}\ cm^{-3}$$
$$N_{VAL} = 8.0x10^{15} T^{\frac{3}{2}}\ cm^{-3} \tag{15.1}$$

15.2 Hexagonal Close Packed Structure (HCP Wurtzite) Gallium Nitride [1–33]

Unlike FCC III-V semiconductors (e.g., GaAs, InP etc.,), the thermodynamically stable phase of InN, GaN and AlN (i.e., indium nitride, gallium nitride and aluminium nitride) is the hexagonal close packed (wurtzite) structure [17]. **The wurtzite structure** (Fig. 15.2a) **consists of alternating biatomic close packed (0001) planes of gallium and nitrogen atom pairs stacked in an ABABAB sequence. Atoms in the first and third layers are directly aligned with each other**. In this phase III-nitride materials form a continuous alloy system indium gallium nitride (InGaN), indium aluminum nitride (InAlN), aluminum gallium nitride (AlGaN) whose direct optical bandgaps range from 0.7 eV for α- InN 1 and 3.4 for α-GaN to 6.2 eV for α- AlN.

The hexagonal crystal structure of group III-nitrides is characterized by three length parameters $a_0(hexagonbaselength)$, $c_0(hexagonalprismheight)$ and an internal parameter u defined as the *anion cation bond length* along the (0001) axis. Owing to the different cation and ionic radii (Al $3+$: 0.39 Å, Ga $3+$: 0.47 Å, In $3+$: 0.79 Å) InN, GaN and AlN have different lattice constants, bandgaps and binding energies. The bonds in the 0001 direction for wurtzite and 111 direction for zincblende are all faced by nitrogen in the same direction and by the cation in the opposite.

As the basis material for heterostructure growth, GaN is the most investigated nitride compound. The most common growth direction of hexagonal GaN is normal to the {0001} basal plane, where the atoms are arranged in bilayers consisting of two closely spaced hexagonal layers, one with cations and the other with anions. Bilayers have polar faces. Lack of inversion symmetry means that group III-nitrides do not have any inversion plane perpendicular to the c-axis. So crystal surfaces have either a group III element (Al, Ga, or In) polarity (designated 0001) or a N-polarity. *Gallium faced* means gallium on the top position of the {0001} bilayer, corresponding to [0001] polarity. **Gallium faced does not mean gallium terminated.** As termination describes a surface property, gallium faced material can be gallium terminated or nitrogen terminated. *The gallium, nitrogen (0001) surfaces of GaN are* ***not equivalent*** *(by convention, the [0001] direction is given by a vector pointing from a Ga to a nearest neighbor N-atom) along the longitudinal bond.* The properties of GaN that make it very attractive for device applications are:

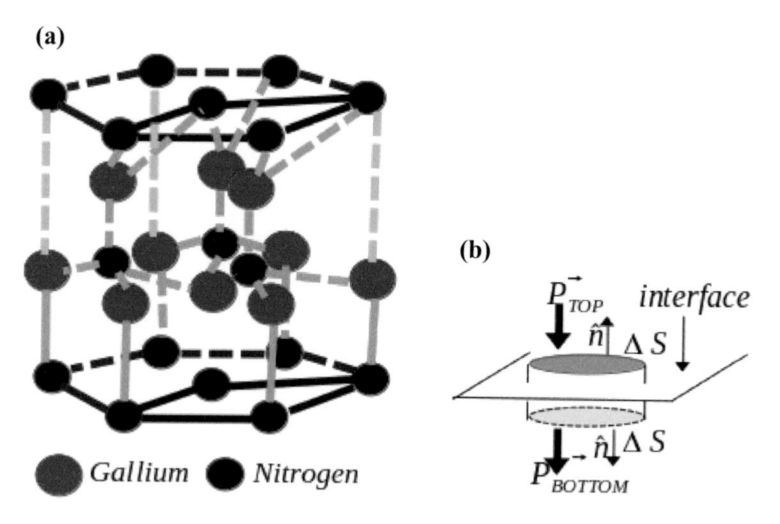

Fig. 15.2 **a** The hexagonal close packed (HCP wurtzite) lattice structure. **b** The closed surface enclosing a volume V, used to measure polarization charge density at the interface

- High thermal conductivity (4 × that of GaAs) and high breakdown field allow this material to withstand high power levels.
- A high electron saturation velocity enables GaN to operate at high frequencies. The peak drift velocity for AlGaN/GaN is $2.7x10^7$ cm/s as compared to AlGaAs/InGaAs.with $2.1x10^7$ cm/s since an electric field is able to accelerate the electrons up to higher field values before optical phonon scattering—the optical phonon energy—is much higher in GaN (91.2 meV) than in GaAs (33.2 meV).

AlGaAs/GaAs Based Devices Are Doped, AlGaN/GaN Are not.

15.3 Piezoelectric and Spontaneous Polarization and Symmetry

No AlGaN|GaN based device is doped because of two crystalline properties—piezoelectric and spontaneous polarization [17–32]. Certain crystals exhibit an electrical polarization ***spontaneous polarization*** [17–32], **characterized by a non-zero value in absence of an external electric field.** *This polarization is quantified by the spontaneous polarization vector,* P_{SP}. Spontaneous polarization is different from induced polarization which occurs only when a dielectric is placed in an electric field.

If a stress σ is applied to certain crystals they develop an electric dipole moment whose magnitude is proportional to the applied stress—direct piezoelectric effect. *The general relationship between the piezoelectric polarization, the vector* $\vec{P}_{PE}$, *and the second rank stress tensor,* σ_{ij}, *is given by* $\vec{P}^{i}_{jk} = d_{ijk}\sigma_{jk}$, *where* d_{ijk} *are elements of the the third rank piezoelectric moduli tensor*. Polarization, as a physical property of a crystal, is related to the symmetry properties of the semiconductor crystal structures examined here.

Unlike most III-V semiconductor compounds with FCC(zincblende) crystal structure, the nitride equilibrium crystal structure is the hexagonal wurtzite type. *Neumann's principle states that the symmetry elements of any physical property of a crystal must include the symmetry elements of the point group of the crystal.*

Spontaneous polarization, is described by a vector with a fixed orientation in the crystal. This form of polarization can occur only in those crystals (also called pyroelectric crystals) for which ***there is at least one direction (a vector) that remains invariant under all the symmetry operations of the crystal****. This is the case only for rotation through any angle about the vector and mirrored in any plane containing the vector. Any kind of inversion, through a centre of symmetry or through the presence of a n-fold inversion axis (rotation followed by inversion through a given point on the axis) as well as multiple rotation axis are symmetric elements not compatible with a spontaneous polarization in the crystal.* The key facts about pyroelectric crystals are:

- Pyroelectric crystals do not have a center of symmetry

- Pyroelectric crystals either have no rotation axis, or have a single rotation axis that is not an inversion axis.

Point group symmetry rules dictate that both zincblende and wurtzite are non-centrosymmetrical. **The wurtzite lattice, with its unique sixfold symmetry axis and mirror planes satisfies both requirements and can therefore have a spontaneous polarization which is parallel to the polar axis (c-axis).** *This is strictly not true for the FCC(zincblende) lattice, which has both four polar three fold rotation axis (the [111] equivalent direction) and four fold inversion axis (the [001] equivalent directions). Thus spontaneous polarization is absent in FCC(zincblende) crystals.*

The piezoelectric effect is described by a third rank tensor and is restricted to non-centrosymmetric classes. From the form of the non-zero components of piezoelectric moduli, FCC(zincblende) symmetry crystals can develop piezoelectric polarization. *Therefore, all pyroelectric substances are also piezoelectric, but not vice-versa.*

The macroscopic electric polarization of a crystal is defined as the dipole of a unit cell. *In the lack of a no clear scheme to select the unit cell, taking into account contributions that result from charge transfer between unit cells across individual cell boundaries.* For example by referring to the two possible choices of zincblende unit cell combined with the rudimentary approximation of assigning equal positive and negative point charges to the cation and anion lattice sites, it is seen that the calculated dipoles of the two unit cells are different.

The spontaneous polarization $\vec{P}_{SP}$ of a pyroelectric material can not be measured as an intrinsic equilibrium property, as the physical observables are only the variations of the polarization ΔP, measured as bulk material properties. **In the wurtzite lattice atoms of opposite electronegativity lie above each other along the symmetry axis and the charge displacement generates a dipole along the same axis.** *An atomic relaxation in the wurtzite structure (non ideal wurtzite) due to Coulomb forces acting differently along c-axis on the different tetrahedral bonds enhances spontaneous polarization.* On the other hand, the symmetry of the zincblende lattice cancels these contributions along the four [111] equivalent directions.

The HCP lattice (wurtzite) has the highest symmetry needed for the existence of spontaneous polarization. **Also, the piezoelectric tensor of wurtzite has three independent non-vanishing components. So polarization in a III-nitride system will have both a spontaneous and a piezoelectric component. In the absence of external electric fields, the total macroscopic polarization P of a solid is the sum of the spontaneous polarization $\vec{P}_{SP}$ in the equilibrium lattice and the strain-induced or piezoelectric polarization $\vec{P}_{PE}$.**

The spontaneous polarization in the hexagonal close packed (HCP|wurtzite) GaN lattice occurs along the [0001] axis. *All subsequent analysis will be with reference to this axis.* The sign of the spontaneous polarization is determined by the polarity and is opposite to the [0001] direction. The piezoelectric polarization along the c axis is calculated using the piezoelectric coefficients e_{13}, e_{33}, equilibrium lattice parameters a_0, c_0, in-plane strain parameters $\varepsilon_1 = \varepsilon_2 = \frac{a-a_0}{a_0}$ and out of plane (along c axis) strain parameter $\varepsilon_3 = \frac{c-c_0}{c_0}$:

$$P_{PE_3} = e_{33}\varepsilon_3 + e_{31}(\varepsilon_1 + \varepsilon_3) = e_{33}\left(\frac{c-c_0}{c_0}\right) + 2e_{31}\left(\frac{a-a_0}{a_0}\right) \tag{15.2}$$

The third independent component of the piezoelectric tensor is related to the polarization induced by shear strain and is ignored for heterogeneous epitaxial layers grown in the [0001] direction, in absence of external applied forces. For the HCP lattice, the lattice parameters are related via the elastic constants (C_{13}, C_{33}) and then (15.2) can be re-written as:

$$\frac{c-c_0}{c_0} = \frac{-2C_{13}}{C_{33}}\left(\frac{a-a_0}{a_0}\right) \quad P_{PE_3} = 2\frac{a-a_0}{a_0}\left(e_{31} - \frac{e_{33}C_{13}}{C_{33}}\right) \tag{15.3}$$

This equation is valid in the linear regime for small strain values. *It defines the piezoelectric tensor through the change in polarization induced by variations of lattice constants a, c only.* **Inside the HCP lattice, a strain parallel or perpendicular to the c axis produces an internal displacement of the metal sub-lattice with respect to the nitrogen ones, i.e., a variation of the parameter u of the wurtzite structure. Therefore the spatial distribution of the polarization charges changes in comparison with the unstrained state. The piezoelectric contribution P_{PE} to the total polarization P becomes large enough so that it cannot be ignored any more.** The calculated values of the piezoelectric constants in GaN, InN, and AlN are up to ten times larger than in GaAs based crystals and the sign is opposite to other III-V compounds. *The value of the* ***piezoelectric polarization increases with the strain*** *and, for crystals or epitaxial layers under the same strain, piezoelectric polarization increases from GaN - > InN - > AlN.*

Experimentally measured spontaneous polarization is negative for each of aluminum nitride (AlN), gallium nitride (GaN) and indium nitride (InN): meaning that for alloys like $Al_xGa_{1-x}N$ $0 < x < 1$ $\left(e_{31} - \frac{e_{33}C_{13}}{C_{33}}\right) < 0$.

Therefore piezoelectric polarization is negative for tensile and positive for compressive strained films, $Al_xGa_{1-x}N$. **So the orientation of the piezoelectric polarization is parallel to the spontaneous polarization in the case of tensile strain and antiparallel in the case of compressively strained layers of $Al_xGa_{1-x}N$.** The remainder of the analysis is based on both the piezoelectric and spontaneous polarizations parallel to the c-axis.

Inside GaN bulk, for homogeneous top|bottom pairs layer the total polarization is constant in the bulk and has a discontinuity at the interface with fixed polarization charge density σ (Fig. 15.2b). The polarization induced charge density and the fixed polarization charge density(using theorem of divergence) is:

$$\rho_P = -\nabla \cdot \vec{P} \quad \sigma\Delta S = -\int \rho_p dP = -\int \vec{P} \cdot \hat{n} ds = (|P_{TOP}| - |P_{BOTTOM}|)dS \tag{15.4}$$

where the cylindrical surface enclosing the volume V, is formed by two surfaces ΔS just above and below the interface—*'x' represents the mole fraction.*

To estimate the polarization induced sheet charge located at $Al_xGa_{1-x}N/GaN$ interface, the linearized versions of piezoelectric constants, lattice constants, elastic constants and spontaneous polarization respectively used:

$$\begin{aligned} e_{ij}(x) =& xe_{ij}(AlN) + (1-x)e_{ij}(GaN) \quad a_0(x) = 3.189 - 0.077x \\ c_0(x) =& 5.189 - 0.203x \quad C_{13}(x) = 103.0 + 5.0x \; GPa \\ C_{33}(x) =& 405.0 - 32.0x \; GPa \quad P_{Spontaneous\ Polarization}(x) = -(0.052 + 0.029) \end{aligned} \tag{15.5 a,b,c,d,e,f}$$

where Gpa is GigaPascals and the units of the spontaneous polarization are $\frac{Coulomb}{m^2}$.

The piezoelectric polarization for partially relaxed barriers can be determined either by using the measured lattice constants or from the measured degree of relaxation:

$$P_{PE}(x) = 2(r(x) - 1)\left(\frac{a_0(x) - a(GaN)}{a_0(x)}\right)\left(e_{31}(x) - \frac{e_{33}(x)C_{13}(x)}{C_{33}(x)}\right) \tag{15.6}$$

where r(x) is the degree of polarization relaxation as a function of the mole fraction x.

15.4 Two Degree Electron Gas (2DEG) GaN-AlGaN–GaN Pseudomorphic Heterogeneous Junction Properties

A junction between two different semiconductor materials on the two sides of the junction is a heterogeneous semiconductor junction. Moreover, the material on one side has a narrow bandgap (type I e.g., $Al_xGa_{1-x}As$, $Al_xGa_{1-x}N$) and as expected, the material on the other side is wide bandgap (type II GaAs, GaN). Therefore at the junction there are differences in the conduction and valence bands of the two sides [17]. **When equilibrium is achieved following band bending and corresponding equilibrium Fermi level creation, a potential well is created at the narrow and wide bandgap material interface junction. For electrons to collect inside the potential well, the equilibrium Fermi level must lie inside the potential well.** *When these conditions are satisfied, a two dimensional electron gas(2DEG) is produced inside the potential well—key to the operation of a high electron mobility transistor (HEMT).* Figure. 15.3a, b show the band structure of a lightly doped n type narrow bandgap semiconductor (I) and a heavily n doped wide bandgap semiconductor (II) before and after contact. In thermal equilibrium, the electrons are confined in the triangular quantum well of the narrow bandgap semiconductor (I) and form a two dimensional electron gas (2DEG). $E_{COND}, E_{VAL}, E_F, \phi_S, \chi_S, V_{BI}$ are respectively the conduction band minimum, the valence band maximum, the equilibrium Fermi level, the semiconductor work function, the electron affinity and the built-in voltage.

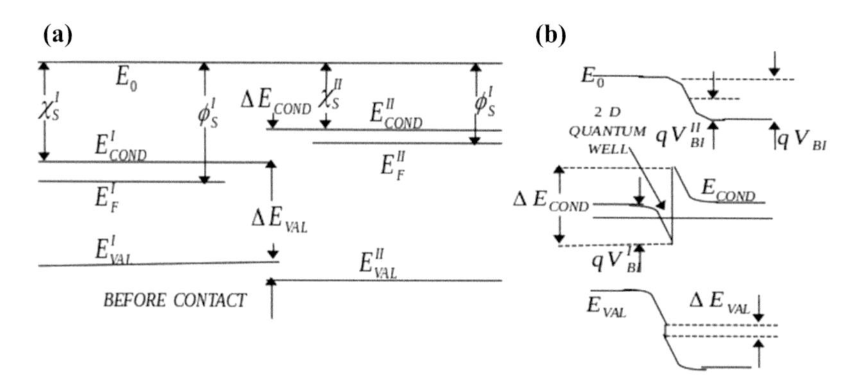

Fig. 15.3 **a**, **b** Energy band diagram and band bending for lightly n doped Type I semiconductor in contact with heavily doped n type Type II semiconductor to form a heterogeneous junction before and after contact

The band discontinuity is determined by high electrostatic potential gradients acting on the carriers on the length scale of some atomic interplanar spacing. The electric fields are are of the order of the atomic fields $10^8 \frac{V}{cm}$. Since semiconductor (I) has a smaller gap than semiconductor (II), there are regions in the gap of (II) where the continuum of bulk valence and conduction band states of (I) leak into the gap of semiconductor (II). Thus, in limited energy range in the upper and lower parts of the gap of semiconductor (II) there exist a number of **V**irtual **I**nduced **G**ap **S**tates (VIGS) derived from the bands (wave functions) of both semiconductors (I) and (II)—Fig. 15.4a, b. These two diagrams explain how a semiconductor heterostructure is formed, using the Virtual Induced Gap States (VIGS) model. *The band schemes of two semiconductors are plotted with conduction and valence band band edges.* ***The matching of the two band schemes (band offsets), are controlled by charge neutrality within the VIGS—i.e., equilibrium is achieved only when the branching energies on the two sides E^I_{BR}, E^{II}_{BR} coincide.***

Any electronic state in the gap of a semiconductor, including VIGS, is a mixture of conduction and valence band states—*the corresponding wavefunction is a superposition*

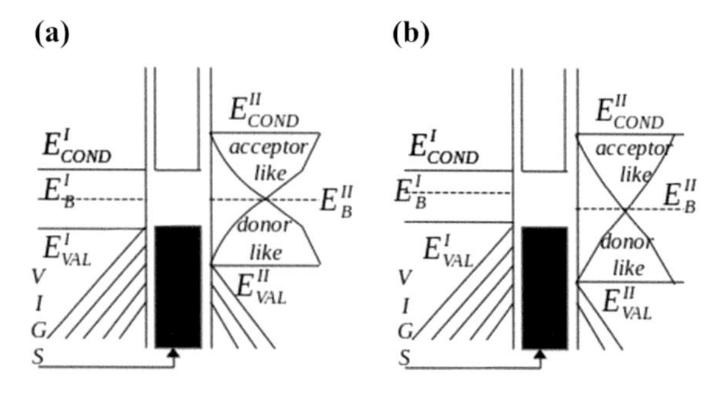

Fig. 15.4 **a**, **b** Creation of virtual induced gap states

of conduction and valence band wavefunctions. The wavefunction of the state closer to the valence band edge will have a larger contribution from valence band only wavefunctions, as compared to contribution from conduction band wavefunctions. **The crossover point (also branching point or neutrality point) E_B occurs when there is an equal contribution from both conduction band only and valence band only wavefunctions.**

When the Fermi level is close to the branching point of the VIGS, overall charge compensation occurs. When the branching points (E_B^I, E_B^{II}) of the two semiconductors do not match, the negative charge carried by the VIGS below E_B^{II} exceeds (small conduction band wavefunction contribution) the tiny positive charge in the interface states in the upper half of the gap. This positive charge is a result of these predominantly acceptor type states having a small contribution from valence band only wavefunctions. Both the positive and the negative interface charges in the VIGS are compensated when the branching points in the two materials are aligned. From energy balance arguments, the condition of zero interface dipole therefore requires alignment of the branching energies. For an ideal semiconductor heterostructure the alignment of the branching points in the two semiconductors yields the valence-band offset:

$$\left(E_B^I - E_{VAL}^{II}\right) - \left(E_B^{II} - E_{VAL}^{II}\right) = \Delta E_{VAL} \tag{15.7}$$

Built-in potential in a material are from free charges, which accumulate in the lowest energetic states. Free charges, in combination with the ionized dopants create a space charge zone. The built-in potential overlaps the effective crystal potential. *Its value is determined by the interface position of the Fermi level and extends on a length scale that depends on the bulk doping of the two semiconductors, i.e., of the order of the Debye length.* It could be as long as thousand Å for doping of the order of $10^{16}/cm^3$, or as short as 10– 100 Å for doping up to $10^{20}/cm^3$. The electric fields are similar to the those which are spread over space charge in a np junction (10^5 V/cm). Figure 15.5 shows a depletion region of fixed ionized donor atoms spread in the wide gap semiconductor near the interface due to the accumulation of donor dopant electrons in the narrow-gap semiconductor, which form a two dimensional electron gas (2DEG).

The pseudomorphic $GaN - Al_xGa_{1-x}N - GaN$ structure (Fig. 15.6) illustrates how the concepts presented above apply to real world GaN based semiconductor devices. This semiconductor heterostructure consists of a thick GaN layer deposited on a thin aluminum gallium nitride ($Al_xGa_{1-x}N$) layer, which itself is grown on a GaN buffer layer. **The middle thin AlGaN layer is under tensile strain—the piezoelectric and spontaneous polarizations point in the *same* direction. Since the values of the piezoelectric constants and spontaneous polarization increase from GaN to AlN, the total polarization of a strained (or even unstrained) $Al_xGa_{1-x}N$ layer is larger than that of a relaxed GaN buffer layer. Consequently, a positive polarization charge is present both at the lower AlGaN|GaN interface for the gallium face structure and at the upper GaN|AlGaN interface for the nitrogen face structure.** *Electrons neutralize this positive polarization*

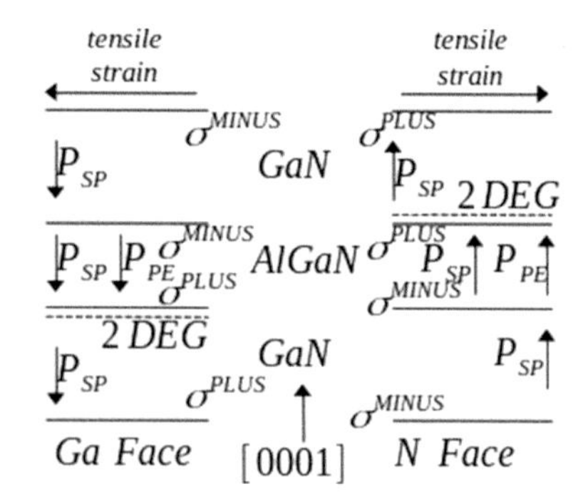

Fig. 15.5 Spontaneous, piezoelectric polarizations, two dimensional electron gas(2DEG) and tensile strain at pseudomorphic GaN|AlGaN|GaN heterogeneous junction

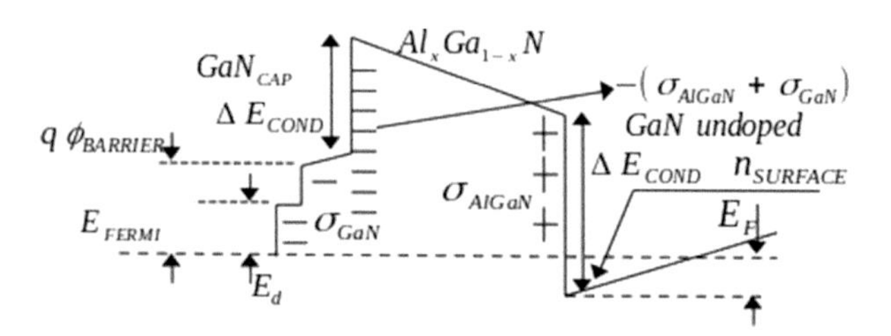

Fig. 15.6 Energy band diagram for GaN-AlGaN-GaN heterogeneous junction

charge resulting in the formation of a two dimensional electron gas(2DEG), when the triangular quantum well at the AlGaN|GaN interface is deeper than the Fermi level. Analogously, negative polarization sheet charge density results in accumulation of holes at the interface, if the valence band edge of the AlGaN|GaN heterostructure crosses the Fermi level. **For AlGaN|GaN interfaces, the spontaneous and piezoelectric polarization are large enough to generate 2DEGs with high electron concentration *without* doping the barrier.** *In contrast, the 2DEG in AlGaAs|GaAs modulation doped heterogeneous junction is due to remote doping* (Fig. 15.5).

15.4.1 Surface Charge Concentration AlGaN|GaN Interface 2DEG

Exploiting and extending Eqs. 15.2–15.6, the surface (sheet) carrier (charge) concentration at a AlGaN|GaN interface 2DEG can be estimated [18]. Free electrons compensate a positive polarization induced sheet charge which is bound at the lower AlGaN|GaN interface for Ga(Al) face or at the upper GaN|AlGaN interface for N-face GaN|AlGaN|GaN structures. The value of the total polarization induced sheet charge is the same in heterogeneous structures of different polarities for a given Al concentration and strain of the barrier. For undoped Ga-face AlGaN|GaN or GaN|AlGaN|GaN heterogeneous structures, the sheet electron concentration $n_{SURFACE}(x)$ can be calculated by using the total bound sheet charge $\sigma(x)$. Equation 15.6 expresses the piezoelectric polarization and Eq. 15.5**f** expresses the spontaneous polarization, both as functions of the aluminum mole fraction

x. The total bound sheet charge density is:

$$q\sigma(x) = P_{SPONTANEOUS} + P_{PIEZOELECTRICEFFECT} \tag{15.8a}$$

The sheet carrier concentration and carrier distribution profile in doped and undoped GaN|AlGaN|GaN and AlGaN|GaN heterogeneous junctions are computed numerically using a Poisson-Schrodinger solver. Contributions from both spontaneous and piezoelectric polarizations are included in the solver using thin layers (few Angstroms thick) of charge at the heterogeneous structure interface equivalent to the bound sheet charge density $\frac{\sigma}{q}$. Appropriate boundary conditions at the correct surface and substrate interfaces need to be added. The aluminum mole fraction based linearized equations for the various intermediate parameters as bandgap (AlGaN), dielectric constant, Schottky barrier height (metal contact at the surface), conduction band offset (for the two dissimilar materials at the heterogeneous junction) are:

$$\begin{aligned} \Delta E_{COND} &= 10.4 - 0.3x \quad q\phi_B = 1.3x + 0.84\ NickelContact \\ \Delta E_{COND} &= 0.7(E_{GAP}(x) - E_{GAP}(0)) \\ E_{GAP}(x) &= 6.13 + 3.42(1.0 - x) - x(1 - x) \end{aligned} \tag{15.8 b,c,d,e}$$

Then the interface sheet electron concentration is:

$$n_{SURFACE}(x) = \frac{\sigma(x)}{q} - \frac{\varepsilon_0 \varepsilon(x)}{d_{AlGaN}}(q\varepsilon_B(x) + E_F - \Delta E_{COND}(x)) \tag{15.8e}$$

where a_{AlGaN} is the thickness of the AlGaN layer.

15.4.2 Surface Charge Concentration GaN-AlGaN–GaN Interface 2DEG

The 2DEG in AlGaN|GaN HEMTs is due to the presence of the surface donor states at the AlGaN top. It is assumed that the donor states are present at the top of the GaN cap layer, once deposited. Both the GaN cap thickness $d_{cap,GaN}$ and AlGaN thickness d_{AlGaN} in combination with the aluminum mole fraction x influence the 2DEG. This analysis uses the following parameters:

- AlGaN piezoelectric charge density $(q\sigma)$
- Dielectric permittivity (ε)
- 2DEG electron concentration $(n_{SURFACE})$
- Fermi level and conduction band minimum at the GaN heterogeneous junction interface (E_F).
- Conduction band offset between the AlGaN and GaN at the junction (ΔE_{COND})
- Surface donor level at the top (E_d)
- Constant surface donor density (n_0)

- Surface barrier height $(q\phi_{BARRIER})$

Both AlGaN, GaN with Wurtzite crystal structure, and asymmetric bonding generates strong spontaneous polarization charges. The lattice mismatch between the GaN substrate and the AlGaN barrier layer, generates strong piezoelectric polarization charge, depending on the value of the Al mole fraction x. Denoting the electric fields in the GaN cap layer and the AlGaN layer as E_1, E_2 and ignoring E_F (of the order of a few multiples of k_BT) the electric field in the AlGaN layer is:

$$E_2 = \frac{q\phi_{BARRIER} + d_{cap,GaN}E_1}{d_{AlGaN}} \tag{15.9a}$$

To maintain the continuity of displacement at the GaN cap—AlGaN interface, E_1, E_2 must satisfy:

$$\varepsilon E_2 = q\sigma - \varepsilon E_1 \quad E_2 = \frac{q(\sigma - n_{SURFACE})}{\varepsilon} \tag{15.9b}$$

Combining Eqs. 15.9a, 15.9b gives the expression for the surface free electron charge density of the 2DEG:

$$n_{SURFACE} = \frac{n_0\left(\sigma - \frac{\varepsilon q\varepsilon_{BARRIER}}{qd_{AlGaN}}\right)}{1 + \frac{d_{cap,GaN}}{d_{AlGaN}}} \tag{15.9c}$$

The surface barrier height is:

$$q\varepsilon_{BARRIER} = \frac{n_0E_dd_{cap,GaN} + d_{AlGaN}(\sigma + n_0E_d)}{\frac{\varepsilon}{q} + n_0\left(d_{cap,GaN} + d_{alGaN}\right)} \tag{15.9d}$$

The 2DEG density must monotonically decrease with increasing GaN cap layer thickness. Experimentally it is observed that the 2DEG charge value saturates at higher values of the GaN cap layer thickness. It has been postulated that this might be because the valence band and any trapped charge effect are ignored in this analysis. The free electron 2DEG charge density increases with the height of the AlGaN layer even in the presence of the GaN cap layer. Then the expression for the surface charge density can be re-written as:

$$n_{SURFACE} = \frac{n_0\left(\frac{\sigma d_{AlGaN}}{\varepsilon} - \frac{E_d}{q}\right)}{\frac{1}{q} + \frac{n_0\left(d_{cap,GaN} + d_{ALGaN}\right)}{\varepsilon}} \tag{15.9e}$$

The energy band diagram for this stacked GaN|AlGaN|GaN system is in Fig. 15.6.

15.5 High Electron Mobility Transistor Properties

State-of-art molecular beam epitaxy (MBE) and chemical vapor deposition (CVD) techniques enable the fabrication of ultra high purity semiconductor layers essential for constructing High Electron Mobility Transistors (HEMTs). These transistors are essential for ultra high frequency (10 s of GHz) and power (10 s of Amperes—100 s of Volts) applications as wireless communication.

The epitaxial growth of wide bandgap AlGaN semiconductor on a thick GaN undoped buffer layer creates a heterostructure essential for HEMTs. Figure 15.6 shows the schematic structure and electronic band scheme perpendicular to the wafer surface underneath the gate electrode of an AlGaN|GaN *normally on* (depletion mode) HEMT device. The conductive channel is formed by the 2DEG. The band schemes are under zero gate voltage, in thermal equilibrium, under negative gate bias. E_{COND}, E_{VAL}, $E_{F,metal}$, $E_{F,semi}$, $q\phi_{SCHOTTKYBARRIER}$, V_{GATE} are respectively the conduction band minimum, the valence band maximum, the Fermi level in metal, the Fermi level in the semiconductor, the Schottky barrier height and applied the gate voltage. The n channel is built by a 2DEG at the AlGaN|GaN interface with typical electron concentrations of $10^{13}\ cm^{-2}$. Similar transistors based on AlGaAs|GaAs layer structure have an electron concentration in the 2DEG of $2x10^{12}cm^{-2}$. Ultra high electron mobility and low noise operation is a result of high electron concentration (separated from the ionized donors). **Drain and source contacts under the metal overlayers must make a good ohmic contact to the 2DEG at the interface**. I*n thermal equilibrium, the conduction, valence band structure along an intersection normal to the layers below the gate electrode appears in* Fig. 15.7. *When positive drain voltage is applied, the potential drop along the source-drain connection perturbs the equilibrium band structure, parallel to the AlGaN|GaN interface.* Depending on the local potential, the accumulation layer does not contain any more electrons—the position of the Fermi level with respect to the band edges varies along the current channel.

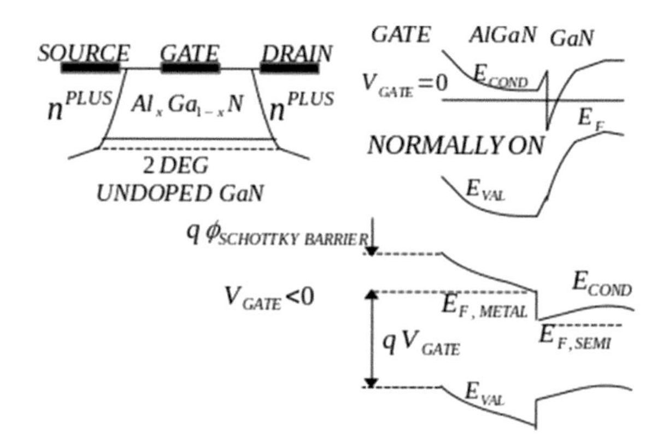

Fig. 15.7 Layered structure and energy band diagrams in equilibrium ("normally on") and switched off conditions

Externally applied gate voltage shifts the Fermi level in the gate metal with respect to its value deep in the undoped GaN layer. Then due to the Schottky depletion layer just below the metal gate electrode (donors in the AlGaN layer having been emptied), most of the voltage drop occurs across this AlGaN layer, thus establishing a quasi-insulating barrier between gate electrode and 2DEG. Depending on the gate voltage, the triangular potential well at the interface is raised or lowered in energy and the accumulation layer is emptied or filled. This changes the carrier density of the 2DEG and switches the drain source current. For large enough gate bias, the depletion region penetrates into the 2DEG region, the electron concentration is negligible and the current channel is pinched off. The corresponding relative gate voltage is called the threshold voltage.

As a depletion mode (normally on) field effect transistor, the amplification is based on a low voltage gate bias that cuts off the drain source current. A simplified model with the gate length, width L, W respectively enables the gate capacitance to be estimated:

$$C_{GATE} = \frac{\varepsilon_{SEMICONDUCTOR}\varepsilon_0 LW}{d_{AlGaN}} \tag{15.10a}$$

where ε_s, ε_0, d_{AlGaN} are respectively the dielectric constant of the semiconductor the dielectric constant of the vacuum and the thickness of the AlGaN layer below the gate electrode. The charge density $qn_{SURFACE}$ in the 2DEG is induced by the gate voltage V_{GATE}, and so the gate capacitance can be re-written as:

$$C_{GATE} = \frac{dQ}{dV_{GATE}} \approx \frac{qn_{SURFACE}LW}{V_{GATE} - V_{BI}} \tag{15.10b}$$

where V_{BI} is the built-in voltage at zero external gate bias. For normal operation the drain source voltage is so high that the electrons in the channel move with saturation velocity $v_{SAT}(v_{SAT} \approx 10^7 \frac{cm}{s})$ independent of drain voltage. The drain source current can then be written as:

$$I_{DS} \approx qn_{SURFACE}v_{SATURATION}W \approx \frac{C_{GATE}v_{SATURATION}(V_{GATE} - V_{BI})}{L} \tag{15.10c}$$

The key HEMT parameter describing the performance of HEMT is the **transconductance** which defines the change of drain source current with the change of the gate voltage at constant drain source voltage:

$$g_m = \left(\frac{\partial I_{DS}}{\partial V_{GATE,SOURCE}}\right)_{V_{DS}} \approx \frac{qv_{SATURATION}Wdn_{SURFACE}}{dV_{GATE}} \tag{15.10d}$$

The transconductance depends on the influence of electron concentration on the gate voltage. The relation between the transconductance and the gate capacitance defines the the transit time τ for an electron to pass under the gate.

$$f_{MAX} = \frac{1}{\tau} = \frac{v_{SATURATION}}{L} = \frac{g_m}{C_{GATE}} \quad (15.10e)$$

where f_{MAX} is the maximum frequency. For gate lengths in the order of 1 μm and saturation velocities around $10^7 \frac{cm}{s}$ transit times of about 10 picosecond are reached. This makes the GaN HEMT ideal for circuits operating at the microwave frequency range. *To improve the high frequency properties, gate capacitance should be minimized and the transconductance maximized.*

15.6 Scattering Processes in GaN HEMTs [33–52]

The quantum well at the AlGaN|GaN interface results from a large conduction band discontinuity (as high as 2.4 eV for AlN/GaN), in combination with band bending induced by charge transfer across the junction leads (a growth in the [0001] direction). **As long as the equilibrium Fermi level lies within the quantum well, free electrons are confined to the quantum well—2DEG**.

The eigenvalues of electrons along the normal to the surface are quantized due to their confinement in the quantum well. Parallel to the interface the assumption of a free electron gas, controlled by Bloch waves holds. The eigenvalues of the electrons in a 2DEG are:

$$E_i^{2D} = \frac{\hbar^2 k_x^2}{2m_x^P} + \frac{\hbar^2 k_y^2}{2m_y^P} + E_l \quad (15.11a)$$

The 2DEG controls the electronic transport along the interface. In a 3 dimensional non degenerate semiconductor the transport properties are calculated by solving the Boltzmann equation for elastic scattering under the assumption that the relaxation time τ_m is averaged through the temperature dependent energy distribution. A 2DEG can be treated as a degenerate semiconductor: *only the electrons near the Fermi edge contribute to the transport and so the relaxation time is valid only for* $E = E_F$. This different energy dependency leads to a different temperature behavior in the 3 dimensional case. The Fermi impulse is dependent on the carrier concentration $k_F = \sqrt{2\pi n_{SURFACE}}$, that determines the quantity of initial and final states, which are available for scattering processes.

The different scattering mechanisms limit the electron mobility in different ways, but their effects can be treated as independent of each other. The effective mobility is expressed in first approximation by the Matthiensen rule:

$$\frac{1}{\mu_{TOTAL}} = \sum \left(\frac{1}{\mu i}\right) \quad (15.11b)$$

15.6.1 Polar Optical Phonon Scattering

For the optical phonon the atoms in the elementary cell move in opposite directions. In a polar crystal this behavior is similar to that of an oscillating dipole moment. Since polar optical phonons in GaN have large energy ($E_0 \approx 91meV$) compared to the energy separation between sub-bands, the effect of a large number of sub-bands must be included when calculating their effects, and the problem changes from two-dimensional to a three dimensional one. **That is electrons absorbing optical photon(s) gain so much energy that they can be scattered completely out of the confining potential and into the bulk.** Thus an analytical solution to the optical phonon limited mobility uses a variational principle method. This scattering mechanism is dominant in degrading electron mobility for temperatures above 200 K.

15.6.2 Acoustic Phonon Scattering

Acoustic phonons are generated when the crystal lattice atoms move in the same direction, all together. In modulation doped heterogeneous junction structures, although the movement of the electrons is confined to a thin layer (~100 Å) near the interface, it is assumed that acoustic phonons can propagate freely in all three dimensions. **The electrons in a polar crystal can interact either electrostatically through the piezoelectric interaction or through the deformation potential.** *The temperature dependency of mobility of confined electrons with three-dimensional acoustic phonons due to screened deformation potential scattering has inverse power law dependency on temperature, just as mobility in the screened piezoelectric mode.* The acoustic phonon scattering rates are linear functions of temperature. This approximation is true at temperatures at which the thermal energy is greater than the acoustic phonon energy, but is not true for lower temperatures. Since temperature independent processes (e.g., Coulomb scattering) dominate the low temperature mobility, the deviations of the acoustic phonon scattering rate from linearity will have little effect on the total mobility.

15.6.3 Ionized Impurity Scattering

Variations in the perfect crystal lattice structure result in potential variations, and so scatter electrons. For example, the interaction with the Coulomb potential of the ionized impurities is very strong at low temperatures. In an AlGaN|GaN modulation doped heterogeneous junction structure, there are two types of ionized impurity scattering. The first is scattering by residual ionized impurities in the GaN. In bulk semiconductors, the ionized impurities occupy the same region of space as the conduction electrons, making a Coulomb scattering a very efficient process. The electrostatic interaction between an ionized donor and a conduction electron is screened by other conduction electrons. However to achieve the high electron concentrations needed for efficient screening, the crystal must be highly doped, leading to higher concentrations of ionized impurity centers and counteracts any beneficial screening effects. The second type is scattering by the ionized donors in the AlGaN barrier left behind the conduction electrons. Since the electric field of ionized centers drop off as the distance squared, this type of scattering is much less effective in limiting the electron mobility, and can be neglected for concentrations up to $\frac{10^{15}}{cm^3}$ for spacers in the AlGaN barrier less than few hundred Å. *Assuming the impurities screening (Thomas–Fermi screening) both two and three dimensional mobilities are inversely proportional to the impurity concentration. Three dimensional mobility is power law dependent on the temperature, and two dimensional mobility is power law dependent on the surface charge concentration.* Coulomb scattering rates are often estimated using a temperature independent approximation for 2DEGs, assuming that all scattering events involve electrons at the Fermi level. At temperatures above 100 K, as the Fermi level starts to shift upward, the approximation is invalid, and the mobility is limited by phonon scattering.

15.6.4 Alloy Disorder Scattering

AlGaN is an alloy of aluminum and GaN. In the alloys the statistical distribution of the elements leads to local potential fluctuations. The height of the scattering potential can be assumed as the difference of the energy gaps of the two different binary compounds. *The effect of this scattering mechanism depends on the degree of disorder in the crystal, i.e. on the frequency of the appearance of the single constituents. This is taken in consideration through the parameter x(1-x) where x is the mole fraction of the alloy material.* The three dimensional mobility is inversely proportional to both this parameter and the temperature, while the two dimensional mobility is inversely proportional to this parameter and the surface charge concentration.

15.6.5 Dislocation Scattering

Common defects in GaN(0001) include stacking disorder|faults, Shockley|Frank partial dislocations, inversion domains and *threading dislocations* (TDs). The stacking disorder and partial dislocations occur in regions immediately adjacent to the substrate and are associated with the growth of a disordered low temperature nucleation layer. Inversion domains occur in nitrogen polar domains that have grown either through the free surface of a Ga polar film or are overgrown by Ga polar material. The TDs have typical total densities in the range $\frac{10^8-10^{10}}{cm^2}$ as a result of the substantial GaN film substrate chemical and lattice mismatch. There are two different predominantly observed TDs:

- pure edge, with Burgers vectors in the family 13 < 2110 > and [0001] line directions
- mixed character, with Burgers vectors in the family 31 < 2113 > and line directions inclined $\sim 10^o$ from [0001] towards the Burger vector.

Pure screw TDs, with line direction [0001], represent a small fraction ($\sim 0.1 - 1\%$) of the total density of TDs. *TDs in the group III-nitrides behave as nonradiative recombination centers, with energy levels in the forbidden energy gap.* **These TDs act as charged scattering centers in doped materials, providing leakage current pathways.** *The highly dislocated wurtzite crystal can be considered as consisting of hexagonal columns rotated relatively to each other by a small angle, with inserted atomic planes to fill the space between the columns—resulting from coalescence of slightly disoriented GaN high temperature islands.* Grain boundaries between prisms as in poly crystalline material would require arrays of dislocations along the interface between two prisms.

Pure screw and mixed dislocations decrease with distance from the substrate buffer interface. Edge dislocations with mainly vertical orientation thread to the epitaxial layer surface. The charged vertical dislocation lines form space charge regions scattering electrons and reducing mobility.

Mobility reduction also occurs in the buffer region due to scattering at screw dislocations, point defects and stacking faults. Empty traps are electrically neutral, but each filled trap carries one electronic charge—negatively charged dislocation lines act as Coulomb scattering centers. The effect of scattering at charged dislocation lines on mobility, as compared to effect of lattice and ionized impurity scattering on mobility, becomes significant at threading dislocation densities above $\frac{10^9}{cm^2}$.

At low growth temperatures "V-defect"s are formed, *consisting of six {1011} family planes and form an inverted hexagonal pyramid*. They form mostly at mixed character TDs. V-defects are speculated to be due to a kinetically limited growth process such that the surface depression associated with a TD assists in the formation of {1011} facets. For HEMTs the V-defects form during the last grown AlGaN layer and concentrate electric fields at ohmic source drain and Schottky gate contacts. The two dimensional electron mobility is inversely proportional to the number of dislocations and directly power law proportional to the surface charge density.

15.6.6 Interface Scattering

In a quasi two dimensional electron gas the charge transport takes place along the interface between two semiconductors. As expected, interface roughness produces an additional deviation from the periodical lattice and can substantially reduce the mobility of a 2DEG. Theoretical calculations performed on AlGaN|GaN wurtzite and zincblende heterogeneous junction structures(to determine the scattering processes that limit the electron mobility) indicate that:

- At room temperature, the mobility is dominated by polar optical phonon scattering.
- For higher sheet electron densities in the 2DEG the remote donor (impurity) and piezoacoustic scattering processes (which dominate at lower sheet densities) are screened and the electrons are pushed closer to the interface.

For poor interface quality, the mobility is significantly reduced due to increase in interface roughness scattering. For ideal defect free interfaces, the mobility is $\frac{2000cm^2}{Vs}$ at room temperature, *which does not agree with either theoretically estimated or experimentally measured values.* Depleting the electrons from the 2DEG channel by applying negative gate voltage the electron screening is less effective and Coulomb scattering dominates. *Increasing the negative gate bias, the maximum mobility shifts towards higher temperatures along with decrease of the maximum mobility.*

15.6.7 Dipole Scattering

Dipole scattering is dominant scattering mechanism at low temperatures. The 2DEGs in III-V nitride modulation doped heterogeneous junction structures are polarization induced, compared to the 2DEG in modulation doped heterogeneous junction structures in AlGaAs|GaAs and gate induced inversion in Si-MOSFETs. Spontaneous and piezoelectric polarization in III-V nitride modulation doped heterogeneous junctions is large enough to produce 2DEGs without doping the barrier. Strong polarization along the c-axis of the wurtzite nitride compounds and fluctuations of a perfectly periodic structure in the AlGaN alloy, a random distribution of microscopic dipoles in AlN and GaN regions of the alloy scatters the electrons in 2DEG.

15.7 Basic and Advanced Physical Structures of HV, RF HEMTs

While semiconductor device research groups and semiconductor device manufacturers have improvised a variety of physical structures for the GaN based HEMTs, the common property shared by each of them is that the power HEMTs are less complicated than the ultra high frequency counterparts. Typical layered structure of these two devices is shown in Fig. 15.8.

To satisfy the ever increasing stringent performance characteristics being placed on GaN HEMTs for wireless communications (e.g., K-, W-band wireless communications) and high power switching (e.g., electric vehicle battery charging) a number of innovative features have been added to GaN HEMTs.

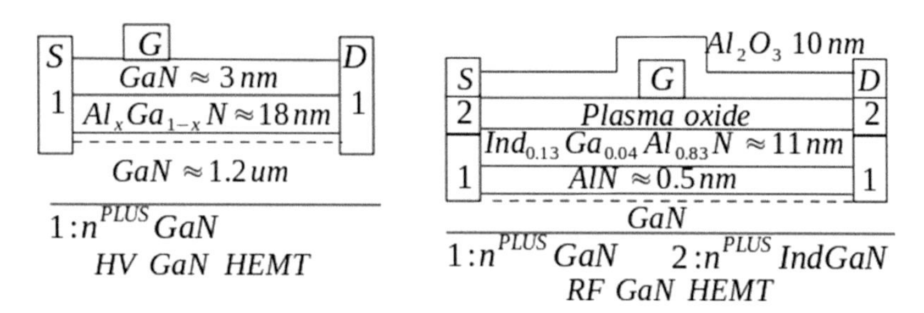

Fig. 15.8 Typical layered structure of power and RF GaN HEMTs. Dashed lines indicate 2DEG

15.7.1 Gate and Source Connected Field Plates [53]

The cross sections of a gate and a source connected field plate structures in GaN HEMTs are in Fig. 15.9a, b respectively. **The field plate (FP) lowers the electric field peak value, thereby changing the electric field profile**. *Unwanted trapping effects are minimized and breakdown voltages are increased*. Initial FPs were either constructed as part of the gate or tied to the gate externally. This improves large signal|power performance as well as high voltage operation. Up to a critical value for the FP length, increasing its physical length increases output power. But in this configuration the capacitance between the FP and drain becomes gate to drain capacitance C_{GD}, causing negative Miller feedback, which in turn reduces both current gain and power gain cutoff frequencies(f_T, f_{MAX}).

As the voltage swing across the gate and source is only 4–8 V for a typical GaN HEMT, much less than the dynamic output swing up to 230 V, terminating the FP to the source also satisfies the electrostatics for it to be functional. With the source tied FP, the FP-to-channel capacitance becomes the drain-source capacitance, which is absorbed in the output tuning network. The key drawback of the gate tied FP, i.e., additional gate drain capacitance, is eliminated. Depending on the implementation, the source connected field plate can add parasitic capacitance to the device input. This parasitic capacitance is absorbed into the input impedance tuning circuit, for narrow band signals.

15.7.2 Deep Recessed Gate HEMTs [53]

Silicon nitride(SiN_x) passivation is used to reduce the dispersion, but reproducibility of breakdown voltage, gate leakage, and effectiveness of dispersion elimination is strongly fabrication|process related. A solution, based on molecular beam exitaxy (MBE) is the deep recessed GaN HEMT with a thick cap layer to eliminate dispersion (Fig. 15.9c, d).

The effect of the surface on the channel is inversely proportional to the distance between surface and channel. The thick AlGaN or GaN cap layers in the deep recessed HEMTs

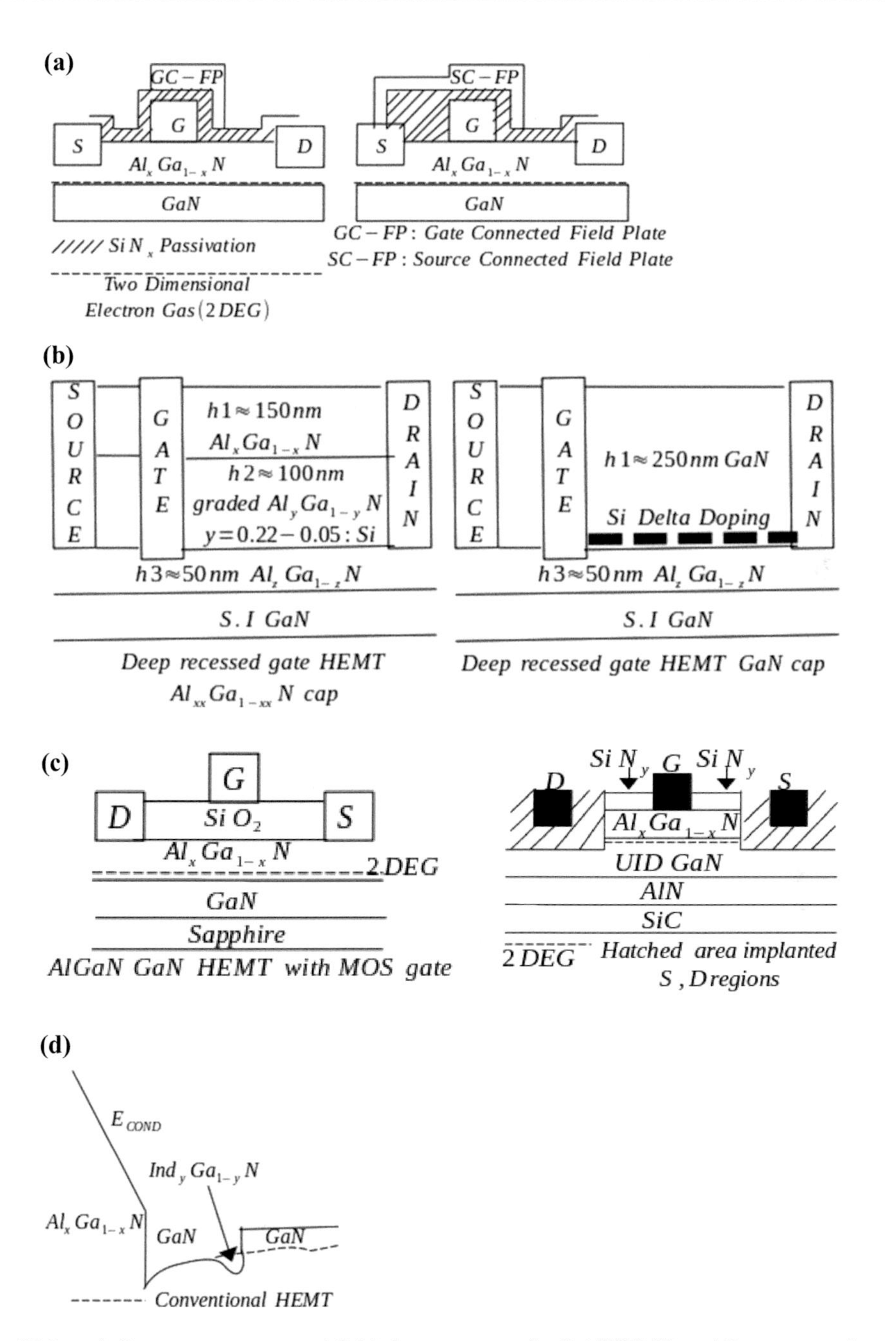

Fig. 15.9 **a**, **b** Gate, source connected field plate structures for GaN HEMTs c, d Deep recessed gate structures with AlGaN and GaN caps. e, f MOSHEMT and ion implanted drain source AlGaN|GaN HEMT. g Conduction band of back InGaN plane HEMT

increase the surface-to-channel distance, minimizing surface trap induced dispersion. Surface passivation is not necessary, as only a smaller portion of the channel charge is affected compared to the original AlGaN|GaN HEMT structure. The graded AlGaN layer is Si-doped to compensate the negative polarization charge and prevent hole accumulation.

The device fabrication processing flow is similar to that of the standard HEMT except that to create the deep ohmic and gate recess. A processing trick that reduces gate leakage by two orders of magnitude and also increases the breakdown voltage is to treat the recessed surface with fluorine plasma before gate metallization.

Accurate control of recess depth is achieved with a selective dry etch using a mixture of boron trichloride and silicon hexafluoride (BCl_3 SF_6). Fluorine decreases the etch rate of AlGaN due to the formation of a non-volatile aluminum trifluoride (AlF_3) residue on the AlGaN surface. The deep-recessed structure has a GaN cap(>200 nm) and an abrupt GaN|AlGaN interface to clearly define the etch-stop position.

15.7.3 Metal Oxide Semiconductor(MOS) HEMT [53]

The MOSHEMT (Fig. 15.9e) exploits the best features of both the MOSFET and the GaN HEMT. **Specifically, the MOS structure minimizes gate leakage current and the AlGaN/GaN heterogeneous junction provides high density high mobility 2DEG channel**.

Just like in a regular AlGaN|GaN HEMT, the MOSHEMT's built-in channel is formed by the high density 2DEG at the AlGaN|GaN heterogeneous junction. *The key difference between a AlGaN|GaN based MOSHEMT and the corresponding regular HEMT, is that the* ***gate metal is isolated from the AlGaN barrier layer by a thin dielectric film*** (e.g., silicon dioxide, aluminum nitride|oxide, zirconium oxide, niobium oxide, hafnium dioxide). So, the MOSHEMT gate behaves like the gate in a regular MOSFET. Because a properly designed AlGaN barrier layer is fully depleted by electron transfer to the adjacent GaN layer, the gate insulator in the MOSHEMT consists of two sequential layers: dielectric film (e.g., silicon dioxide) and AlGaN epitaxial layer. The advantages of the double layer are:

- Extremely low gate leakage current–density.
- Large positive to negative gate voltage swing.

The maximum DC saturation drain current at positive gate voltages is a key parameter controlling maximum output RF power. For standard AlGaN|GaN HEMTs, gate voltages above 1.2 V result in excessive forward current. In contrast, for a MOSHEMT, the gate voltages as high as 10 V cause no problems, resulting in significant increase in maximum channel current.

The gate-to-source spacing of a RF (e.g., millimeter wave) HEMT must be minimized, to keep the source access resistance low. *Conventional alloyed ohmic contacts have rough edges, which prevents the source gate spacing from being reduced below a limit.* Therefore, a non-alloyed ohmic contact is preferred for the high frequency devices. Ion implantation has been used in the GaN device fabrication to form non-alloyed ohmic contacts (Fig. 15.9f).

To boost the performance of GaN HEMTs beyond the Ka-band, the standard HEMT features as shorter gate length, multiple fingers to reduce gate resistance and Γ-shaped.

gate to decrease gate-to-drain capacitance, are being augmented with the back barrier structure, whose energy band diagram is shown in Fig. 15.9g.

References

1. https://www.ioffe.ru/SVA/NSM/Semicond/GaN/bandstr.html
2. Rammohan, R. L. (2001). *Physical Review., 89*, 11.
3. Johnson, W. C., Parsons, J. B., & Crew, M. C. (1932–1992). *Journal of Physical Chemistry, 36*, 2561; *Science Letter, 11*, 261.
4. Bloom, S., Harbeke, G., Meier, E., & Ortenburger, I. B. (1974). *Physica Status Solidi, B, 66*, 161.
5. Wright, A. F., & Nelson, J. S. (1994). *Physical Review, B, 50*, 2159.
6. Strite, S., & Morkoc, H. (1992). *Journal of Vacuum Science and Technology, B, 10*, 1237.
7. Kung, P., & Razegui, M. (2000). *Optical Review, 8*, 201.
8. Pankove, J. I. (1987–1990). *Proceedings of Material Research Society Symposium, 97*(409), 162, 515.
9. Mizuta, M., Fujied, S., Matsumoto, Y., & Kawamara, T. (1986). *Japan Journal Applied Physics, 25*, L945.
10. Scanlon, A. P. (1991). *Letters, 95*, 944.
11. Paisley, M. J., Sitar, Z., Posthill, J. B., & Davis, R. F. (1989). *Journal Vacuum Science and Technology, A, 7*, 701.
12. Sitar, Z., Paisley, M. J., Yan, B., & Davis, R. F. (1990). *Material Research. Society Symposium. Proceedings, 162*, 537.
13. Humphreys, T. P., Sukow, C. A., Nemanich, R. J., Posthill, J. B., Radder, R. A., Hattangady, S. V., & Markunas, R. J. (1990). *Material Research Society Symposium Proceedings, 162*, 53.
14. Lambrecht, W. R. L., & Segall, B. (1991). *Physical Review, B, 43*, 7070.
15. Phillips, J. C. (1973). *Bonds and bands in semiconductors.* Academic Press.
16. Brust, D. (1964). *Physical Review, A114*, 1357.
17. https://d-nb.info/968925243/34
18. Ambacher, O., Foutz, B., Smart, J., Shealy, J. R., Weimann, N. G., Chu, K., Murphy, M., Sierakowski, A. J., Schaff, W. J., Eastman, L. F., Dimitrov, F., Mitchell, A., & Stutzmann, M. (2000). Two dimensional electron gases induced by spontaneous and piezoelectric polarization in undoped and doped AlGaN/GaN heterostructures. *Journal of Applied Physics, 87*(1).
19. Ambacher, O., et al. (1999). Two-dimensional electron gases induced by spontaneous and piezoelectric polarization charges in N- and Ga-face AlGaN/GaN heterostructures. *Journal Applied Physics., 85*, 3222.
20. Ibbetson, J. P., et al. (2000). Polarization effects, surface states, and the source of electrons in AlGaN/GaN heterostructure field effect transistors. *Applied Physics Letters, 77*, 250.

21. Koley, G., & Spencer, M. G. (2005). On the origin of the two-dimensional electron gas at the AlGaN/GaN heterostructure interface. *Applied Physics Letters, 86*, 042107.
22. Gordon, L., et al. (2010). Distributed surface donor states and the two-dimensional electron gas at AlGaN/GaN heterojunctions. *Journal of Physics D: Applied Physics, 43*, 505501.
23. Higashiwaki, M., et al. (2010). Distribution of donor states on etched surface of AlGaN/GaN heterostructures. *Journal Applied Physics, 108*, 063719.
24. Heikman, S., et al. (2003). Polarization effects in AlGaN/GaN and GaN/AlGaN/GaN heterostructures. *Journal Applied Physics, 93*(12), 10114.
25. Smorchkova, I. P., et al. (2001). AlN/GaN and (Al,Ga)N/AlN/GaN two-dimensional electron gas structures grown by plasma-assisted molecular beam epitaxy. *Journal of Applied Physics, 90*(10), 5196.
26. Goyal, N., et al. (2012). Analytical modeling of bare surface barrier height and charge density in AlGaN/GaN Heterostructures. *Applied Physics Letters, 101*, 103505.
27. Shin, J. S., et al. (2015). Investigating gate metal induced reduction of surface donor density in GaN/AlGaN/GaN heterostructure by electroreflectance spectroscopy, curent. *Applied Physics, 15*, 1478.
28. Jogai, B. (2003). Influence of surface states on the two-dimensional electron gas in AlGaN/GaN heterojunction field-effect transistors. *Journal Applied Physics, 93*, 1631.
29. Mizutani, T., et al. (2007). AlGaN/GaN HEMTs with thin InGaN cap layer for normally off operation. *IEEE Electron Device Letters, 28*(7), 549.
30. Green, R. T., et al. (2010). Characterization of gate recessed GaN/AlGaN/GaN high electron mobility transistors fabricated using a SiCl 4 /SF 6 dry etch recipe. *Journal Applied Physics, 108*, 013711.
31. Xu, P., et al. (2012). Analyses of 2-DEG characteristics in GaN HEMT with AlN/GaN superlattice as barrier layer grown by MOCVD. *Nanoscale Research Letters, 7*, 141.
32. Goyal, N., et al. (2013). Barrier layer thicknesses in AlGaN/GaN heterostructure field effect transistors. *AIP Conference Proceedings, 1566*, 393.
33. Resta, R. (1994). *Review of Modern Physics, 66*, 899–915.
34. Posternak, M., Baldereschi, A., Catellani, A., & Resta, R. (1990). *Physical Review Letters, 64*, 1777–1780.
35. Ambacher, O., Foutz, B., & Smart, J. (2000)2DEG induced by spontaneous and piezoelectric polarization in undoped and doped AlGaN/GaN heterostructures. *Journal of Applied Physics, 87*(1), 334–344.
36. Lüth, H. (1993). *Surfaces and interfaces of solids*. Springer.
37. Harris, J. J., Pals, J. A., & Woltjer, R. (1989). *Reports of Progress Physics, 52*, 1217.
38. Nag, B. R. (1980). *Electron transport in compound semiconductor. Solid states science* (Vol. 11). Springer Edition.
39. Walukiewicz, W., Lagowski, L., Jastrzebski, L., Lichtensteiger, M., & Gatos, H. C. (1979). *Journal of Applied Physics, 50*, 899.
40. Shur, M., Gelmont, B., & Asif, K. M. (1996). *Journal of Electronic Materials, 25*, 821.
41. Walukiewicz, W. (1992). *Semiconductor interfaces and microstructures* (Z. C. Feng edition, p. 1). World Scientific.
42. Störmer, H. L., Pfeiffer, L. N., Baldwin, K. W., & West, K. (1990). *Physical Review B, 41*, 1278.
43. Hsu, L., & Walukiewicz, W. (1997). Electron mobility in AlGaN/GaN heterostructures. *Physical Review B, 56*(3), 1520–1528.
44. Wu, X. H., Brown, L. M., Kapolnek, D., Keller, S., Keller, B., DenBaars, S. P., & Speck, J. S. (1996). *Journal of Applied Physics, 80*, 3228.
45. Wu, X. H., Kapolnek, D., Tarsa, E. J., Heying, B., Keller, S., Keller, B. P., Mishra, U. K., DenBaars, S. P., & Speck, J. S. (1996). *Applied Physics Letters, 68*, 1371.

46. Jena, D., Gossard, A. C., & Mishra, U. K. (2000). Dislocation scattering in a two-dimensional electron gas. *Applied Physics Letters, 76*(13), 1707–1709.
47. Read, W. T. (1954). *Philosophical. Magazine, 45*, 775.
48. Ponce, F. A. (1997). *Material Research Bulletin, 22*, 51.
49. Wu, X. H., Fini, P., Tarsa, E. J., Heying, B., Keller, S., Mishra, U. K., DenBaars, S. P., & Speck, J. S. (1998). *Journal of Crystal Growth, 189–190*, 232.
50. Kapolnek, D., Wu, X. H., Heying, B., Keller, S., Keller, B. P., Mishra, U. K., DenBaars, S. P., & Speck, J. S. (1995). *Applied Physics Letters, 67*, 1541.
51. Ng, H. M., Doppalapudi, D., & Moustakas, T. D. (1998). The role of dislocation scattering in n- type GaN films. *Applied Physics Letters, 73*(6), 821–823.
52. Mishra, U., Shen L., Kazior T. E., & Wu, Y.-F. (2008). GaN based RF power devices and amplifiers. *Proceedings of the IEEE, 96*(2).
53. https://engineering.purdue.edu/~yep/Papers/GaN%20MOSHEMT%20International%20Journal%20of%20High%20Speed%20Electronics%20and%20Systems.pdf